Spatial Polarization Characteristics of Radar Antenna

Huanyao Dai · Xuesong Wang
Hong Xie · Shunping Xiao
Jia Luo

Spatial Polarization Characteristics of Radar Antenna

Analysis, Measurement and Anti-jamming Application

National Defense Industry Press

Springer

Huanyao Dai
State Key Laboratory of Complex
  Electromagnetic Environment Effects on
  Electronics and Information System
Luoyang
China

Xuesong Wang
National University of Defense Technology
Changsha
China

Hong Xie
State Key Laboratory of Complex
  Electromagnetic Environment Effects on
  Electronics and Information System
Luoyang
China

Shunping Xiao
National University of Defense Technology
Changsha
China

Jia Luo
Ningbo Institute of Microelectronic
  Applications Chinese Academy of
  Sciences
Beijing
China

ISBN 978-981-10-8793-6   ISBN 978-981-10-8794-3   (eBook)
https://doi.org/10.1007/978-981-10-8794-3

Jointly published with National Defense Industry Press, Beijing, China

The print edition is not for sale in China Mainland. Customers from China Mainland please order the print book from: National Defense Industry Press, Beijing.

Library of Congress Control Number: 2018937334

Printed on acid-free paper

This Springer imprint is published by the registered company Springer Nature Singapore Pte Ltd. part of Springer Nature
The registered company address is: 152 Beach Road, #21-01/04 Gateway East, Singapore 189721, Singapore

Preface

Local wars of high technology in recent years show that the battlefield awareness has gradually become the key factor to decide the outcome of a war. Radars are one of the most widely used sensing systems in battlefields. They play a more and more important role in modern wars. With the electromagnetic environment in battlefields getting increasingly harsh, the anti-jamming capability, intelligent identification capability, and survivability of radar systems such as battlefield search, warning, guidance, and tracking guidance are seriously restricted. For the complex and abominable battlefield environment in future military warfare, it is urgently necessary for the radar systems to improve anti-jamming capability.

The jamming and anti-jamming of existing radars are mostly carried out in time, frequency, and space domains. The role and potential of polarization information in anti-jamming has not yet been fully valued and studied. Moreover, polarization is seldom considered in the design and tactical use of all kinds of radar jamming modes. This also prompted experts, scholars, and other researchers to continuously develop and utilize various polarization information processing technologies to improve the detection capability and perception capability of sensing systems. Besides information of time, frequency, and space domains, polarization is also available important information of electromagnetic waves. As an effective way to improve the anti-jamming and target identification capabilities of radars, fully extracting polarization information makes considerable room for improving the performances of modern radar system. If radars have the capability of polarization measurement and can make full use of the difference between target and jamming polarization information, they are expected to achieve significant anti-jamming effects.

For more than half a century, with the development of radar polarization measurement technology and gradually deepened understanding of the electromagnetic scattering characteristics of targets, fruitful results have been achieved in the radar polarization technology as an important branch of modern radar technology, greatly promoting the development of modern radar technology. More and more polarimetric radars have been developed and put into operation, and a large number of polarimetric radars have been produced in the USA, Canada, Italy, Germany,

France, the UK, Holland, and other countries. Most of the polarimetric radars adopt two orthogonal polarization channels, to transmit multiple orthogonally polarized pulses alternately or simultaneously, and receive signals simultaneously. The polarization state can be estimated by processing several adjacent pulses. However, a common feature between the two polarization measurement methods is that they both need dual-polarization antenna and two polarization channels. This imposes rather stringent requirements on the systematic design, complexity, and R&D funding of radars.

According to the antenna theory, in the far-field region at a given frequency and space orientation, an antenna has a certain fixed polarization mode. The polarization mode of antenna radiation field varies with the operating frequency and space orientation. This means that polarization of the antenna is the function of frequency and spatial orientation. Hence, this book presents a new concept, namely the spatial polarization characteristic of antenna". In the past research, this characteristic was often regarded as an imperfect factor and not taken into account. It was not the main research and development direction in the field of radar polarization technology. However, to the development of radar information processing theory, the spatial polarization characteristic of antenna has contributed a new idea or new direction, reasonably using the "non-ideal orthogonal basis" constituted by antenna spatial polarization variation to perform target measurement and anti-jamming. Taking into consideration the experience gained in the national defense preliminary research projects, the projects supported by the National Natural Science Fund, and the "973" major basic national security projects as carried out in the periods of the Tenth Five-Year Plan and the Eleventh Five-Year Plan, the authors, in recent years, have achieved some research results of high academic and application significance in the following aspects: characterization of spatial polarization characteristic of antenna, polarization characteristic analysis and modeling, antenna polarization measurement and new test method, measurement error correction method, and new anti-jamming method and experiment based on spatial polarization characteristic of antenna. Based mainly on these research results, and taking into account the requirements of information equipment (including radars) upgrading, we wrote this monograph in respect of spatial polarization modulation effects of radar antenna and their processing technologies. Providing a theoretical and technological summary in the field of radar signal processing research is the intention of this monograph for reference by scientific and technological personnel in this field.

Chapter 1 mainly summarizes the theory and application of radar polarization information processing in such aspects as target recognition, target detection, clutter suppression, and anti-jamming. Chapter 1 also points out the shortcomings of existing research on radar polarization technology and illustrates the scientific connotation and research trend of antenna spatial polarization characteristic. In Chap. 2, various characterization methods of antenna spatial polarization characteristic are introduced, and the evolution law of antenna spatial polarization is discussed with a dynamic and statistical point of view. Chapters 3–5 are introduction to the basic theory. The spatial polarization characteristics of wire antenna and aperture antenna are analyzed in Chap. 3 with clear mathematical formula.

Particularly, the parabolic reflector antenna polarization characteristic is analyzed in detail, providing qualitative and quantitative results through electromagnetic calculation and anechoic chamber measurement. Chapter 4 discusses the spatial polarization characteristic of phased array antenna. Similarity and difference in terms of spatial polarization characteristic between phased array antenna and mechanical scan antenna are indicated. Polarization characteristic variation during the azimuth and elevation two-dimensional scan of phased array antenna is described through computer simulation and high-frequency electromagnetic field calculation. A new method for antenna polarization characteristic measurement and a method for error correction are presented in Chap. 5, mainly intended to realize fast and accurate measurement in the field. Without requiring antenna removal, the methods can realize effective measurement by just processing the signals received by the radar. Indicating both steps and results of field measurement, these methods are notable for high practicability, simple algorithm, and convenient use. They can provide valuable data for reference, laying a foundation for making full use of radar antenna polarization characteristic. Chapters 6 and 7 deal with application of antenna spatial polarization characteristic. To solve the problems concerning target identification and anti-jamming as encountered by a large number of military mechanical scan radars, such as warning radars, target designation radars, and guidance radars, new information processing methods are designed. In-depth discussion is performed on application. With the antenna spatial polarization characteristic considered, time series received at each spatial position in the process of antenna scan are processed, to realize target scattering matrix measurement, polarization measurement and suppression of jamming signals, as well as polarization measurement and discrimination of true and false targets. To ensure actual application of the polarization technology, the following aspects are discussed thoroughly: compatibility of the polarization technology with coherent processing, measurement accuracy improvement under the influence of the automatic gain control system, and jamming suppression parallel processing with the existence of elevation angle estimation error. Results of both the simulation experiment and actual test demonstrate that the new methods achieve better anti-jamming effects. This verifies the correctness, feasibility, and superiority of these new methods. All of these methods are implemented on traditional single-polarization radars. In other words, by improving the signal processing software of the existing radars, it is possible to provide these radars with a certain level of polarization information processing capability, to obtain remarkable anti-jamming effects. At the end of this book is a list of nearly 200 references on radar polarization information processing methods. They are useful for those who wish to do research in this field or to explore new polarization information processing technology.

Research results shown in this book have greatly expanded the scope of radar polarization information processing technology and increased the contents of radar polarization theory. Combination of the radar polarization information processing technology with the anti-jamming measures based on time domain, frequency domain, and spatial domain is expected to effectively improve the anti-jamming and target identification capabilities of the existing radar systems, hence their adaptability and survivability in complex battlefield environment. The radar polarization information processing technology has a bright prospect as far as application is concerned.

Authors of this book are Dr. Dai Huanyao (CEMEE Lab), Dr. Luo Jia, Prof. Wang Xuesong (National University of Defense Technology), Prof. Xie Hong(L), and Prof. Xiao Shunping. In the process of compilation, Senior Engineer Kong Depei and Cui Yun Editor of the National Defense Industry Press have provided us with great help. To them, we want to express our sincere gratitude.

Since the compilation time is short and our academic knowledge is limited, it is possible that mistakes exist in this book. Please inform us of such mistakes if you find. Thank you.

Luoyang, China	Huanyao Dai
Changsha, China	Xuesong Wang
Luoyang, China	Hong Xie
Changsha, China	Shunping Xiao
Beijing, China	Jia Luo
October 2017	

Contents

Chapter 1
Introduction

Electromagnetic environment in battlefields turns to be increasingly harsh and complex in future; therefore, various information-processing technologies are hence constantly developed and utilized to enhance the detection or sensing capability of sensor systems. In addition to time domain, frequency domain, and spatial domain, polarization is also important useful information of electromagnetic waves. Acquiring as much polarization information as possible offers great possibility in improving performances of modern radar systems [1, 2]. During the past more than half a century, with the development of radar polarization measurement technology, and the increasing deeper understanding of the scattering characteristics of target electromagnetic wave, we have gained a good many research results in radar polarization science as an important branch of modern radar science, which has greatly drove the growth of radar technology.

Nowadays, research achievements in radar polarization science are mainly in two fields: polarization characterization and polarization characteristics, and acquisition and processing of polarization information. Research in polarization characteristics and characterization provides a basis for the application of radar polarization information. Main aspects of this research are: polarization characterization of electromagnetic waves and radar target scattering electromagnetic waves [1, 2], polarization characteristic of electromagnetic waves and radar target [3–5], instantaneous polarization and the statistical theory of instantaneous polarization [3, 6], characteristics and characterization of antenna polarization [7], and target optimum polarization theory [1, 5], etc. The ultimate goals of research in radar polarization science are to acquire and process polarization information, and to enhance the detection or sensing especially the anti-jamming capability of radar systems. The content included in this research deals with the following issues: polarized radar data measurement and correction [1, 8–10], polarization filtering [11–20], polarization detection [21–29], polarization anti-jamming, extraction of radar target polarization characteristics [1, 2, 30], polarization classification and identification parametric inversion [30–35], and imaging of polarized radars [36–41].

© National Defense Industry Press, Beijing and Springer Nature Singapore Pte Ltd. 2019
H. Dai et al., *Spatial Polarization Characteristics of Radar Antenna*,
https://doi.org/10.1007/978-981-10-8794-3_1

It is due to the development of basic theory of polarization and the progress of application technology of polarization information that more and more polarized radars have been successfully developed and put into use. A good number of polarized radars have come into being in the USA, Canada, Italy, Germany, France, the UK, and Holland [42–45]. Most of these radars are capable of alternatively transmitting two kinds of orthogonal polarization waves and simultaneously receiving two kinds of orthogonal polarized waves to acquire all information on target polarization scattering matrix.

As early as the 1960s polarization measurement radars were invented in foreign countries. During that period, China also spared no efforts in polarized radar development. However, as far as engineering realization is concerned, polarization measurement radar systems are both rather complex and expensive. Just in settling engineering realization issues, we often encounter a lot of difficulties, such as transmission switch-over between polarization channels, channel amplitude-phase consistency correction, orthogonal-polarization isolation, and calibration of measurement systems. Here we take the searching and tracking radars as an example. Improving their polarization measurement capability will probably cause terrible increase in the quantity of equipment in the RF systems, the complexity of the radar systems, and the cost of engineering realization. Therefore, while developing special radars for measuring polarization characteristics of targets, we should seek new technological approaches for solving critical technological problems by fully tapping the potential information of the radar system and the target based on the existing single-polarized radars. Without making considerable modification to the hardware of the existing radar systems and by merely carrying out technical updates, we should effectively improve the capabilities of the radar systems in polarization measurement, tracking, identification and anti-jamming via fully acquiring and using radar antenna and polarization information of the target. Doing so is of great importance, which lays solid theory and technology foundation for future advanced radar. Improving performances and increasing functions of radars at the possibly least cost is not only the pursuit of radar engineering industry but also the trend of radar polarization information processing and utilization.

According to antenna theory, the antenna has a certain fixed polarization style in the far-field region with the given frequency and spatial orientation. The polarization style in antenna radiation field varies with different operating frequency and spatial orientation [46, 47]. It implies that antenna polarization is a function of frequency and spatial orientation, which associated with antenna type. The characteristic variation of antenna spatial polarization is called the spatial polarization characteristics of antenna [48–50]. In traditional research, however, this characteristic is intentionally ignored because of different research emphasis and application requirement, or the cross-polarization of antenna is considered to be harmful and avoided to the greatest extent. Do not blindly suppress the cross-polarization of antenna but thoroughly investigate and effectively use the rule of antenna spatial polarization characteristic variation. Antenna spatial polarization characteristic will be broadly used with great attraction. Nowadays, research in the polarization characteristics of electromagnetic waves and antennas and the application of such characteristics is mostly performed in time domain, frequency domain and joint time-frequency domain. Research in aspects of antenna spatial polarization characteristic variation, its distribution rule and application has not yet been reported.

With the electromagnetic environment in modern battlefields getting increasingly harsh and complex, the following aspects have become the basic topics and compelling tasks currently in the research of radar polarization technology: fully acquiring and utilizing polarization information of electromagnetic waves and radar antennas to study the radar antenna spatial polarization characteristics and further extend and perfect the theory regarding polarization characterization and polarization characteristic; fully acquiring and utilizing electromagnetic information obtained by radar sensor systems; improving capability of radar systems in acquisition and processing of information so that they are adaptive to the complicated and changing electromagnetic environment in battlefields.

Based on the instantaneous polarization theory and in the context of engineering application demand, this book is organized with the research in the theory and application of antenna spatial polarization characteristic as the main topic. We have established the method for representation of antenna spatial polarization characteristic. Our deep research in antenna spatial polarization characteristic has extended and refined the systematic theory on radar polarization. For the mechanical-scan-system radars that are widely used for military purpose, for example, the warning radars, target designation radars and guidance radars, we have carried out the research in application of antenna spatial polarization characteristic in such fields as the measurement of target polarization characteristic, anti-active jamming, and discrimination between true and false targets. Our research has obtained a good many significant results.

1.1 Science Connotation and Progress of Antenna Spatial Polarization Characteristic

Polarization describes the locus of the electric field vector of spatial electromagnetic wave that moves with time in the propagation cross section. Polarization reflects the vector attribute of electromagnetic wave. Polarization is another piece of useful information, in addition to time domain, frequency domain, and spatial domain. For enhancing the performances of modern radars in detection, tracking, anti-jamming, and target identification, deep exploration of polarization information is of great importance. A relatively systematic theory on radar polarization information processing has resulted from more than sixty years' efforts. Besides issues of vector signal processing and issues of radar target identification, the theory introduces also basic issues, such as polarization characterization of electromagnetic waves and radar target, description of radar target polarization characteristic, and polarization data measurement and correction. Polarization filter, polarization detection, polarization optimization and intensification, polarization tracking and correlation, and imaging of polarized radar are examples of issues of vector signal processing, while examples of issues of radar target identification are extraction of polarization characteristic, polarization discrimination, polarization classification, and identification parametric inversion. Most of these issues are associated with basic radar theory, electronic countermeasure and pattern identification. Divided by the content of the research, the radar polarization science can be roughly divided into four levels.

1.1.1 Study Present Situation of Electromagnetic Wave Polarization Property

In the field of electromagnetic wave research, at the beginning of the 1950s, the polarization state and its effects drew attention for the first time. Classic methods for describing electromagnetic wave polarization by using static parameters such as Jones vector, elliptic geometric descriptor, polarization phase descriptor, polarization ratio, and Stokes vector were put forward successively by scholars [2]. Using these methods, abstract polarization phenomena can be described visually in a two- or three-dimensional space. The elliptic geometric descriptor of polarization is defined based on the assumption that the locus of electric field vector tip which moves with time in the electric wave propagation cross section is an ellipse. The definition is physically both clean and direct. The linear polarization ratio and circular polarization ratio, which are defined on the linear polarization basis and circular polarization basis respectively, are considered to be effective means for describing polarization states. They can be used to plot planar graphs, for describing complex polarization states. The Poincare polarization sphere derived from the Stokes vector descriptor establishes one-to-one correspondence between polarization states and points on the sphere. These polarization descriptors of electromagnetic wave are used in a variety of applications, for example, theoretic analysis and engineering design. It is implied in all of the applications that the electromagnetic waves under consideration are "of narrow band type" and "time-harmonic". The locus of spatial movement of electric field vector tips of electromagnetic waves is required to be excellent in both geometric regularity and periodicity.

With the development of wide band electromagnetic theory and polarization measurement technology in recent years, the frequency, amplitude and polarization in the duration of waves no longer remain unchanged for complex modulated wide band electromagnetic waves and transient electromagnetic waves. For time-variant electromagnetic waves, the locus of spatial movement of electric field vector tips usually does not have excellent geometric regularity and long-range repeatability. Classic polarization descriptors, for example, the simple static curves, are far from complete and accurate characterization. In 1999, Wang Xuesong innovatively framed the concept "instantaneous polarization" to characterize electromagnetic wave polarization [3]. For instantaneous polarization descriptors, he built the set of parameters applicable for describing time-variant electromagnetic waves. Included in the set were instantaneous Stokes vector, instantaneous polarization projection vector, polarization cluster center, polarization divergence, polarization measure, and instantaneous polarization state change rate. Next, he studied the equivalence between time-domain and frequency-domain instantaneous polarization information, and formulated the new concept of instantaneous polarization distribution in the joint time-frequency domain of electromagnetic waves. He disclosed the intrinsic link between time-domain and frequency-domain instantaneous polarization descriptors of electromagnetic waves. From the point of view of polarization distribution on the boundaries of time and frequency domains, he made unitary explanation of the two classes of instantaneous polarization descriptors, extending the concept of instantaneous polarization to the joint time-frequency domain, so as to furnish a powerful theoretical tool for polarization characterization of time-variant electromagnetic waves and wide band polarization information processing. In 2004, Li Yongzhen and Zeng Yonghu broadened and improved the theory of statistical distribution of instantaneous polarization of electromagnetic waves [6, 31]. Using the signal distribution in joint time-frequency domain as the theoretical tool, Zeng Yonghu devised the basic method for analysis of instantaneous polarization of electromagnetic waves in joint time-frequency domain. He offered a theoretical fundament and mathematic tool for depiction and analysis of instantaneous polarization characteristic of time-variant electromagnetic waves. Li Yongzhen investigated the statistical characteristics of instantaneous polarization of electromagnetic wave scattering of radar targets under the Gaussian distribution conditions, laying a foundation for subsequent investigation into the statistical characteristics of instantaneous polarization of electromagnetic wave scattering of radar targets under the non-Gaussian distribution conditions.

To be brief, electromagnetic wave polarization characteristic and deterministic characterization, statistical characterization of electromagnetic wave polarization, radar target polarization characteristic and deterministic description, and statistical description of radar target polarization characteristic are the main concerns of the research in electromagnetic wave polarization characteristic. Methods for polarization characterization may be divided into two classes: classic polarization characterization and instantaneous polarization characterization. Figure 1.1 demonstrates the relation among various methods for polarization characterization.

1.1.2 Study Present Situation of Antenna Polarization Characteristic

Since the 1970s, both domestic and international efforts have been devoted to the research in antenna polarization purity. For the research in application of antenna systems, the polarization direction at the main lobe center is generally defined as the

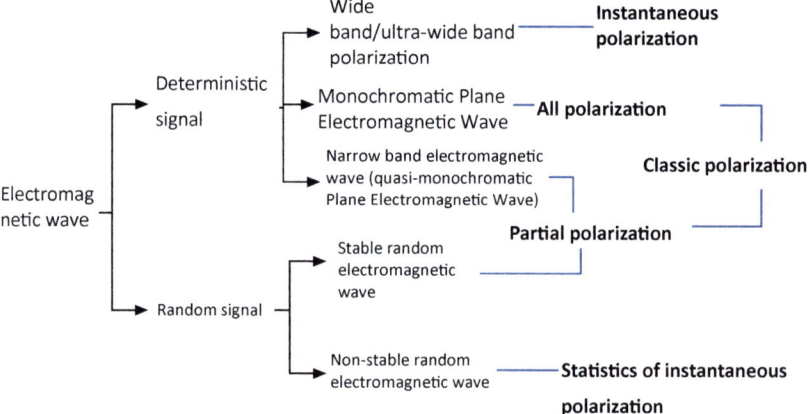

Fig. 1.1 Relation among various methods for electromagnetic wave polarization characterization

co-polarization direction, i.e. the reference polarization direction. Polarization vertical to the co-polarization is cross-polarization, which is often considered as a jamming and negative factor. Thus for developing antennas with high polarization purity, scholars committed themselves to identifying the causes for cross-polarization and the approaches to reducing cross-polarization. As a result of nearly a decade's efforts, the relation between the performances of the components of the antenna feed system and the antenna cross-polarization has been clearly stated theoretically: a variety of factors account for antenna cross-polarization, mainly the non-ideal characteristics of antenna [51–57], including antenna port isolation, antenna static cross-polarization discrimination (XPD), antenna consistency, material and mounting of antenna radome, and flatness of antenna XPD in operating frequency band. With the machining precision of the components of antenna being continually raised, scholars at home and abroad have gained lots of experience in the improvement of antenna polarization purity. They have an increasingly clear understanding about the influence of various parameters of the widely used reflector antenna on cross-polarization, and have established a good number of optimization criteria. Today, in most cases, research in the antenna cross-polarization characteristic focuses on the calculation and suppression of cross-polarization.

On the other hand, it is pointed out by antenna theory that the antenna has a certain fixed polarization style in the far-field region with the given frequency and spatial orientation. The polarization style of antenna radiation field varies somewhat with operating frequency and spatial orientation. It implies that antenna polarization is a function of frequency and spatial orientation, and associated with antenna type and function. Figure 1.2 shows the amplitude graphs of co-polarization and cross-polarization of a typical offset parabolic antenna. Figure 1.3 indicates the rule of antenna polarization purity varying with azimuth angle and pitch angle. The antenna spatial polarization characteristic variation can affect in many respects the

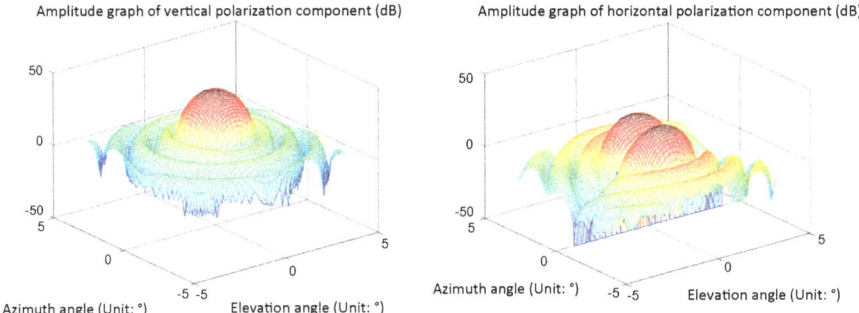

Fig. 1.2 Amplitude graphs of co-polarization and cross-polarization of a single-offset parabolic antenna

Fig. 1.3 Rule of antenna polarization purity variation

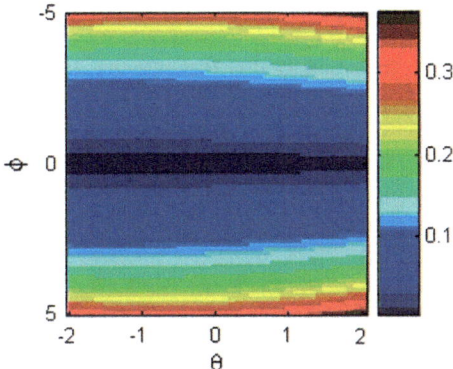

radar polarization information processing and electronic countermeasure. Doctor Luo Jia was the first scholar who named this characteristic "the spatial polarization characteristic" of antenna. For the characterization of spatial polarization characteristic, he established the theoretic frame, offering a theoretical fundament and powerful tool for depiction of antenna spatial polarization characteristic. He performed theoretic analysis and numerical simulation of spatial polarization characteristic of a variety of communication antenna including dipole antenna, helical antenna, and loop antenna. Mainly taking into consideration the offset parabolic antenna, he analyzed the spatial polarization characteristic of typical wire antenna and aperture antenna. He unveiled the rule for antenna spatial polarization characteristic variation, and proved the rule with calculated data and anechoic chamber measured data.

The development of microwave solid-state components and monolithic microwave integrated circuits (MMIC) leads to huge reduction to the volumes and production costs of phased-array radars. The phased array technology is increasingly broadly used in both military and civil fields. Examples are early-warning radars, communication radars, and meteorological radars. A trend in the development of phased-array radar antenna technology is rapid phased-array beam scanning velocity, flexible beam control, space-time signal processing, multi-target information extracting, and self-adaptive anti-jamming. For this special type of antenna, at the given operating frequency, the relation between polarization characteristic variation and main lobe spatial orientation has not yet been reported. The item that requires further investigation is the antenna polarization characteristic variation with radiation unit, polarization style, scan range, feed distribution, and observation direction.

Few papers have been published in the report of the influence of antenna spatial polarization characteristic variation on radar signal processing and electronic countermeasure performance. When the target direction deviates from the normal direction of the array antenna, the polarization state of received electric waves varies with the pitch angle and direction deviating from the electrical bore sight. Hanle in Germany pointed out that when the method of orthogonal circular polarization reception was used to dampen meteorological (rain and cloud) clutter jamming, the actual effect was generally not as strong as expected [58–61]. The primary cause is that the antenna deviation from the electrical bore sight changes polarization characteristic. J. G. Worms in Germany discussed the influence of antenna array element polarization characteristic on array antenna anti-active jamming performance [62]. He argued that attention should be given to antenna array element polarization in selection of self-adaptive spatial filter. Ni Jinlin discussed the influence of unit cross-polarization on the performances of self-adaptive array [63]. To summarize, nowadays at home or abroad there are few papers that examine the influence of antenna spatial polarization characteristic on radar signal processing and electronic countermeasure performance. Without systematic deep theoretical research, these papers just analyzed the in adverse influence of cross-polarization characteristic on radar signal processing. Using the cross-polarization characteristic to improve parametric measurement, characteristic measurement, and anti-jamming capability of radar received no consideration in these papers. Therefore, on the basis of research in the spatial polarization characteristic of typical and array antenna, this book thoroughly discusses the influence of antenna spatial polarization characteristic variation on radar signal processing and electronic countermeasure performance. Accordingly, this book puts forward the corresponding method for compensation and processing.

1.1.3   Study Present Situation of Radar Polarimetry

As one of important means for acquiring radar polarization information, the radar polarization measurement is in essence acquisition of all elements of target polarization scattering matrix by changing polarization states of transmission and reception. Ever since a long time ago, accurate measurement and effective utilization of target polarization characteristic has been a fundamental issue that attracts full attention in such fields as polarization anti-jamming, polarization target identification, and polarization SAR imaging.

All the four elements of polarization scattering matrix are complex numbers, which provides not only amplitude information but phase information as well. In addition to the capability of transmission with variable or orthogonal-polarization, full-polarized radar must also have the coherent processing capability. Due to restriction on development cost and realization technology, a majority of radars produced at the early stage operate in the single-polarization mode, i.e. both transmitted and received waves are single polarized. Measurement result in this operating mode is just a single element of polarization scattering matrix. For example, if both transmission and reception are of H-polarization, the measurement result is just the element S_{HH}. Note that some radar may operate in the mode of dual-polarization reception. In this case, only a column of elements of target scattering matrix can be measured, with half the target information being lost. Only the radars that are capable of accurately acquiring all the four elements of polarization scattering matrix for static, dynamic, and distributed targets can be called full-polarized radars. Described below are the latest research and development of radar polarization measurement system.

1.1.3.1 Non-full-Polarization Doppler System with Single-Polarization Transmission and Dual-Polarization Reception

"Non-full-polarization Doppler system with single-polarization transmission and dual-polarization reception", or dual-polarization system for short, is such that a pair of orthogonal-polarization antenna and the respective receiving channels are added to single-polarized radar to possess to some extent the capability of polarization measurement. Although this kind of measurement systems cannot acquire all polarization information of the target, they can get two elements of polarization scattering matrix. We may assume that the radar constantly transmits H-polarization signals and receives H- and V-polarization signals, and that the measurement sequence relation schema is as shown in Table 1.1. Then the radar transmits successively H-polarization signals in pulse repetition periods, and meanwhile the receiver receives H- and V-polarization signals.

A majority of polarized radars produced in the 1960–1970s adopted the dual-polarization system, due to restriction by the component machining precision and signal processing technology level. These radars were mainly used in such fields as weather observation, air defense and anti-missile, and missile guidance.

Table 1.1 Measurement sequence relation schema of type 1 polarization measurement radar

Sequence	T_p	$2T_p$	…	$K \cdot T_p$
Transmission	H	H	…	H
Reception	$H + V$	$H + V$	…	$H + V$
Measured value	s_{HH} s_{VH}	s_{HH} s_{VH}	…	s_{HH} s_{VH}

In the dual-polarization system, two orthogonal-polarization reception signals are mixed to intensify the target detection performance and anti-jamming capability, with the signal-to-noise ratio increased by several decibels. For the development of polarization filter and self-adaptive cancellation algorithm, the dual-polarization system provides a validation platform. The dual-polarization system has been widely used. In particular, it exhibits obvious advantage in eliminating clutter from rain. However, considering their limited capability of polarization measurement, the radars that adopt non-full-polarization measurement system are not frequently in the fields requiring high capabilities of anti-jamming and target identification. Typical representatives of this radar type were the USA-made Millstone Hill radar, which was used for observing and tracking satellites, and the AMRAD and ALTAIR radars, which were used for research in ballistic missile defense, and the USSR-made "Fan Song" SAM missile guidance radar.

The ballistic missile early-warning radar (Millstone Hill, UHF band) made in 1958 by the USA underwent significant modification in 1962, for example, the radar antenna was replaced, with the transmission frequency changed to L-band (carrier frequency 1295 MHz), and the UHF-band conical scan feed was replaced with the L-band Cassegrain feed. The Millstone Hill radar after modification adopted the polarization diversity technology. Signals transmitted by the radar were with right-hand circular polarization. When the radar received signals, the echoes with left- and right-hand circular polarization were simultaneously processed in the two angular error channels. Thus the radar could obtain a column of elements of target scattering matrix. Application of the polarization diversity technology increased the detection and tracking performance of the radar by several decibels. The target identification capability of the Millstone Hill radar had not been greatly improved because of its incapability of acquiring all information of target scattering matrix.

In the same period, the USA Defense Advanced Research Projects Agency (DARPA) initiated the research and development of the new generation of ballistic missile terminal defense system (ARPAT). The critical radar in the ARPAT system was the ARMRAD radar (L-band), which was designed by Lincoln laboratory and produced by Raytheon Company. Compared with the Millstone Hill and TRADEX (L-band) which were developed earlier, the ARMRAD radar employed a series of new technology, for example, narrow pulse width (minimum pulse width 0.1 μs), the digital signal processor (developed by Raytheon Company), and the RF component newly designed with polarization agility. Application of the new technology provided the ARMRAD radar transmission signals with agility between linear polarization and circular polarization. When receiving signals, the radar was capable of simultaneously processing two mutually-orthogonal echoes with linear polarization. The improvement of the ARMRD radar was carried out through the entire 1960s. In the 1970 shortly after its finalization, it was capable of identifying ballistic missile target with the polarization characteristic of the target.

While developing the TRADEX and ARMRAD radars, the USA detected that the ballistic missile defense radars (Hen House and DogHouse) of the USSR adopted different technological approaches—large VHF and UHF radars.

(a) ALTAIR missile defense radar

(b) "Fan Song" air defense guidance

radar

Fig. 1.4 Typical radar of non-full-polarization measurement system

To analyze the penetration capability of strategic ballistic missiles against the radars of the USSR, in the early 1960s the USA started the research and development of the ALTAIR radar of VHF/UHF-band. The research and development was performed through the entire 1960s. In May 1970, the radar was successfully fabricated. Since then, the radar experienced several times of major modification. To improve its target identification capability, major improvement has been carried out recently. The ALTAIR radar of latest version (VHF-band) has the capability of measuring polarization scattering matrix of ballistic missile target. The radar (Fig. 1.4a) can transmit signals with right-hand circular polarization. It can receive signals with left-hand circular polarization and those with right-hand circular polarization simultaneously.

According to the disclosed data, "Fan Song" A/B model adopted the single-polarization system. As it had only an azimuth angle scan antenna and a pitch angle scan antenna with horizontal polarization, it was easily affected by angle deception jamming and clutter from objects at low altitude or on land. The main parameters of the "Fan Song" D/E model as shown in Fig. 1.4b with updated H-polarization are given in Table 1.2 with the anti-jamming performance greatly improved. Technologically critical addition of an orthogonal-polarization double-feed parabolic antenna near the azimuth angle antenna enabled the "Fan Song" D/E model to adopt the "Irradiation" system in target tracking, being capable of creating both sum-difference signals with horizontal polarization and sum-difference signals with vertical polarization. Application of advanced polarization processing technology greatly enhanced the radar's capability of resisting angle deception jamming. It was reported that, by using the difference in respect of polarization characteristic between target and clutter/repeater jamming/responsive jamming/passive jamming (for example, chaff jamming), the radar has improved its operation under the jamming conditions with the technology of polarization estimation, polarization detection and polarization filter.

Table 1.2 Main parameters of "Fan Song"

"Fan Song" D/E air defense guidance radar	
Carrier frequency	5010–5090 or 4910–4990 MHz
Pulse repetition frequency	828–1440 or 1656–2880 Hz
Pulse width	0.4–1.2 µs
Peak power	1500 kW
Detection range	75–150 km
Beam width	7.5° (pitch), 1.5° (azimuth)
Polarization style	Transmission with H-polarization, reception with H/V-polarization

1.1.3.2 Non-full-Polarization Doppler System of Transmission-Reception with Variable Polarization

In the type-2 dual-polarization operating mode, a transmitting channel and a receiving channel are needed, the polarization style selector switch is used to select polarization style, and the reception is that with co-polarization. This type of polarization measurement system can be called "non-full-polarization Doppler system" of transmission-reception with variable polarization. The sequence relation taken by the system to realize polarization measurement is shown in Table 1.3. At PRI, the radar successively transmits signals with H-polarization and signals with V-polarization, and the receiver employs the reception with co-polarization.

In this measurement mode, two continuous pulses are measured to get the co-polarization component of target polarization scattering matrix. For example, in the $(2k - 1)$th PRI, the radar transmits signals with H-polarization, and simultaneously receives signals with H-polarization, then the measurement result is $s_{HH}[(2k - 1)T_p]$; in the $2k$th PRI, the radar transmits signals with V-polarization and simultaneously receives signals with V-polarization, then the measurement result is $s_{VV}(2kT_p)$.

Radar of this polarization measurement system requires addition of the ferrite phase shifter or the high-frequency changeover switch. This system imposes stringent requirements on pulse repetition interval. It is only capable of measuring the co-polarization component of target polarization scattering matrix. It cannot obtain target depolarization characteristics. Therefore, at the early stage, this polarization measurement system was not advantageous in the fields of

Table 1.3 Measurement sequence relation schema of type-2 polarization measurement radar

Sequence	T_p	$2T_p$	$3T_p$	$4T_p$	$\cdots$	$(2K - 1)T_p$	$2K \cdot T_p$
Transmission	H	V	H	V	$\cdots$	H	V
Reception	H	V	H	V	$\cdots$	H	V
Measured value	S_{HH}	S_{VV}	S_{HH}	S_{VV}		S_{HH}	S_{VV}

meteorological observation and target recognition, and was only adopted by several radars to achieve frequency diversity meeting specific needs, for example, target detection and anti-jamming.

1.1.3.3 Full-Polarization Doppler System with Dual-Polarization Alternative Transmission and Simultaneous Reception

It is impossible for the types 1 and 2 polarization measurement systems to obtain the complete target polarization scattering matrix. To obtain the complete target polarization information, since the 1980s, the international scholars have carried out extensive research in the radar of full-polarization measurement system. With the advancement in polarization theory and polarization measurement technology, the radar of type-3 polarization measurement system has gradually become the mainstream of research in polarized radar. At PRI, radar of this system alternatively transmits signals with orthogonal-polarization (H- and V-polarization), and receives simultaneously signals with H- and V-polarization. Two continuous pulses are measured to obtain complete polarization scattering matrix. Polarization measurement system of this type can be called "full-polarization Doppler system with dual-polarization alternative transmission-reception", or "alternate polarization system" for short. Radar of this system has one transmitting channel with variable polarization and two independent receiving channels with orthogonal polarization. In terms of transmission signal waveform and reception signal processing, no essential difference exists between the radar of this system and the conventional radar. The sequence relation of alternate polarization measurement is indicated in Table 1.4.

In the alternate polarization operating mode, two continuous pulses may be used to obtain once the measurement result of polarization scattering matrix. The measurement result is $s_{HH}\left[(2k-1)T_p\right]$ and $s_{VH}\left[(2k-1)T_p\right]$ in the $(2k-1)$th PRI, and $s_{HV}\left(2kT_p\right)$ and $s_{VV}\left(2kT_p\right)$ in the $(2k)$th PRI. So, the target Sinclair scattering matrix sequence is:

$$\mathbf{S}(k) = \begin{bmatrix} s_{HH}\left[(2k-1)T_p\right] & s_{HV}\left(2kT_p\right) \\ s_{VH}\left[(2k-1)T_p\right] & s_{VV}\left(2kT_p\right) \end{bmatrix} \tag{1.1}$$

Table 1.4 Measurement sequence relation schema of alternate full-polarization measurement radar

Sequence	T_P	$2T_P$	$3T_P$	$4T_P$	$\cdots$	$(2K-1)T_P$	$(2K)T_P$
Transmission	H	V	H	V	$\cdots$	H	V
Reception	$H+V$	$H+V$	$H+V$	$H+V$	$\cdots$	$H+V$	$H+V$
Measured value	S_{HH}	S_{HV}	S_{HH}	S_{HV}	$\cdots$	S_{HH}	S_{HV}
	S_{VH}	S_{VV}	S_{VH}	S_{VV}		S_{VH}	S_{VV}

Since the mid-1980s, in the developed countries such as the USA, Canada, Italy, Germany, France, Russia, Japan, and Holland, emphasis has been placed on the research and development of radar of alternate full-polarization measurement system. Quite a number of polarized radars have been produced, including the surveillance and tracking radars, airborne/satellite-borne polarized SARs, land ISARs, and meteorological radars. Examples were the S-band polarized radar developed by Italy [64], the Convair-580 X/C-band synthetic aperture radar developed by Canada Center for Remote Sensing [65, 66], the airborne Ka-band synthetic aperture radar ADTS developed by Lincoln Laboratory of Massachusetts Institute of Technology in the USA, the P-3 multi-band (X, C, L) polarized SAR imaging system jointly developed by the Environmental Research Institute of Michigan (ERIM) and the Naval Air Warfare Center (NAVC), the AER airborne X-band polarized SAR developed by Germany Research Society of Applied Natural Science/Research Institute for Radio and Mathematics (FGAN/FFM), the DLR multi-band (S, L, P) airborne SAR developed by Germany Aerospace Center, the RAMSES multi-band (P, L, S, C, X, Ku) polarized SAR developed by France National Office for Aerospace Studies and Research (ONERA), the EMISAR dual-band (L, C) synthetic aperture radar developed by Denmark, the PALSAR system developed by Japan, and the satellite-borne TecSAR system developed by Israel. Since the mid-1980s, China has also carried out the research and development of the radar of alternate polarization measurement system. For example, Lanzhou Plateau Atmosphere Physics Research Institute under Chinese Academy of Science has developed China's first 5-cm-wavelength dual-linear-polarization radar, and the No.38 Research Institute under China Electronics Technology Group Corporation has developed the airborne dual-polarization SAR system. See Fig. 1.5a for the configuration of the Convair-580 synthetic aperture radar.

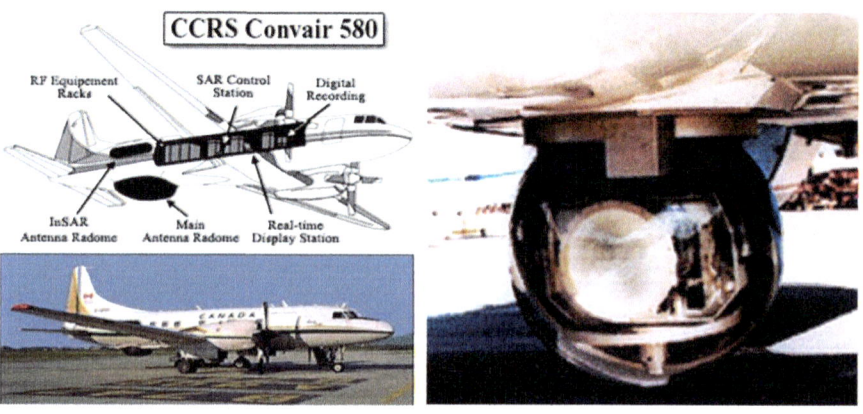

(a) Convair-580 synthetic aperture radar   (b) ADTS airborne synthetic aperture
    radar

Fig. 1.5 Typical radar of alternate full-polarization measurement system

Table 1.5 Main parameters of the ADTS airborne synthetic aperture radar

Carrier frequency	33.56 GHz
Pulse waveform	LFM
Pulse width	32.5 μs
Pulse bandwidth	650 MHz
Scan area width	375 m
Distance to scan center	7.26 km
Pulse repetition frequency	3 kHz
Polarization style	H/V alternate transmission, H/V simultaneous reception

The ADTS airborne synthetic aperture radar is shown in Fig. 1.5b, with its main parameters given in Table 1.5.

Although the radar of alternate polarization measurement system is capable of obtaining the four elements of target polarization scattering matrix, it inherently has the following defects:

(1) For the relatively rapidly moving non-stationary targets, the radar of alternate full-polarization measurement system can cause the decorrelation effect between the two columns of elements of the target polarization scattering matrix;

(2) The Doppler effect of the target can cause phase difference between the measured values of the two columns of elements of target polarization scattering matrix. This will affect the measurement accuracy.

(3) Range ambiguity can affect the normal reception of pulse echoes, thus affecting polarization measurement accuracy.

(4) Radar of alternate full-polarization measurement system requires polarization switchover, while the inherent cross-polarization coupling interference of the polarization switchover component can adversely affect the polarization measurement accuracy.

It is because of the above mentioned defects that the alternate polarization measurement radars encounter difficulty in accurately measuring polarization scattering characteristic of moving targets especially rapidly-moving targets. Therefore, they are often used in such fields as meteorological observation and SAR earth-reconnaissance, to improve the performance of target detection and classification.

1.1.3.4   Full-Polarization Doppler System of Simultaneous Transmission and Reception with Full-Polarization

At the end of 1980s, considering the defects of alternate polarization measurement system, the idea of simultaneous polarization measurement was first put forward in the field of meteorological radar. The scan velocity of polarization metrological radars was hence greatly improved [35]. Based on this idea, Giuli etc. supposed a new type of polarization measurement system that transmitted primary pulses to measure target scattering matrix. This type of polarization measurement system is so-called type-4 polarization measurement system, or full-polarization Doppler system of simultaneous transmission and reception with full-polarization, or "instantaneous/simultaneous polarization measurement system" for short. The core idea of this system is as follows: The radar transmitted signal is obtained through coherent superposition of two modulated signals with a certain bandwidth. The transmission waveforms of the two orthogonal polarization channels are orthogonal as much as possible. That is, the correlation function is as small as possible. For each echo signal, correlated reception of the two orthogonal waveforms is carried out simultaneously. Using the orthogonally of signal modulation, echoes corresponding to different transmission polarization styles are separated from one another, so that the estimates of the four elements of target polarization scattering matrix can be obtained using a pulse interval. The radar having the capability of polarization measurement is also known as "full-polarization radar" in the industry Fully Polarimetric. Throughout the 1990s to the present, instantaneous polarization measurement has already become the main development trend of full-polarization measurement system. The instantaneous polarization measurement waveforms such as reverse-slope linear FM waveform pair, digital phase coded waveform, and frequency-shift vector pulse are signal wave-forms generally adopted by instantaneous polarization measurement radars nowadays. For the frequency-shift pulse waveform, some scholars thought it was impossible to restore the phase difference between the two columns of elements of scattering matrix, and it was hence impossible to obtain accurate scattering matrix. However, this problem was effectively resolved by X. S. Wang, by using the reciprocity of target scattering matrix. This further improved the performance of instantaneous polarization measurement of full-polarization radar.

Table 1.6 shows the sequence logic for the instantaneous full-polarization measurement radar to realize polarization measurement. This radar type has two independent modulation orthogonal-polarization transmitting channels, and two orthogonal-polarization receiving channels. Orthogonal modulation wave-forms are transmitted simultaneously through the two polarization channels. Correlated reception of the orthogonal wave-forms from the two channels is carried out simultaneously too. Matching filter is performed separately. The complete information of target full-polarization scattering characteristic is obtained via the four matching filter outputs.

Table 1.6 Measurement sequence relation schema of instantaneous full-polarization measurement radar

Sequence	T_P	$2T_P$	$\cdots$	$(2K)T_P$
Transmission	$H + V$	$H + V$	$\cdots$	$H + V$
Reception	$H + V$	$H + V$	$\cdots$	$H + V$
Measured value	$[S]$	$[S]$	$[S]$	$[S]$

In the full-polarization Doppler system of simultaneous transmission and reception with full-polarization, the wave-forms of transmission signals from the H- and V-polarization channels are $g_H(t)$ and $g_V(t)$, respectively. Correlation between them approximates zero:

$$\int_0^T g_H(t)g_V^*(t)dt \approx 0 \tag{1.2}$$

The full-polarization Doppler system is capable of obtaining the target coherent Sinclair scattering matrix at any moment, i.e. $S(k) = \begin{bmatrix} s_{HH}(kT_p) & s_{HV}(kT_p) \\ s_{VH}(kT_p) & s_{VV}(kT_p) \end{bmatrix}$, whereby the estimates of elements of polarization covariance matrix can be obtained.

This new radar type obviates the expensive polarization switch-over component, reduces decorrelation between measurement pulses, and eliminates influence of Doppler frequency on phase measurement. However, addition of multiple RF channels increases both the quantity of equipment and the complexity of the system. Realization cost is hence relatively high. Furthermore, in respect of transmission waveform design and signal processing, considerable difference exists between radar of simultaneous full-polarization measurement system and traditional radar. Consequently, the transmission waveform design and signal processing is always a hot topic for the research of radars using this system.

As the newest polarization measurement system nowadays, the instantaneous full-polarization measurement system is capable of realizing the coherent measurement of the four elements of target polarization scattering matrix. Having also the capability of acquiring full-polarization scattering characteristic of dynamic targets, thus providing dynamic and coherent full-polarization information sources for radar polarization information processing, the system promotes the development of the time-variant and dynamic polarization information processing technology. Combination of the simultaneous full-polarization measurement system and the advanced polarization information processing technology can acquire, process, and utilize time-variant and dynamic full-polarization information, greatly improving the capabilities of the radar in target detection, anti-interference, target classification and recognition, and target parametric inversion. The simultaneous full-polarization measurement system is enormously valuable in these fields: target classification of meteorological radar, high-velocity moving target detection and tracking of air defense anti-missile radar, and earth reconnaissance of polarized SAR. From 2005 to present, radars of simultaneous full-polarization measurement system have been successfully developed in foreign countries. Examples are the C-band ARMOR meteorological radar developed by the USA University of Alabama Huntsville/ National Space Science and Technology Center (UAH-NSSTC), the S-band S-Pol metrological radar developed by the USA National Center of Atmospheric Research (NCAR), the CSU-CHILL meteorological radar developed by the USA Colorado

State University, the X-band PARSAX meteorological radar developed by Holland Delft University of Technology, and the X-band air target instantaneous polarization measurement radar MERIC developed by France ONERA. The MERIC air defense radar and PARSAX meteorological radar are shown in Fig. 1.6a, b, respectively. Their parameters are given in Tables 1.7 and 1.8, respectively.

(a) MERIC air defense radar   (b) PARSAX meteorological radar

Fig. 1.6 Typical radar of alternate full-polarization measurement system

Table 1.7 Main parameters of PARSAX meteorological radar

Carrier frequency	9.6–10 GHz
Pulse waveform	FMCW
FM scan duration	≥ 0.655 ms
Modulation frequency scan range	2–50 MHz
Receiving antenna	Diameter 2.12 m
	Beam width 4.6°
	Gain 32.75 dB
Transmitting antenna	Diameter 4.28 m
	Beam width 1.8°
	Gain 40.0 dB
Polarization style	H/V simultaneous transmission, H/V simultaneous reception

Table 1.8 Main parameters of MERIC radar

X-band	Modulation bandwidth: 300 MHz Resolution: 0.5 m Pulse width(PW): 30 μs
Transmitter	Solid-state power amplifier
Transmission power	10 W continuous
Antenna	Parabolic reflector antenna Polarization coupling > 30 dB
Waveform	Orthogonal code, linear FM

1.1.3.5 Compact Polarization Measurement System

The fifth type of polarization measurement system is a new system proposed in recent years. Today, it is still at the stage of conceptual demonstration. It is called the Compact-polarity or Hybrid-polarity system, i.e. the compact-polarization measurement system. The system may be used in space SAR detection. Two operating modes of the system are $\pi/4$ and hybrid-polarity. In the $\pi/4$ mode, the radar transmits signals with 45° linear polarization, and simultaneously receives signals with H/V polarization. In the hybrid-polarity mode, the radar transmits signals with circular polarization, and simultaneously receives signals with H/V-polarization. The sequence logic of the compact full-polarization measurement is shown in Table 1.9, taking the DTLR mode as an example.

Note that when transmission polarization is circular polarization, the four quantities of the Stokes parameter of scattering matrix are invariable quantities with regard to target orientation. Advantages of this system are relative simplicity of manufacturing process, self-calibration of characteristic, and little possibility of being affected by noise or cross-polarization channel. The polarization decomposition method is the $m - \delta$ method. Actual experimental results demonstrate that the CL-Pol method is the best polarization configuration for outer space and moon detection.

In this mode, the measurement result is combination of elements s_{HH} and s_{HV} for the H-polarization channel, or combination of elements s_{VH} and s_{VV} for the V-polarization channel. With the assumptive condition $\langle s_{HH}s_{HV}^* \rangle = \langle s_{VV}s_{HV}^* \rangle = 0$ satisfied, the full-polarization measurement result may be restored using the compact measurement result.

Table 1.9 Measurement sequence relation schema of compact full-polarization measurement radar

Sequence	T_P	$2T_P$	$\cdots$	KT_P
Transmission	R	R	$\cdots$	R
Reception	$H + V$	$H + V$	$\cdots$	$H + V$
Measured value	$\frac{1}{\sqrt{2}}(S_{HH} - jS_{HV})$	$\frac{1}{\sqrt{2}}(S_{HH} - jS_{HV})$	$\frac{1}{\sqrt{2}}(S_{HH} - jS_{HV})$	$\frac{1}{\sqrt{2}}(S_{HH} - jS_{HV})$
	$\frac{1}{\sqrt{2}}(S_{VH} - jS_{VV})$	$\frac{1}{\sqrt{2}}(S_{VH} - jS_{VV})$	$\frac{1}{\sqrt{2}}(S_{VH} - jS_{VV})$	$\frac{1}{\sqrt{2}}(S_{VH} - jS_{VV})$

1.1.3.6 Polarization Measurement System Based on Non-ideal Orthogonal Basis and Sequence Analysis

For an excellently-designed antenna, the azimuth polarity purity is often the highest at the center and descends in the direction off the center. That is to say, the antenna polarization is a "spatial slowly-variable" quantity. This characteristic can be called "spatial polarization characteristic" of antenna. By using the inherent attribute of "non-ideal orthogonal basis" constructed with spatial distribution "impurity" of antenna polarization, and adopting special means for processing echo series received in the antenna scan process, Luo Jia etc. designed a processing algorithm that is capable of obtaining estimates of target polarization scattering matrix [88–90]. This kind of measurement method is temporarily called "polarization measurement system based on non-ideal orthogonal basis and sequence".

In terms of performance and efficiency this type of polarization measurement system is not as good as the above mentioned five system types. Moreover, this system type encounters restriction in application. For example, while ensuring measurement accuracy, the antenna scan rate, spatial sampling interval, and polarization measurement performance become mutually restrictive factors; this to some extent reduces the applicability of polarization measurement to the moving or extended targets. However, this system type has innovative contribution to design idea. Probably in future it will be used in equipment reconstruction.

Conclusions drawn from the comprehensive comparison among the mentioned polarization measurement systems are as follows. Radars of types 1 and 2 polarization measurement systems have the minimum system complexity, the lowest costs, and not-very-strict requirements on antenna polarization isolation and signal processing capability. But they are weakest in polarization measurement capability. They are incapable of obtaining complete target polarization scattering matrix. The radars of alternate polarization measurement system (type 3 radars) have not-very-strict requirements on antenna polarization isolation. Compared with types 1 and 2 radars, type-3 radars have stricter requirements on signal processing capability, as well as higher system complexity, costs, and polarization measurement capability. They are capable of obtaining complete polarization scattering matrix. However, for high-velocity moving targets or undulating targets, it is difficult for type-3 radars to ensure ideal measurement accuracy. Therefore, strictly speaking they cannot be called full-polarization measurement radars. Radars of type-4 polarization measurement system have relatively high requirements on antenna polarization isolation and signal processing capability. Compared with radars of alternate polarization measurement system, type-4 radars have higher system complexity, costs, and polarization measurement capability. They are capable of realizing full-polarization scattering characteristic measurement of dynamic, static, and distributed targets. In near and relatively distant future, they will be the main stream of the development of polarization measurement system. Radars of type-5 polarization measurement system are simpler in respect of manufacturing process, compared with types 1 to 4. For polarization measurement of some targets of special shapes, type-5 radars have a big advantage over the others. They could

Table 1.10 Comparison between polarization measurement systems

Technical indicator measurement system	Required antenna polarization isolation	Signal processing capability	System complexity	Cost	Polarization measurement capability
Type 1	Low	Low	Low	Low	Low
Type 2	Low	Low	Medium	Medium	Low
Type 3	Low	Low	Medium	Medium	Medium
Type 4	High	High	High	High	High
Type 5	Low	High	Low	Low	High
Type 6	Low	High	Low	Low	Low

play an important role in the field of polarization measurement in future. Radars of type-6 polarization measurement system impose relatively strict requirements on antenna cross-polarization characteristic, target movement condition, and signal processing means. Currently they are still at the stage of theoretical research. Probably in near future new ideas will be created and breakthrough can be made. Table 1.10 presents comparison among the 6 system types in five respects.

Development of polarization measurement system boosts the advancement of polarization information processing technology. Research, development, and application of the radars of wide band and simultaneous full-polarization measurement system facilitate proposition and development of the instantaneous polarization theory. Taking into consideration the time-variant and dynamic characteristic of electromagnetic wave polarization, the instantaneous polarization theory as proposed for wide band high-resolution polarized radars not only fully covers the classical polarization theory but also removes the restrictions of classical polarization such as "of time harmony" and "of narrow band". For the polarization characterization of time-variant electromagnetic waves as well as the description of full-polarization scattering characteristic of dynamic targets, the instantaneous polarization theory provides a kind of theoretical tool. By using such means as time-frequency transformation and statistical analysis, the polarization information processing technology based on the theory improves the capability of processing time-variant and dynamic polarization information. Related theoretic and algorithm research in polarization filter of wide band electromagnetic signals, radar target intensification, radar target detection, and radar target recognition will be gradually carried out. The results will be demonstrated and applied in actual systems.

1.2   Present Situation of Radar Polarization Information Processing and Application

In principle, polarization information may be combined with various radar systems to improve capabilities of radar in detection, tracking, anti-jamming, and target recognition. Applicable fields of polarization information processing technology include polarization filter, polarization recognition, polarization detection, polarization tracking, and target polarization recognition.

1.2.1   Polarization Target Classify and Identification

Processing the polarization information acquired by airborne and satellite-borne full-polarization remotely-sensing radars is able to classify and identify natural targets, for example, aperture features, land-forms, forests, and vegetation. According to particular imaging mechanism and imaging environment complexity, as much target information as possible is extracted, among which useful information is further extracted or obtained via inversion. Accurate information of natural environment so obtained is useful in the utilization, estimation, exploration, map-making, and detection of natural environment.

Processing the polarization information acquired by full-polarization Doppler meteorological radar can obtain polarization parameters of metrological targets (cloud, rain, fog, hail, etc.), for example, reflectivity factor (Z_H, Z_V) of orthogonal-polarization wave, reflectivity factor (A_{HV}, K_{HV}) of cross-polarization wave, differential reflectivity factor Z_{dr}, specific differential phase shift of propagation K_{dp}, zero lag correlation coefficient $|\rho_{hv}(0)|$, and linear depolarization ratio LDR_{vh}. Physical attributes of various metrological targets, such as size, density, shape, intensity, and thickness can be obtained through parametric inversion [67–75], providing highly accurate criteria information for weather forecast, and disaster pre-warning.

Battlefield surveillance full-polarization radars are capable of classifying, identifying, and marking various artificial military targets of interest in the battlefield environment. They can provide convenience for the selection of favorable military land-forms and paths; they can provide assurance for the smooth movement, strategic deployment and rapid attack of forces; they can provide basis for further military attack.

Polarization characteristics currently in target classification and recognition are roughly classified into two types: based on simple combination and transformation of measurement data, and based on target decomposition. The first type is mainly used for the classification of polarized SAR images, and the second for classification of scattering characteristics.

It shall be particularly pointed out that the target polarization decomposition theory has been widely applied in the PolSAR target classification and recognition.

Basic idea of target polarization decomposition is to decompose the target polarization scattering into several combinations of basic scattering mechanisms. These combinations may be used to characterize target scattering or geometric information. Classification is then performed according to the similarity between classification unit and basic scattering mechanism, or by directly using extracted new characteristics. Merit is that classification results are able to relatively accurately reveal the scattering mechanism of aperture features. This helps us understand images. Moreover, no training data are needed in classification, so the theory can be widely applied. Methods for polarization target decomposition may be divided roughly into two types: intended for target scattering matrix, and intended for polarization covariance matrix, polarization coherent matrix, Muller matrix and Kennaugh matrix. If the first type is used, target scattering characteristic shall be determinate (or stable), and scattering echo shall be coherent. Hence this type is also called coherent target decomposition (CTD). Methods for CTD include decomposition based on Pauli basis, Cameron decomposition, and SDH decomposition. If the second type is used, target scattering may be time-variant, and echo shall be partially coherent. Therefore, this type is also known as partially coherent target decomposition (PCTD). Methods for PCTD include Huynen decomposition, odd order-even order-diffuse reflection put forward by Freeman etc., $H/\alpha/A$ decomposition put forward by Cloude etc., Holm & Barnes decomposition, and least square decomposition based on Kennauth matrix.

Among these methods, the $H/\alpha/A$ decomposition based on characteristic value/characteristic vector analysis put forward by Cloude and Pottier has been most frequently used in the classification of images of polarized SAR. A series of physically clear characteristics can be obtained by using the $H/\alpha/A$ decomposition of coherent matrix. Also based on characteristic decomposition, Qong [80] extracted two new characteristics: rotation angle φ and offset angle e. Xu Junyi etc. [81–82] extracted new characteristics describing target mean scattering and scattering randomness, and introduced discrimination information as measure of difference between targets. Jin Yaqiu and Chen Fei [83, 84] derived direct relation between the measured values of co-polarization and cross-polarization indexes, and the characteristic value and information entropy of coherent matrix. In addition to the method put forward by Cloude and Pottier, the method of decomposition based on model put forward by Freeman and Durden has also brilliantly succeeded in the classification of images of polarized SAR. A group of new characteristics extracted in the theory of deorientation based on polarization scattering target supposed by Jin Yaqiu and Xu Fengji have been ideally used in actual classification of aperture features.

1.2.2 Polarization Detection

The same target at different attitudes exhibits different degrees of sensitivity to different polarized wave. The polarization difference can be up to 10 dB. Therefore, the use of polarization diversity technology can significantly improve detection performance of radar system. From the end of 1980s to the early 1990s, the polarization detection technology received extensive attention, and relevant research was carried out by a good number of scholars. Remarkable achievements were from L. M. Novak etc., who worked with the Lincoln Laboratory in Massachusetts Institute of Technology (MIT). Concerning concept of optimal polarization detector, they performed relatively systematic research. They also performed research in quasi-optimal polarization detectors suitable for engineering use, for example, identicalness likelihood ratio detector, polarization whitening filter, polarization constant false alarm detector, polarization formation detector, and power maximization synthesis detector.

In the coherent radar system, the target is required to be detected from the clutter. Echoes to medium- and low-resolution radars contain superposition of scattered waves from different scattering objects. Both target echo vector and clutter vector can be considered to be approximately compliant with zero-mean complex Gaussian distribution. For coherent radar detection in Gaussian clutter in unknown clutter covariance matrix, E. Pottier performed research in optimal polarization detection in the slowly undulating clutter environment [121]. He proposed the Stokes vector estimation polarization detector. Kelly proposed the general likelihood ratio testing (GLRT) algorithm. Using a series of auxiliary data that have the same distribution clutter as the detection unit and do not contain signals, the detection algorithm tests the likelihood ratio of the detection unit, which has the nature of constant false alarm to the clutter covariance matrix.

For target adaptive polarization detection in non-Gaussian noise and clutter environment, the corresponding constant false alarm detector has been designed [85, 86], and Park etc. proposed the polarization GLRT algorithm that combines two polarization channels for unknown target and clutter polarization characteristics [87, 88]. Pastina etc. extended the results given in Refs. [89, 90] to the situations having three or more polarization channels. Research results demonstrated that using echo signals from multiple polarization channels can further enhance the target detection capability.

In the high-resolution radar system, due to increase in transmission signal bandwidth, the range resolution unit will diminish, and the echo of the complex target will continuously occupy multiple resolution units, and the strong scattering center of target can be segregated. High-resolution target detection can be equivalent to scattering center detection. There have been lots of achievements in the research in high-resolution radar target detection. In the Ref. [91], for the radar detection of targets distributed in Gaussian noise of known power spectrum, two detection ideas were put forward, and comparison and analysis were also performed. In the Ref. [92], assuming both the echo amplitude and clutter covariance were unknown, research in using the array antenna to detect radially distributed targets in multiple range units was performed, and a so-called general likelihood ration detection method was proposed. In the Ref. [93], research in the method for multiple-pulse combination detection in time domain was performed, assuming that the time correlation of clutter was produced by white noise via an auto regression model. In the Ref. [94], research in adaptive polarization detection of distributed radar targets was performed, considering polarization information of echo signal. In the Ref. [25], for the high-resolution radar system, the concept of transverse polarization filter was put forward, and used for research in radar signal polarization detection. Research results were excellent. In the Ref. [26], the method for high-resolution polarization target detection based on strong scattering point radial accumulation was put forward taking into consideration the energy accumulation of multiple polarization channels. In the Ref. [95], a new method for target detection in the background of strong aperture feature clutter, based on one-dimensional range image in a wide band millimeter wave radar system, i.e. adaptive range unit accumulation detection, was put forward. In the Ref. [96], a polarization detection method was put forward, first the polarization whitener was used to mix echoes from the three polarization channels, then detection of scattering centers was carried out, finally the number of detected scattering centers was used to judge whether existed any radar targets. In the Ref. [97, 98], considering the amplitude of Stokes vector in the resolution unit interval occupied by the target, and the amplitude of Stokes vector in the resolution unit relative to clutter, it was seen that the two polarization states differed remarkably from each other. Based on this fact, a method, so-called double-threshold non-parametric detection, was put forward.

1.2.3 Polarization Anti-jamming

Development of the polarization anti-jamming technology has been undertaken for many years. Clear development ideas have come into being. In terms of anti-jamming object, the technology may be classified into four types: polarization anti-clutter jamming, polarization anti-active suppression jamming, polarization anti-active deception jamming, and polarization anti-passive jamming. Most jamming currently encountered may be suppressed using proper polarization anti-jamming technology, provided that the jamming types are known. In terms of technological realization approach, anti-jamming technology may be classified into three types: polarization filter technology, polarization intensification technology, and target polarization discrimination technology.

Polarization filter and target polarization intensification are intended to increase Signal to Interference plus Noise Ratio (SINR). The former places emphasis on suppressing jamming or clutter via polarization selection, while the latter emphasizes maximizing target reception power via polarization selection. In essence, the latter belongs also to polarization filter. The SINR polarization filter criteria as optimization with two degrees of freedom take account of the two kinds of technology at the same time.

In 1975, in the research of rain clutter cancellation, Nathanson supposed the block diagram for realization of adaptive polarization canceller (APC). Essentially, the realization of the APC was using the correlation between signals of orthogonal polarization channels to calculate weighing coefficients of the two channels. This filter system was notable for simple construction. It was capable of automatically compensating for amplitude-phase unbalance between channels. For polarized fixed or slowly-variable clutter and jamming, it exhibited excellent suppression performance. In 1984, A. J. Poelman set forth the multi-notch polarization filter used for suppressing partially-polarized clutter or jamming [15–17]. By combining APC and multi-notch H-polarization (MLP), Italian scholars Giuli and Gherardelli proposed MLP-APC and MLP-SAPC, respectively, in 1985 and 1990, respectively. They were able to improve adaptive capability of MLP. Because MLP is a kind of non-linear processing, polarization filters of this type can destroy signal coherence hence their use is seriously restricted in reality. Chinese scholars also made contribution in this regard. Doctor Zhang Guoyi in Harbin Institute of Technology, using the characteristic that in frequency domain and polarization domain difference existed among multiple sky-wave radio jamming, put forward several methods of interference suppression in combination of frequency domain and polarization domain. The experimental results were favorable [18].

In 1985, regarding polarization intensification of non-time-variant target echo in no-noise-clutter environment, Kostinski and Boemer put forward the famous three-step solution for optimum polarization. In 1987, they proposed the extended three-step solution in case of clutter. In 1995, D. P. Stapor investigated the optimization according to the criterion of maximum Signal to Interference plus Noise Ratio (SINR) in the case of single signal source, jamming source, and full polarization [103]. In recent years, Wang Xuesong and Xu Zhenhai etc. who worked for China National University of Defense Technology issued also several related papers. Using the idea of polarization track restriction, they changed the issue of optimization at two degrees of freedom into the issue of search at one degree of freedom. They investigated the optimization of target functions such as SINR and PDSI (Power-Difference-between-Signal-and-Interference), and carried out the performance analysis of the corresponding filter.

To summarize, at present research in polarization filter and polarization intensification is aimed, in most cases, at optimization of reception polarization, and seldom at optimization of transmission polarization. If the target, jamming and clutter have similar polarization states, the above mentioned algorithms will become invalid. Therefore, target echo polarization must be changed by changing transmission polarization so that sufficient difference exists between them to smoothly realize polarization filter. Santalla V. and Yang J., etc. issued quite a few papers regarding polarization comparison and intensification [104–106]. Both reception polarization and transmission polarization are objects of optimization, but these algorithms are on the condition that target scattering matrix must be known. However, it is generally difficult to know target scattering matrix as it is rather sensitive to frequency and attitude. Hence, actual application of the algorithms is confronted with many problems.

In recent years, the work of identifying active false target jamming, passive chaff jamming, and passive corner reflector jamming by extracting and using polarization domain characteristic gradually attracts scholars' attention. In the Refs. [107, 108], difference in respect of scattering characteristic between several simple-shape targets and fixed-polarization false targets was analyzed, and characteristics of scattering matrix were extracted for discrimination purpose. In the Ref. [109], the rules for inter-pulse variation of instantaneous polarization projection vectors of jamming signals and target echoes were investigated, and whereupon the idea was proposed of using the undulation characteristic of instantaneous polarization projection vector for discrimination. In the Ref. [110], mainly difference in respect of matrix singularity between true and false targets was used to discriminate fixed polarization from inter-pulse variable-polarization false target. In the Ref. [111], the difference in respect of polarization characteristic between target echo and false target echo was considered. Target echo polarization is linearly related to radar transmission polarization; while fixed polarization of false target is irrelevant to radar transmission polarization (i.e. false target polarization ratio is fixed). Received by polarization vector tensor product, characteristic vectors of target echoes converge at one point; while characteristic vectors of polarized modulated false target echoes disperse in the complex space. By designing new polarization processing method and transmission polarization combination, it is possible to very effectively discriminate target from false target (including fixed-polarization and polarized modulated). The polarization scattering mechanism of chaff passive jamming differs greatly from that of reflector passive jamming; but they are simpler than that of true target [112–115]. By using the difference in respect of statistical characteristic between chaff and target, discrimination and comparison were performed to obtain their polarization ratio statistical distribution and depolarization coefficient distribution. By using the relation between polarization ratio and polarization angle, variable substitution was performed to obtain their polarization angle statistical distribution. The results demonstrated that dual-polarization statistical characteristic may be used as reliable characteristic for target identification. Using the polarization process technique to extend target echo characteristic to polarization domain provides an effective technical approach for discrimination of passive jamming. It is expected to greatly improve the anti-jamming capability of radar.

Based on antenna spatial polarization characteristic, Luo Jia proposed a polarization estimator [116, 117]. Considering the case where received incoming waves were noise blanket jamming, he designed an adaptive filter and theoretically analyzed and practically simulated and demonstrated its jamming suppression effect. As the open-loop model of jamming suppression polarization filter, the polarization estimator was the front end of polarization filter. In the paper, Luo Jia took into consideration only the polarization estimation aspect, other than the effect of polarization estimation error, or the compatibility between polarization information processing and coherent processing. And the polarization filter information processing still employed orthogonal dual channel.

In summary, there are a great number of achievements in the research of radar polarization anti-jamming. However, nowadays only several technologically mature jamming suppression methods can be used in actual engineering applications. Research in polarization anti-jamming shall be closely connected with engineering actuality. Polarization information shall be integrated with the traditional processing frame in time domain, frequency domain, and spatial domain. All useful information on target and jamming shall be used to solve the problems that could not be solved with traditional anti-jamming methods.

Chapter 2
Connotation and Representation of Spatial Polarization Characteristic

The concept of polarization is first defined in the field of optics, used to describe the phenomenon of directional oscillation of light. Therefore, it is also called directional oscillation. In fact, as a common nature of vector waves, polarization receives wide attention in the infrared, optical, radar and other fields. As far as radar is concerned, polarization describes the variation of the electric field vector of electromagnetic wave with time on the propagation cross section [1]. It reflects the vector characteristic of electromagnetic waves. Besides information of time domain, frequency domain and spatial, polarization is also available important information of electromagnetic waves.

In the classical radar polarization theory [3], the research on electromagnetic wave polarization basically focuses on the monochromatic or quasi-monochromatic waves based on the "narrow-band" or "time-harmonic" assumption. Based on the fact that the trajectory of the electric field vector tip changes with time in the cross section of electromagnetic wave propagation is an ellipse, the scholars have proposed static parameters such as Jones vector, elliptic geometry descriptor, polarization phase descriptor, polarization ratio and the Stokes vector as the methods for depicting electromagnetic wave polarization. It is hard for these classical depiction methods to reveal dynamic polarization information of the time-variant electromagnetic waves. The theory of instantaneous polarization proposed in the Ref. [3] regards, in essence, the polarization of electromagnetic waves as a dynamic parameter rather than a non-static parameter. The theory provides a powerful tool for the characterization of dynamic polarization characteristics of electromagnetic waves and target electromagnetic wave scatters. In the theoretical system of instantaneous polarization [3], based on the concept of instantaneous polarization, the set of descriptors was established applicable for describing the instantaneous polarization of time-variant electromagnetic waves; the new concept of distribution of electromagnetic wave instantaneous polarization in the joint time-frequency domain was formulated, to reveal the intrinsic relationship between electromagnetic wave time-domain and frequency-domain instantaneous polarization descriptors. Unitary explanation of the two classes of instantaneous polarization descriptors was

© National Defense Industry Press, Beijing and Springer Nature Singapore Pte Ltd. 2019
H. Dai et al., *Spatial Polarization Characteristics of Radar Antenna*,
https://doi.org/10.1007/978-981-10-8794-3_2

made from the point of view of polarization distribution on the boundaries of time and frequency domains. The concept of instantaneous polarization was thus extended to the joint time-frequency domain. In the Ref. [31] the concept of distribution of electromagnetic wave instantaneous polarization in the joint time-frequency domain was extended and perfected, devising the basic method for analyzing electromagnetic wave instantaneous polarization in the joint time-frequency domain. The method was used in target identification of polarization radar. In the Ref. [6], the time-variant polarization characteristics of electromagnetic waves and radar target electromagnetic wave scatters was re-examined and studied statistically and dynamically, with a basic framework for instantaneous polarization statistics theory established. However, most of current researches focus on the polarization characteristics of electromagnetic waves and antenna in time domain and frequency domain and their application. No research is reported on the law of electromagnetic wave and antenna polarization variation in space domain, especially, on spatial instantaneous polarization characteristics.

This chapter first briefly reviews the classical description of electromagnetic wave polarization and instantaneous polarization characterization method, then discusses in detail the connotation and characterization of antenna spatial polarization characteristic, including the concept and classical description of antenna spatial polarization characteristic, and the characterization of antenna spatial instantaneous polarization characteristic. This chapter provides a theoretical basis and a powerful tool for the subsequent analysis and application of antenna polarization characteristic.

2.1 Polarization Characterization of Electromagnetic Waves

As the theoretical basis of this chapter and the necessary mathematical preparation, this section gives the classical description of electromagnetic wave polarization, and the concept and method of instantaneous polarization characterization.

2.1.1 Classic Description of the Electromagnetic Waves Polarization

The monochromatic wave is a completely polarized wave, or purely polarized wave. At any point in space, its electric field vector tip depicts a polarization ellipse of constant ellipticity and tilt angle. This polarization ellipse does not vary with time, that is to say, it is time invariant. Thus the polarization nature of monochromatic waves can be completely described by polarization ellipse or all equivalent parameters derived.

Assuming that there is a monochromatic TEM plane wave propagates along $+z$ direction in the Cartesian coordinate system, with the horizontal and vertical polarization basis $(\hat{h}, \hat{v})$, the vector field of wave can be expressed by:

$$e_{HV} = \begin{bmatrix} E_H(z,t) \\ E_V(z,t) \end{bmatrix} = \begin{bmatrix} a_H e^{j(\omega t - kz + \varphi_H)} \\ a_V e^{j(\omega t - kz + \varphi_V)} \end{bmatrix}, \quad t \in \mathbf{T} \qquad (2.1.1)$$

where, $k = 2\pi/\lambda$ is the wave number, $\omega = 2\pi c/\lambda$ is angular frequency, λ is wave length, φ_H and φ_V are phases of horizontal and vertical polarization components of electromagnetic wave, respectively, a_H and a_V are amplitudes of horizontal and vertical polarization components of electromagnetic wave, respectively, and $\mathbf{T}$ is the time domain support set of electromagnetic wave.

Therefore, the polarization information of electromagnetic wave depends mainly on the amplitude ratio and phase difference of signals in the two orthogonal directions. The concepts and characterization methods of main classical characterization parameters are briefly described below, providing a basis for the subsequent research on antenna spatial polarization characteristic.

(1) Jones vector

For a monochromatic wave, the Jones vector is expressed as

$$e_{HV} = \begin{bmatrix} E_H \\ E_V \end{bmatrix} = \begin{bmatrix} a_H e^{j\varphi H} \\ a_V e^{j\varphi V} \end{bmatrix} = \begin{bmatrix} x_H + j y_H \\ x_V + j y_V \end{bmatrix} \qquad (2.1.2)$$

Therefore, the Jones electric field vector characterization contains not only the electromagnetic wave polarization information, but also electromagnetic wave intensity and phase information. The value space of the electric vector is a 2 dimensional complex space.

(2) Polarization ratio

Polarization ratio is the most commonly used classical polarization descriptor. On the horizontal and vertical polarization basis $(\hat{h}, \hat{v})$, polarization ratio of electromagnetic wave is expressed as

$$\rho_{HV} = \frac{E_V}{E_H} = tg\gamma e^{j\phi}, \ (\gamma, \phi) \in [0, \pi] \times [0, 2\pi] \qquad (2.1.3)$$

In which, $\gamma = \tan^{-1}\frac{a_V}{a_H}$, $\phi = \varphi_V - \varphi_H$. Polarization ratio contains only the electromagnetic wave polarization information. Its value space is a complex plane containing infinite point (∞).

(3) Polarization phase descriptor

The polarization information is embodied by the amplitude ratio and phase difference between the electric fields in two orthogonal directions. According to the formula (2.2.3), we know that

$$(\gamma, \phi) \in [0, \pi/2] \times [0, 2\pi] \tag{2.1.4}$$

From the above we can see that, the polarization phase descriptor is fully equivalent to polarization ratio. Its value space is a rectangular subset of 2-dimensional real plane.

The polarization characteristic characterization methods are considered above for monochromatic waves. In addition to the Jones vector, polarization ratio and polarization phase descriptor mentioned above, the polarization ellipse descriptor and the Stokes vector are also commonly-used methods for polarization characteristic characterization. But in the electromagnetic engineering field, the signals generated by a radiation source are absolutely not monochromatic waves, but signals with a certain bandwidth, in most cases the narrow-band signals. The signals have a limited bandwidth far less than the center frequency. The electric field vector movement becomes relatively complex at the moment. The trajectory depicted by the tip of the electric field vector at a given point in space is no longer a constant ellipse, but an ellipse-like curve whose shape and direction change slowly with time. For the quasi-monochromatic waves, the classical polarization science proposes the concept of "partial polarization" to describe their polarization characteristic. Essentially, the quasi-monochromatic waves are deemed a stationary stable random process. Replacing time average with set average for the quasi-monochromatic waves obtains a group of partial-polarization descriptors of statistical significance. The polarization ellipse geometric descriptor is defined based on the elliptical shape, for the trajectory of the electric field vector tip changing with time, in the cross section of electromagnetic wave propagation. It is difficult to extend the characterization when electromagnetic wave polarization changes with time, i.e. the trajectory is no longer an ellipse. In contrast, other polarization characterization methods such as the Jones vector, polarization ratio, polarization phase descriptor, and the Stokes vector can all in principle be extended to describe the polarization information of time-variant electromagnetic waves [1].

2.1.2 Instantaneous Polarization Characterization of Electromagnetic Waves

It can be seen from the analysis in Sect. 2.1.2 that, for the electromagnetic wave with time-variant polarization, the classical polarization characterization method is no longer applicable. The concept of instantaneous polarization comes into being accordingly [3]. The theory of instantaneous polarization has been continuously perfected and developed [6, 38, 71]. In this section, the core contents of instantaneous polarization characterization of electromagnetic waves are briefly reviewed, providing a theoretical basis for subsequent research on antenna spatial polarization characteristic.

The electromagnetic waves of space propagation in the horizontal and vertical polarization basis $(\hat{h}, \hat{v})$ can be expressed as

$$e_{HV}(t) = \begin{bmatrix} a_H(t)e^{j\varphi H(t)} \\ a_V(t)e^{j\varphi V(t)} \end{bmatrix}, \quad t \in \mathbf{T} \tag{2.1.5}$$

where, $a_H(t)$ and $a_V(t)$ represent the amplitudes of time-variant horizontal and vertical polarization components of electromagnetic waves, respectively, and $\varphi_H(t)$ and $\varphi_V(t)$ represent the phases of the horizontal and vertical polarization components of electromagnetic waves, respectively.

Based on the above formula, the time domain instantaneous Stokes vector and instantaneous polarization projection vector (abbreviated as IPPV) of electromagnetic waves are defined as

$$\dot{J}_{HV}(t) = \begin{bmatrix} g_{HV0}(t) \\ g_{HV}(t) \end{bmatrix} = \mathbf{R}e_{HV}(t) \otimes e^*_{HV}(t), \quad t \in \mathbf{T} \tag{2.1.6}$$

and

$$\tilde{g}_{HV}(t) = [\tilde{g}_{HV1}(t), \tilde{g}_{HV2}(t), \tilde{g}_{HV3}(t)]^T = \frac{g_{HV}(t)}{g_{HV0}(t)}, \quad t \in \mathbf{T} \tag{2.1.7}$$

In formula (2.1.6), "$\otimes$" and "*" represent the Kronecker product and conjugate, respectively; the superscript "T" represents vector transposition, $g_{HV}(t)$ is called the instantaneous Stokes sub-vector, and $\mathbf{R}$ is the four order quasi-unitary matrix

$$\mathbf{R} = \begin{bmatrix} 1 & 0 & 0 & 1 \\ 1 & 0 & 0 & -1 \\ 0 & 1 & 1 & 0 \\ 0 & j & -j & 0 \end{bmatrix} \tag{2.1.8}$$

The definitions of polarization ratio and phase descriptor given in the formula (2.1.3) can be extended, to define instantaneous polarization ratio and instantaneous polarization phase descriptor, for characterizing the dynamic polarization information of time-variant electromagnetic waves. From the perspective of characterization of polarization information, instantaneous polarization ratio and instantaneous polarization phase descriptor are fully equivalent to IPPV [6].

The time domain instantaneous Stokes vector of electromagnetic waves contains their intensity information and polarization information, while IPPV mainly depicts the polarization characteristic of electromagnetic waves. On the Poincare unit sphere, the IPPV of electromagnetic waves forms a 3-dimensional vector ordered set with time as the order parameter. The 3-dimensional vector ordered set, namely the instantaneous polarization projection set, describes the evolution of instantaneous polarization characteristic of electromagnetic waves with time. For the definition and properties of time domain instantaneous polarization descriptor of electromagnetic waves, refer to the Ref. [3], which are not given here.

2.2 Primary Meaning of Antenna Spatial Polarization Characteristic

2.2.1 Antenna Polarization

In the antenna theory, assuming that the antenna center is the center of the sphere and the radiation pattern of the antenna is the radial direction of the sphere, both the radiation electric field vector and the magnetic field vector are perpendicular to the direction of radiation. If the radius of the sphere is very large, in the local area of the sphere, the radiation field of the antenna can be regarded as plane wave. In a plane perpendicular to the direction of radiation, i.e. the plane where the radiation electric field and magnetic field exist, the trajectory of the electric field vector changing with time is defined as the polarization of radiation wave. The antenna radiation field may have different polarization modes. But all of them can be decomposed into a linear combination of two orthogonal polarizations, and the two orthogonal polarization components of the antenna have their own patterns. Generally speaking, the polarization mode of antenna radiation field is not fixed, but in fact closely related to the location of the measured radiation field. In different observation directions, the polarization state of antenna radiated electromagnetic waves may be different, which is a spatial variable of antenna polarization. For the radiation field of any antenna, selecting the antenna aperture center as the origin of the spherical coordinate system, the radiation field of the antenna can be written as [1]:

$$\boldsymbol{E}(r, \theta, \varphi) = \frac{\mathrm{j}Z_0 I}{2\lambda r} \exp(\mathrm{j}kr)\boldsymbol{h}(\theta, \varphi) \tag{2.2.1}$$

where Z_0 is the intrinsic impedance in free space, I is the strength of antenna feed time-harmonic current; $\boldsymbol{h}(\theta, \varphi)$ is called the effective length of antenna (also known as the effective height of antenna in some references), which is associated with the space angular coordinates (θ, φ) of the measuring point.

For example, for the dipole antenna in free space, $\boldsymbol{h}(\theta, \varphi) = l \cdot \sin \theta \cdot \hat{\boldsymbol{u}}_\theta$, in which, l and θ are the length and elevation angle of the dipole antenna, respectively, and $\hat{\boldsymbol{u}}_\theta$ is the elevation direction unit vector. It can be easily seen that the polarization mode of radiation field of the dipole antenna changes with elevation angle, as shown in Fig. 2.1.

Fig. 2.1 Radiation field
of dipole antenna

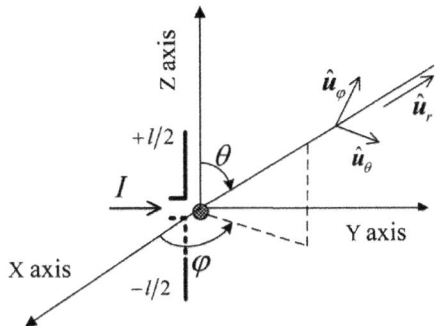

For a given antenna, when the feed current is given, the radiation field is only related to the space location of the measuring point, and the effective length of the antenna is only related to the space angular coordinate of the measuring point. It implies that, when the position of the measuring point relative to the antenna is given the polarization state of the antenna can be defined using the radiation field of the antenna at the point [1]. Moreover, since the Jones vector of antenna polarization differs from the effective length of antenna by merely a scalar factor, usually they can be expressed with the same mark h, and they can be considered identical with each other when it is unnecessary to consider antenna gain or receiving power. Virtually, the antenna polarization definition is based on the polarization state of the antenna radiated electromagnetic wave in a given direction. Hence all electromagnetic wave polarization descriptors discussed in Sect. 2.2 are also perfectly suitable for antenna polarization description [7, 85, 93].

The antenna polarization discussed above is defined in monochromatic wave conditions, that is to say, only considering the situation where the antenna is excited by time-harmonic signals. This definition consideration applies also to most of the narrowband systems. According to the theory of linear system, the antenna can be viewed as a linear filter. When the feed current is no longer a time-harmonic or narrowband signal, the input non-time-harmonic current is converted by the antenna into a non-harmonic radiation electromagnetic wave. The spectrum of the radiation wave is the product of the spectrum of feed current and the frequency-domain response of antenna system. Therefore, where an antenna is excited by non-time-harmonic signals, the instantaneous polarization of antenna in a given propagation direction can be defined using the antenna impulse response vector function $h(t)$ [3]. In physics, it is the function of time-variant electric field of radiation wave when impulse current is fed into the antenna. It is defined using precisely the same method for defining non-time-harmonic electromagnetic wave instantaneous polarization [3].

2.2.2   Antenna Cross Polarization

To discuss the spatial polarization characteristic of antenna, we have to mention the "cross-polarization" concept. In the design of antenna, the required radiated or received electromagnetic wave polarization is often referred to as "expected polarization". But due to various actual factors, for the transmitting antenna, the radiated electromagnetic wave polarization is not so "pure", always mixed with some polarization components that we don't want. Similarly, for the receiving antenna, it is impossible receiving only co-polarization waves without receiving any orthogonal-polarization waves.

2.2.2.1   Definition of Cross-Polarization of Antenna

According to the definition given by IEEE, "in a specific plane containing the reference polarization ellipse, the polarization orthogonal to this reference polarization is called cross-polarization". This reference polarization is called co-polarization [55, 56]. The field component parallel to the field of reference source is called the co-polarization field or main-polarization field; the field component vertical to the field of reference source is called cross-polarization field. For example: An antenna is designed to serve the following purpose: it should radiate horizontal linear polarized waves, but the electromagnetic waves it radiates also contain the vertical linear polarization component. In this case, the horizontal linear polarization component can be regarded as the co-polarization component, while the vertical linear polarization as the cross-polarization component. Furthermore, the smaller the vertical polarization component is than the horizontal polarization component, the purer the polarization is.

A monochromatic wave can be decomposed into two orthogonal polarization components. The two orthogonal components may be linear polarization, such as horizontal and vertical linear polarization, or circular polarization, such as the left and right circular polarization, or general elliptical polarization. Hence, you can choose a pair of orthogonal polarization electric field components with unit power density as the polarization basis, which we may record as $\left(\hat{A}, \hat{B}\right)$. On this polarization bases, the electric field can be written as

$$\boldsymbol{E}(AB) = E_A\hat{\boldsymbol{A}} + E_B\hat{\boldsymbol{B}} \qquad (2.2.2)$$

Here, E_A and E_B represent the complex coordinates of electric field $a = \frac{\delta_\varphi \delta_\theta}{2}\boldsymbol{E}$ on the two polarization bases. With the polarization bases given, the electric field $\boldsymbol{E}$ can be uniquely characterized by them.

Sometimes, in the design of antenna, the designer hopes antenna polarization to be $\hat{A}$ polarization (for example, $\hat{A}$ is the horizontal polarization direction), but actual antenna polarization is $\hat{C}$ (for example, $\hat{C}$ is the 5° linear polarization direction), due to the existence of parasitic polarization $\hat{B}$ (for example, $\hat{B}$ is vertical polarization direction). Naturally, the electric field vector E can be decomposed into $\left(\vec{E}_A, \vec{E}_B\right)$ on the polarization bases $\left(\hat{A}, \hat{B}\right)$; E_A and E_B are defined as the "co-polarization component" and "the cross-polarization component", respectively. "Expected polarization" is essentially a subjective quantity. The definition of reference polarization direction is not unique.

For the antenna cross-polarization, A. C. Ludwig once gave definitions. Using three kinds of reference polarization as the co-polarization, he set the third definition of cross-polarization [55]. The radiation electric field of antenna is represented by $E(\theta, \varphi)$, and $E \cdot \hat{u}_{co}$ is recorded as the reference polarization component of the electric field E. $E \cdot \hat{u}_{cross}$ is the cross-polarization component of the electric field E. The three definitions put forward by Ludwig are shown below:

Definition 2.1

$$\begin{cases} \hat{u}_{co} = \hat{u}_y \\ \hat{u}_{cross} = \hat{u}_x \end{cases} \tag{2.2.3}$$

where $\hat{u}_x$ and $\hat{u}_y$ are the unit vectors in the x and y directions in the rectangular coordinate system, respectively. Definition 2.1 is often used to describe the plane wave in the rectangular coordinate system. The cross-polarization is defined as the linear polarized wave perpendicular to the plane wave.

Definition 2.2

$$\begin{cases} \hat{u}_{co} = \hat{u}_\theta \\ \hat{u}_{cross} = -\hat{u}_\varphi \end{cases} \tag{2.2.4}$$

where, $\hat{u}_\theta$ and $\hat{u}_\varphi$ are the unit vectors in the elevation and azimuth direction in the spherical coordinate system, respectively. Definition 2.2 is often used to describe the polarized wave generated by the electric dipole. Cross-polarization is a polarized wave generated by the magnetic dipole which and the electric dipole are on the same axis.

Definition 2.3

$$\begin{cases} \hat{u}_{co} = \sin \varphi \hat{u}_\theta + \cos \varphi \hat{u}_\varphi \\ \hat{u}_{cross} = \cos \varphi \hat{u}_\theta - \sin \varphi \hat{u}_\varphi \end{cases} \tag{2.2.5}$$

Definition 2.3 is used to describe the polarized wave generated by the Huyghens source. Cross-polarization is the polarized wave generated by the same Huyghens source rotating by 90° in the antenna aperture. Definition 2.3 is most close to the actual situation of antenna pattern measurement. Cross-polarization can be measured under the general conditions for antenna pattern measurement. According to this definition, with the measured antenna and the beacon antenna aimed at each other in the maximum beam directions, let their polarizations be parallel to each other, the pattern measured by rotating the measured antenna is the co-polarization pattern E_p. With the two antenna aimed at each other in the maximum beam directions, let their polarizations be perpendicular to each other, the pattern measured by rotating the measured antenna is the cross-polarization pattern E_q. See the formula:

$$\begin{cases} E_p(\theta, \varphi) = E_\theta(\theta, \varphi) \sin \varphi \hat{\boldsymbol{u}}_\theta + E_\varphi(\theta, \varphi) \cos \varphi \hat{\boldsymbol{u}}_\varphi \\ E_q(\theta, \varphi) = E_\theta(\theta, \varphi) \cos \varphi \hat{\boldsymbol{u}}_\theta - E_\varphi(\theta, \varphi) \sin \varphi \hat{\boldsymbol{u}}_\varphi \end{cases} \qquad (2.2.6)$$

where E_θ and E_φ are the electric field components in the $\hat{\boldsymbol{u}}_\theta$ and $\hat{\boldsymbol{u}}_\varphi$ directions, respectively. With the reference source rotating by 90°, the co-polarization field and cross-polarization field measured in any direction are also interchangeable with each other according to the Definition 2.3. Therefore, this definition has been widely used [91–93, 95]. Figure 2.2 is schematic of the measurement system.

2.2.2.2 Characterization of Antenna Polarization Purity

"Cross-polarization isolation (XPI)" or "cross-polarization discrimination (XPD)" is a commonly-used evaluation index for cross-polarization jamming in dual-polarization channel. Here, we first briefly introduce the commonly-used evaluation index of cross-polarization of dual-polarization system, then discuss the evaluation index of cross-polarization component of single antenna.

Fig. 2.2 Schematic of the antenna pattern measurement system

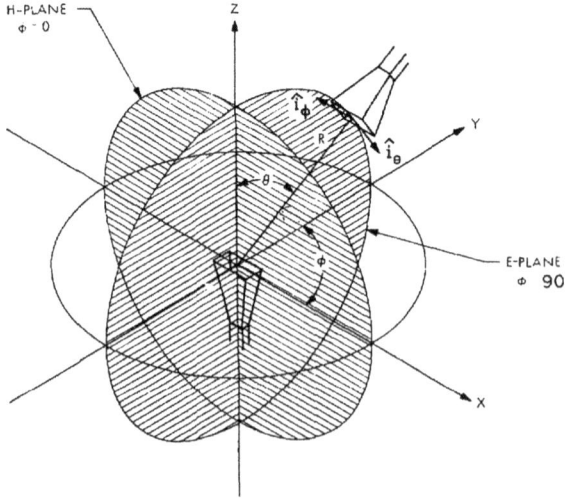

1. Measurement of cross-polarization in the dual-polarization system

In the dual-polarization frequency-multiplexing communication system, inter-channel jamming results between the two orthogonal channels, due to the polarization impurity of the system itself, or the de-polarization effect of the transmission path. The linearly-polarized wave can be decomposed into a co-polarization component, which has the same direction as the original polarized wave, and an orthogonal cross-polarization component. An ideal circular polarized wave becomes an elliptically polarized wave. Consequently, a circularly-polarized wave in the same direction as original rotation, and a circularly-polarized component in the opposite direction to original rotation are generated; the amplitude ratio between them is called counter rotation coefficient b.

Figure 2.3 shows the schematic of a dual-polarization communication system, where, E_{1T} and E_{2T} represent a pair of orthogonal transmission signals, E_{1R} and E_{2R} represent the corresponding pair of orthogonal reception signals, E_{12R} and E_{21R} are the corresponding cross-polarization signals. Cross-polarization isolation (XPI) is defined as the ratio of the cross-polarization component E_{12R} (or E_{21R}) produced in the other channel by the present signal to the co-polarization component E_{1R} (or E_{2R}) produced in the present channel:

$$\mathrm{XPI} = 20\lg b_1 \,(\mathrm{dB}), \ \mathrm{XPI} = 20\lg b_2 \,(\mathrm{dB}) \qquad (2.2.7)$$

$$\mathrm{XPI} = 20\lg \frac{E_{12R}}{E_{1R}} \,(\mathrm{dB}), \ \mathrm{XPI} = 20\lg \frac{E_{21R}}{E_{2R}} \,(\mathrm{dB}) \qquad (2.2.8)$$

Formula (2.2.7) is intended for the linear-polarization mode. Formula (2.2.8) is intended for the elliptical-polarization mode, in which b_1 and b_2 represent the counter rotation coefficients of left-hand and right-hand circular polarizations, respectively. Counter rotation coefficient is equal to the ratio of the counter rotation polarization component to the co-polarization component.

Cross-polarization discrimination (XPD) is defined as the ratio of the cross-polarization component E_{21R} (or E_{12R}) produced in the present channel by the other signal to the co-polarization component E_{1R} (or E_{2R}) produced in the present channel:

$$\mathrm{XPD} = 20\lg \frac{E_{21R}}{E_{1R}} \,(\mathrm{dB}), \ \mathrm{XPD} = 20\lg \frac{E_{12R}}{E_{2R}} \,(\mathrm{dB}) \qquad (2.2.9)$$

Fig. 2.3 Schematic of dual-polarization system

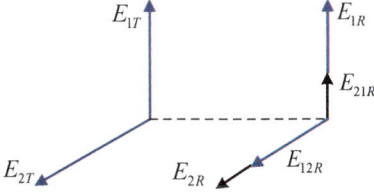

$$\text{XPD} = 20\lg\frac{b_2 E_{2\text{T}}}{E_{1\text{T}}}\,(\text{dB}),\ \text{XPD} = 20\lg\frac{b_1 E_{1\text{T}}}{E_{2\text{T}}}\,(\text{dB}) \qquad (2.2.10)$$

Formula (2.2.9) is intended for the linear-polarization mode. Formula (2.2.10) is intended for the elliptical-polarization mode.

It can be seen from their definitions that both XPI and XPD are used to measure the extent of the jamming between the two channels caused due to the presence of cross-polarization. XPI is used to measure the effect of the cross-polarization component produced in the present channel on the other channel; and XPD is used to measure the effect of the cross-polarization component produced in the other orthogonal channel on the present channel. XPI exists in both the single-polarization system and the dual-polarization system, while XPD can only exist in the dual-polarization system. XPI is usually used to measure the jamming at the transmission end; while XPD is used to measure the jamming at the reception end. In the dual-polarization system, generally, $E_{1\text{T}} \approx E_{2\text{T}}$, $E_{12\text{R}} \approx E_{21\text{R}}$, $b_1 \approx b_2$, and the transmission media are reciprocal, thus, in value, XPI $\approx$ XPD. In other words, in the dual-polarization system, if the signals in the two channels have an equal amplitude and the same cross-polarization effect, XPI = XPD.

2. Measurement of cross-polarization of single antenna

For the antenna itself, for engineering purpose the concept of "cross-polarization discrimination (XPD)" is often also employed to describe the polarization purity of single antenna [61]. It is usually referred to as "cross-polarization discrimination quantity", i.e. the ratio of the power of parallel polarization to that of the expected polarization component:

$$\text{XPD} = 10\lg\!\left(\frac{P_{\text{X}}}{P}\right) = 20\lg\frac{E_{\text{X}}}{E}\,(\text{dB}) \qquad (2.2.11)$$

where, P is the power of antenna co-polarization component; P_{X} is the power of antenna cross-polarization component; E is the electrical level of antenna co-polarization component; E_{X} is the electrical level of antenna cross-polarization component.

Two polarization modes usually adopted by the radar system are linear polarization and circular polarization. In case of linear polarization, if the expected polarization of antenna is the vertical polarization, the cross-polarization discrimination quantity (XPD_{V}) can be determined with the formula:

$$\text{XPD}_{\text{V}} = 20\lg(E_{\text{V-H}}/E_{\text{V-V}}) \qquad (2.2.12)$$

where, $E_{\text{V-V}}$ represents the vertical polarization field, i.e. the co-polarization field received by the antenna when transmission polarization is vertical polarization; $E_{\text{V-H}}$ represents the horizontal polarization field, i.e. the cross-polarization field received by the antenna when transmission polarization is vertical polarization.

If the expected polarization of antenna is horizontal polarization, the cross-polarization discrimination (XPD_H) is:

$$XPD_H = 20lg(E_{H-V}/E_{H-H}) \tag{2.2.13}$$

where, E_{H-H} is the horizontal polarization field received by the antenna when the transmission polarization is horizontal polarization; E_{H-V} is the vertical polarization field received by the antenna when the transmission polarization is horizontal polarization.

Traditionally, the cross-polarization discrimination of the circularly polarized antenna is described with the circular-polarization voltage axis ratio. Circular polarization is a special case of elliptical polarization. By definition, the elliptical polarization wave voltage axis ratio is:

$$AR = \frac{1+b}{1-b} \tag{2.2.14}$$

where, b represents counter rotation coefficient, which is equal to the ratio of counter rotation polarization component to co-polarization component.

The cross-polarization discrimination of elliptically polarized wave can be expressed using the formula below:

$$XPD = 20lg = \frac{AR-1}{AR+1}(dB) \tag{2.2.15}$$

Formulas (2.3.11)–(2.3.15) show that, the smaller the value of XPD is, the less the parasitic polarization component of the antenna is, and the "purer" the polarization of the antenna is.

2.2.3 Antenna Spatial Polarization Characteristic Connotation

It can be seen from the analysis in Sect. 2.3.2 that, due to restriction and influence by various actual factors, there is not any antenna of "pure" polarization. In other words, there is parasitic polarization. In the traditional research, the existence of parasitic polarization is often overlooked because of research focus and application need. In the traditional research, it is considered that the antenna polarization characteristic remains relatively constant on the main lobe. The main lobe polarization, usually known as "co polarization", is used to describe antenna polarization. It is also considered that the side lobe radiation polarization differs greatly from the main lobe polarization [59, 63]. In fact, the variation of antenna polarization in space occurs on not only the main lobe but the side lobes as well, and is associated with the antenna space orientation.

As discussed in Sect. 2.3.1, with the feed current of an antenna given, the radiation field of the antenna is only associated with where the measuring point is located in the space, and the effective length of the antenna is only associated with the space angular coordinates of the measuring point. In other words, the polarization of antenna radiated wave varies with direction. Antenna polarization is a spatial variable. The polarization state of antenna radiated electromagnetic wave may be different in different observation directions. In the traditional research, the factor has been ignored, or the cross polarization of antenna is deemed a harmful quantity and hence avoided to the greatest extent. However, we can treat this issue otherwise. Don't just blindly suppress antenna cross-polarization, but use it effectively, via dedicated research on the rule of antenna polarization characteristic variation. This methodology is of great significance for enhancing radar capability of acquiring and processing polarization information.

For the two field components vertical to the antenna propagation direction, their space distribution functions are different from each other. Consequently, the antenna polarization is a function of space. The characteristic of evolution and distribution in space for the polarization of antenna radiated electromagnetic wave is called as the "antenna spatial polarization characteristic", which indicates the rule for evolution of antenna polarization in the spatial, or the relation between the deviation of actual polarization from expected polarization and the space angular coordinates. Variation of antenna polarization characteristic in space consists of two parts: (1) variation of antenna polarization characteristic caused due to the change of the position of the point of interest relative to the antenna orientation. It can be understood from two aspects: (a) with the antenna stationary, the polarization mode varies with position for the electromagnetic wave radiated; (b) the polarization mode of antenna radiated electromagnetic wave received at the same position varies, with the antenna scanning in the spatial. In fact, these two kinds of ways of understanding is essentially the same, is to investigate the changes of interest and spatial point antenna relative space position change of antenna polarization, just from the goal position change and antenna scanning these two different angles to describe. Spatial polarization characteristic is an inherent characteristic of antenna. Different types of antenna, or antenna of the same type but different structural parameters, observe different rules for variation of spatial polarization characteristic. (2) In practice, due to restriction and influence by such factors as geometrical, dimensional, and machining errors, antenna defocusing, and wave diffraction, the antenna polarization always deviates to a certain extent from the design value (or "expected polarization"). This deviation (referred to as "parasitic polarization"), related usually to operating frequency and space orientation, reduces the polarization purity of antenna.

Antenna spatial polarization characteristic includes the classical spatial polarization characteristic and the spatial instantaneous polarization characteristic. In most cases, traditional radar operates with continuous waves or narrowband conditions. The classical spatial polarization characteristic of antenna is defined based on the fact that the space movement trajectory of the electric field vector tip of monochromatic plane electromagnetic wave is an ellipse. Description of antenna spatial polarization characteristic may also adopt the classical polarization description methods. With the development of wideband and ultra-wideband non-monochromatic wave radars, in case of non-time-harmonic signals, the classical polarization description methods are subject to restriction for the description of spatial polarization characteristic of antenna. Traditional descriptors, for example, "antenna spatial elliptical polarization descriptor" and "polarization axis ratio" have become incapable of effectively describing the evolution of antenna polarization in space. So, nowadays the use of new tools and new methods is necessary for describing antenna polarization characteristic, especially the antenna spatial polarization characteristic.

2.3   Antenna Spatial Polarization Characterization

In the Ref. [1] were listed a variety of methods for the characterization of electromagnetic wave polarization state. In the Ref. [3] were presented the time-domain instantaneous polarization descriptors and frequency-domain instantaneous polarization descriptors of electromagnetic waves. Based on the knowledge, the methods for characterization of antenna spatial polarization characteristic are proposed in this section in the context of the actual application.

The radiation field of antenna is denoted by $E(P, P')$. The space angular coordinates of the measuring point are represented by $P(\theta, \varphi)$, or P for short, which is a 2-dimensional vector of azimuth φ and elevation θ. Moreover, $P \in \Omega$, where Ω is the spatial domain of interest. $P'(\theta', \varphi')$ denotes the deviations from the central orientation in the azimuth and elevation directions during antenna scanning in space. $P' \in \Omega'$, where Ω is the range of antenna scanning. In the discussion of antenna spatial polarization characteristic, P and P' are considered to be consistent with each other in essence. For the sake of convenience, record $E(P, P')$ as $E(P)$, where P is the spatial parameter, i.e. the space angular coordinates of the orientation of the spatial to be measured relative to the orientation of the antenna, during the antenna space scanning. Research on the antenna spatial polarization characteristic means the discussion as to the rule of antenna radiation field polarization mode variation with the angular coordinates $P(\theta, \varphi)$.

Suppose the region of interest is a 3-dimensional space that includes the azimuth $\varphi_1 \sim \varphi_2$ and elevation $\theta_1 \sim \theta_2$. As shown in Fig. 2.4, the "spatial solid angle" can be defined as [118]:

$$V = \frac{S}{R^2} = \frac{1}{R^2} \iint ds = \frac{1}{R^2} \int_{\varphi_1}^{\varphi_2} \int_{\theta_1}^{\theta_2} R^2 \sin\theta d\varphi d\theta$$

$$= (\varphi_2 - \varphi_1)(\cos\theta_1 - \cos\theta_2)$$

(2.3.1)

Fig. 2.4 Definition of spatial
solid angle

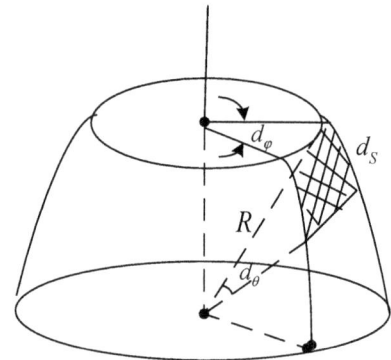

Fig. 2.5 Definition of beam
solid angle

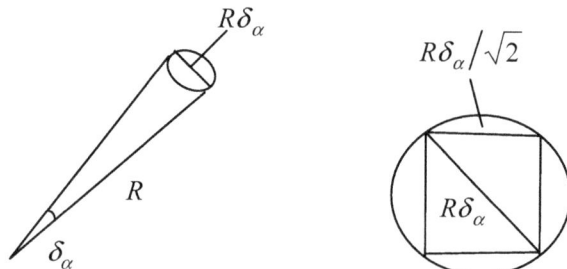

where S is the area of the sphere having an radius of R and cut in the spatial to be measured, $ds = Rd\theta \cdot R \sin\theta d\varphi$.

The solid angle of a certain piece of area on the aperture of a sphere relative to the center of the sphere is defined as the area of this area divided by the square of the radius. Assuming that the beam of antenna has the same width δ_α in the two planes as shown in Fig. 2.5, the beam cuts out a circle, the radius of which is the distance R, on the spherical aperture. The inscribed square of the circle can be viewed as a basic unit of antenna beam scanning. This inscribed square is defined as the "beam solid angle". As shown in the figure, the square has an area of $(R\delta_\alpha/\sqrt{2})^2$, so the beam solid angle is:

$$\alpha = (R\delta_\alpha/\sqrt{2})^2/R^2 = \frac{\delta_\alpha{}^2}{2} \qquad (2.3.2)$$

For the antenna mentioned in formula (2.3.2), the beam has the same width in the azimuth and elevation directions. If the widths in the two directions δ_φ and δ_θ are not the same, the beam solid angle can be expressed as:

$$\alpha = \frac{\delta_\varphi \delta_\theta}{2} \qquad (2.3.3)$$

Then, the spatial solid angle V can be understood as spatial domain of interest Ω Discussion of antenna polarization characteristic variation in a 3-dimensional spatial is essentially the discussion of the rule of antenna polarization characteristic variation during the antenna beam scanning in the azimuth and elevation directions.

In fact, not in every case is it necessary to discuss antenna spatial polarization characteristic in the 3-dimensional spatial. In many cases, dimensionality reduction may be performed according to actual conditions. This will greatly reduce the complexity of the problem being discussed. For example, warning radars (also known as surveillance radars), are mainly used for ground-to-air or coastal defense systems, to provide information on long-distance spatial or maritime targets. Due to the long distance and small elevation of the target, traditionally, most of the warning radars adopt the two-coordinate mechanical scanning system. In other words, they employ the mechanical movement of the whole or part of the antenna system to realize beam scanning. A typical scanning method is circular scanning with a sector beam. In this case, the beam in the horizontal plane is fairly narrow. The azimuth resolution can attain a fraction of degree so that it is possible to accurately measure the azimuth angle of the target. Nonetheless, the beam in the vertical plane is rather wide. Result is scanning rough, so that it is impossible to accurately measure the elevation angle of the target. In a word, the two-coordinate air warning radar often is only capable of measuring the distance and azimuth angle of the target. Generally, a "nodding" type of height-measuring radar is additionally required to provide elevation information. In this case, in the discussion of antenna spatial polarization characteristic, it is permitted to simplify the 3-dimensional issue into a 2-dimensional one. The rule of antenna spatial polarization characteristic variation with the space angular coordinate $P(\theta_0, \varphi)$ is sought at different elevations θ_0. Figure 2.6 illustrates the sector beam scanning.

2.3.1 Classic Spatial Polarization Characteristics Description of Antenna

The electric field radiated by an antenna is usually defined by the spherical coordinate $\left(\hat{r}, \hat{\theta}, \hat{\varphi}\right)$ as shown in Fig. 2.4. In the far-field region, the radiation field E has no radial component. It can be expressed as:

Fig. 2.6 Schematic of sector beam scanning

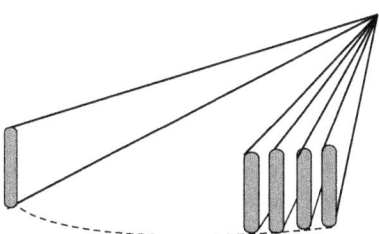

$$E = E_\theta \hat{u}_\theta + E_\varphi \hat{u}_\varphi \tag{2.3.4}$$

where $\hat{u}_\theta$ and $\hat{u}_\varphi$ are the unit vectors in the pitch and azimuth directions, respectively; E_θ and E_φ are the polarization components in the $\hat{u}_\theta$ and $\hat{u}_\varphi$ directions, respectively.

If the xy plane is parallel to the ground, then E_φ is the horizontal component of the wave (it is always parallel to the xy plane thus the ground). $-E_\theta$ is the vertical component. $(\hat{u}_\varphi, -\hat{u}_\theta, \hat{u}_r)$ constitutes a right-hand coordinate system. $\hat{\varphi}$ and $-\hat{\theta}$ can be defined as the horizontal and vertical polarization bases, respectively:

$$\begin{cases} \hat{h} = \hat{\varphi} \\ \hat{v} = -\hat{\theta} \end{cases} \tag{2.3.5}$$

With the position of the measuring point relative to the antenna determined, we can define polarization state of the antenna using the radiation field at this point. The normalized Jones vector of the antenna is expressed as:

$$h = \begin{bmatrix} E_H \\ E_V \end{bmatrix} = \frac{1}{\|E\|} \begin{bmatrix} E_\varphi \\ -E_\theta \end{bmatrix} \tag{2.3.6}$$

The spatial polarization ratio of antenna radiated electromagnetic wave is expressed as:

$$\rho(P) = \frac{E_V}{E_H} = \frac{-E_\theta}{E_\varphi} \tag{2.3.7}$$

The spatial Jones vector and spatial polarization ratio of antenna are related by the formula below:

$$h(P) = \frac{1}{\sqrt{1 + |\rho(P)|^2}} \begin{bmatrix} 1 \\ \rho(P) \end{bmatrix} \tag{2.3.8}$$

The spatial polarization ratio is the most-direct characterization of antenna spatial polarization characteristic. Other description methods may also be defined, for example, spatial polarization phase descriptor $(\gamma(P), \phi(P))$, spatial polarization ellipse descriptor $(\varepsilon(P), \tau(P))$, and spatial Stokes vector.

On the horizontal and vertical polarization basis $(\hat{h}, \hat{v})$, the normalized Jones vector of antenna can be expressed as:

$$h = \begin{bmatrix} E_H \\ E_V \end{bmatrix} = \begin{bmatrix} a_H e^{j\varphi_H} \\ a_V e^{j\varphi_V} \end{bmatrix} \tag{2.3.9}$$

The spatial polarization ratio of antenna can be expressed as:

$$\rho(\boldsymbol{P}) = \frac{E_V}{E_H} = \text{tg}\gamma(\boldsymbol{P})e^{j\phi(\boldsymbol{P})}, \ (\gamma, \phi) \in [0, \pi] \times [0, 2\pi] \qquad (2.3.10)$$

In which, $\gamma(\boldsymbol{P}) = \tan^{-1}\frac{a_V}{a_H}$ and $\phi(\boldsymbol{P}) = \varphi_V - \varphi_H$ are spatial polarization phase descriptors of the antenna.

The antenna spatial polarization ratio and spatial polarization phase descriptors are the most commonly used quantities for describing antenna spatial polarization characteristic. In addition, "antenna cross-polarization XPD (P)" may be used to characterize the deviation of actual polarization from expected polarization varying with spatial angular position. XPD (P) can be calculated using the formula below:

$$\text{XPD}(\boldsymbol{P}) = 20\text{lg}(|\rho(\boldsymbol{P})|)\text{dB} \qquad (2.3.11)$$

As discussed in Sect. 2.3.2, decomposition of the electric field can be made on any pair of orthogonal polarization bases. Therefore, the far-field field of antenna radiation can also be decomposed as follows:

$$\boldsymbol{E}_{\text{far}} = E_1\hat{\boldsymbol{e}}_1 + E_2\hat{\boldsymbol{e}}_2 \qquad (2.3.12)$$

Theoretically, the orthogonal polarization basis $(\hat{\boldsymbol{e}}_1, \hat{\boldsymbol{e}}_2)$ can be arbitrarily chosen. For example, the polarization ratio in formula (2.3.7) is defined on the polarization basis $\left(\hat{\boldsymbol{\varphi}}, -\hat{\boldsymbol{\theta}}\right)$. In the spherical coordinate system, it is the most natural and widely used decomposition method. However, the cross-polarization discrimination XPD obtained using the polarization ratio ρ defined in formula (2.3.7) cannot directly characterize the deviation of antenna polarization from expected polarization at different spatial angular positions. Therefore, different polarization bases may be chosen depending on actual situation. On the polarization basis of $(\hat{\boldsymbol{e}}_{\text{co}}, \hat{\boldsymbol{e}}_{\text{cross}})$, the far field $\boldsymbol{E}_{\text{far}}$ can be expressed as

$$\boldsymbol{E}_{\text{far}} = E_{\text{co}}\hat{\boldsymbol{e}}_{\text{co}} + E_{\text{cross}}\hat{\boldsymbol{e}}_{\text{cross}} \qquad (2.3.13)$$

where, $\hat{\boldsymbol{e}}_{\text{co}}$ represents the expected polarization direction, $\hat{\boldsymbol{e}}_{\text{cross}}$ represents the polarization orthogonal to the direction (also known as parasitic polarization).

The spatial cross-polarization XPD(P) can be then expressed using the formula below:

$$\text{XPD}(\boldsymbol{P}) = 10\text{lg}\left(\frac{P_{\text{cross}}}{P_{\text{co}}}\right) = 20\text{lg}\left(\left|\frac{E_{\text{cross}}}{E_{\text{co}}}\right|\right)\text{dB} \qquad (2.3.14)$$

where, P_{cross} and P_{co} represent the power of parasitic polarization and that of expected polarization, respectively, E_{cross} and E_{co} represent the voltage amplitudes of parasitic polarization and expected polarization, respectively.

It can be seen from the above formula that, the smaller the value of XPD is, the smaller the cross-polarization component is, and the higher the "polarization purity" is. In this sense, the above formula may also be used to define the "spatial polarization purity" *Purity* (*P*) of antenna. Hence, the spatial polarization purity *Purity* (*P*), in agreement with the spatial polarization differential quantity XPD (P), is capable of fairly directly describing antenna cross-polarization characteristic.

With the expected polarization being horizontal polarization $E_H = E_\varphi$, and the parasitic polarization being vertical polarization $E_V = -E_\theta$, the cross-polarization discrimination defined in formula (2.3.14) is the same as that defined in formula (2.3.11). If the expected polarization of antenna is at different polarization state, the corresponding polarization basis may be chosen to define the spatial cross-polarization discrimination XPD(*P*). For example, the polarization basis $\left(\hat{h}, \hat{v}\right)$ may be chosen if the expected polarization is horizontal/vertical linear polarization; the polarization basis $(\hat{e}_{45°}, \hat{e}_{135°})$ may be chosen if the expected polarization is 45°/135° linear polarization; the polarization basis $\left(\hat{l}, \hat{r}\right)$ may be chosen if the expected polarization is left-hand/right-hand circular polarization.

2.3.2 *Instantaneous Polarization Characterization Description of Antenna*

Therefore, in the case of antenna excitation by non-time-harmonic signal, the instantaneous polarization in a given direction of propagation can be defined using the function of electric field in a particular direction of propagation of the antenna radiated wave when the impulse current is fed into the antenna. The amplitude and phase characteristics of antenna are different at different frequency points. The polarization state of antenna is also different at different spatial points. The antenna spatial instantaneous polarization characteristic essentially describes the distribution situation of antenna instantaneous polarization states at different spatial points, including the overall distribution situation of polarization states, central position of polarization distribution, density of distribution of spatial points, and spatial polarization state variation rate. Therefore, the definition method for spatial instantaneous polarization characteristic of antenna agrees completely with that for spatial instantaneous polarization characteristic of non-time-harmonic electromagnetic wave. Parameters, such as spatial instantaneous polarization projection set, spatial polarization cluster center and spatial polarization divergence are used for characterization of spatial instantaneous polarization characteristic of antenna [49, 119].

I. Antenna spatial instantaneous Stokes vector

Supposing that the electric field of antenna radiation wave is $E(P)$, $P \in \Omega$, the spatial coherent vector is defined as:

$$c(P_1, P_2) = E(P_1) \otimes E^*(P_2), \ P_1, P_2 \in \Omega \tag{2.3.15}$$

By definition, the spatial coherent vector of electromagnetic wave is obtained by finding the Kronecker product of the field vectors in two different spatial positions (which also contains a conjugate operation). The spatial coherent vector is a 4-dimensional complex vector, having the exchange property below:

$$c(P_1, P_2) = Q_4 c^*(P_2, P_1) \tag{2.3.16}$$

where, $Q_4 = \begin{bmatrix} 1 & 0 & 0 & 0 \\ 0 & 0 & 1 & 0 \\ 0 & 1 & 0 & 0 \\ 0 & 0 & 0 & 1 \end{bmatrix}$ it is a fourth-order permutation matrix.

Then, the spatial instantaneous Stokes vector of electromagnetic wave is defined as:

$$\dot{J}(P_1, P_2) = R \cdot c(P_1, P_2) \tag{2.3.17}$$

where, R is the quasi unitary matrix as defined in formula (2.2.8).

In particular, with $P_1 = P_2 = P$, $\dot{J}(P_1, P_2)$ is called the spatial instantaneous Stokes vector of electromagnetic wave. This vector can be abbreviated as $\dot{J}(P)$, provided that no confusion results.

$$\dot{J}(P) = Rc(P, P) = RE(P) \otimes E^*(P) \tag{2.3.18}$$

On the horizontal and vertical polarization basis $(\hat{h}, \hat{v})$, the plane electromagnetic wave of spatial propagation can be expressed as

$$e_{HV}(P) = \begin{bmatrix} a_H(P)e^{j\varphi H(P)} \\ a_V(P)e^{j\varphi V(P)} \end{bmatrix}, \ P \in \Omega \tag{2.3.19}$$

where, $a_H(P)$ and $a_V(P)$ are the amplitudes of horizontal and vertical polarization components of electromagnetic wave varying with spatial position, respectively, and $\varphi_H(P)$ and $a_V(P)$ are the phases of horizontal and vertical polarization components of electromagnetic wave, respectively. All of them are the function of the spatial angular position P.

Record $\phi(P) = \varphi_V(P) - \varphi_H(P)$ as the phase difference between the vertical and horizontal components of electromagnetic wave, then, in the horizontal and vertical polarization basis, the instantaneous Stokes vector of space-variant electromagnetic wave can be expressed as

$$g_{HV0}(\boldsymbol{P}) = |E_H(\boldsymbol{P})|^2 + |E_V(\boldsymbol{P})|^2 = a_H^2(\boldsymbol{P}) + a_V^2(\boldsymbol{P})$$
$$g_{HV1}(\boldsymbol{P}) = |E_H(\boldsymbol{P})|^2 - |E_V(\boldsymbol{P})|^2 = a_H^2(\boldsymbol{P}) - a_V^2(\boldsymbol{P})$$
$$g_{HV2}(\boldsymbol{P}) = 2\mathrm{Re}\big(E_H(\boldsymbol{P})E_V^*(\boldsymbol{P})\big) = 2a_H(\boldsymbol{P})a_V(\boldsymbol{P})\cos(\phi(\boldsymbol{P}))$$
$$g_{HV3}(\boldsymbol{P}) = -2\mathrm{Im}\big(E_H(\boldsymbol{P})E_V^*(\boldsymbol{P})\big) = 2a_H(\boldsymbol{P})a_V(\boldsymbol{P})\sin(\phi(\boldsymbol{P}))$$

$$, \boldsymbol{\psi} \in \boldsymbol{\Omega} \qquad (2.3.20)$$

Obviously, the spatial instantaneous Stokes vector of electromagnetic wave contains information on the intensity and polarization of electromagnetic wave varying with spatial position. From formula (2.3.20), the physical implications of the components can be gotten as follows: $g_{HV0}(\boldsymbol{P})$ is the power sum of the two orthogonal components of electromagnetic wave on the horizontal and vertical polarization bases; $g_{HV1}(\boldsymbol{P})$ is the power difference between the two orthogonal components on the horizontal and vertical polarization bases; $g_{HV2}(\boldsymbol{P})$ is the power difference between the two orthogonal components on the $45°$ and $135°$ orthogonal polarization bases; $g_{HV3}(\boldsymbol{P})$ is the power difference between the two orthogonal components on the left-hand and right-hand circular polarization bases.

II. Antenna spatial instantaneous polarization projection set (spatial IPPV)

Record the spatial instantaneous Stokes vector of electromagnetic wave $\boldsymbol{E}(\boldsymbol{P})$ as:

$$\dot{\boldsymbol{J}}(\boldsymbol{P}) = [g_0(\boldsymbol{P}), \boldsymbol{g}^{\mathrm{T}}(\boldsymbol{P})]^{\mathrm{T}} \qquad (2.3.21)$$

then $g_0(\boldsymbol{P}) = \|\boldsymbol{g}(\boldsymbol{P})\|$, finding the normal number of this vector.

Definition 2.4

$$\boldsymbol{g}_{\mathrm{norm}}(\boldsymbol{P}) = \boldsymbol{g}(\boldsymbol{P})/g_0(\boldsymbol{P}) \qquad (2.3.22)$$

then $\|\boldsymbol{g}_{\mathrm{norm}}(\boldsymbol{P})\| = 1$, that is to say, $\boldsymbol{g}_{\mathrm{norm}}(\boldsymbol{P})$ is located on the unit Poincare sphere.

Instantaneous polarization state of antenna is usually different in different spatial positions. Its projection on the Poincare sphere forms an ordered set of 3-dimensional unit vector, called spatial instantaneous polarization projection set, denoted as spatial IPPV:

$$\Pi_P = \{\boldsymbol{g}_{\mathrm{norm}}(\boldsymbol{P})|\boldsymbol{P} \in \boldsymbol{\Omega}\} \qquad (2.3.23)$$

For antenna radiated electromagnetic waves, in evidence, Π_P can describe completely not only the characteristic and rule of polarization state distribution in different spatial positions but also the rule of instantaneous polarization variation in spatial domain. Moreover, it can be seen from formulas (2.3.20) and (2.3.22) that, the instantaneous Stokes vector contains information on the intensity and polarization of electromagnetic wave, while IPPV mainly depicts the polarization characteristic of electromagnetic wave.

III. Antenna spatial instantaneous polarization cluster center and polarization divergence

For general instantaneous electromagnetic waves or complex modulated wide-band electromagnetic waves, their spatial polarization projection sets are usually a point set with a certain spatial distribution. The spatial polarization characteristic of the electromagnetic waves is mainly reflected as the cluster and distribution of polarization states on the polarization sphere. Accordingly, the spatial polarization cluster center and polarization divergence may be defined to quantitatively describe this spatial dispersion characteristic of the spatial polarization projection set. The spatial position of the point set can be roughly determined by the polarization cluster center. The spatial density of the point set distribution can be described using the concept of polarization dispersion.

1. Spatial polarization cluster center

On the Poincare unit sphere, the IPPV of electromagnetic waves forms a 3-dimensional unit vector ordered set, namely the spatial instantaneous polarization projection set. It describes the evolution of instantaneous polarization characteristic of electromagnetic waves with time. The instantaneous polarization projection set is a spatial point set distributed on the unit sphere. The distribution state reflects the overall polarization characteristic of antenna radiated electromagnetic wave.

Set $A = \{a(P), P \in \Omega\}$ as a weighting factor support set on the Ω support. Then

$$a(P) \geq 0 (\forall P \in \Omega), \quad \int_{\Omega} a(P) \mathrm{d}P = 1 \qquad (2.3.24)$$

With the weighting factor set A given, the spatial polarization weighted cluster center $G_P[A]$ of the spatial instantaneous polarization projection set Π_P of electromagnetic wave is defined as:

$$G_P[A] = A \circ \Pi_P = \int_{\Omega} a(P) g_{\mathrm{norm}}(P) \mathrm{d}P \qquad (2.3.25)$$

It can be seen from formula (2.3.25) that $\|G_P[A]\| \leq 1$. The looser the spatial distribution of the polarization projection set is, the closer the polarization cluster center is to the origin. To put it another way, the denser the spatial distribution of polarization projection set is, the closer the polarization cluster center is to the unit sphere aperture.

2. Spatial polarization divergence

The spatial position of the instantaneous polarization projection set of electromagnetic wave can be roughly determined by the polarization cluster center. The spatial density of the point set distribution can be described using the concept of polarization divergence:

$$\boldsymbol{Div}_{(P)}^{(k)}[A] = \int_{\Omega} a(\boldsymbol{P}) \|\boldsymbol{g}_{\text{norm}}(\boldsymbol{P}) - \boldsymbol{G}_P[A]\|^k \mathrm{d}\boldsymbol{P} \qquad (2.3.26)$$

It can be seen from the above formula that $0 \le \boldsymbol{Div}_{(P)}^{(k)}[A] \le 1$, where k is a positive integer, called the order of polarization divergence. In actuality, $k = 1$ or 2 is most commonly used. From the definition of visible, polarization divergence can be interpreted as the polarization projection set with respect to the polarization space weighted average distance of cluster center. For a given full factor set, a greater polarization divergence of electromagnetic wave indicates a looser spatial distribution of the polarization projection set, i.e. an intenser variation of polarization state of electromagnetic wave. In other words, a smaller polarization divergence of electromagnetic wave indicates a denser spatial distribution of the polarization projection set. In particular, with polarization state of electromagnetic wave constant, $\boldsymbol{Div}_{(P)}^{(k)}$ equals zero.

If $\boldsymbol{\Omega}$ is a non-zero measurable set and its measure is a finite value, then $0 < m(\boldsymbol{\Omega}) < +\infty$, where $m(\cdot)$ represents the Lebesgue measure. Let $a(\boldsymbol{P}) < 1/m(\boldsymbol{\Omega})$, then

$$\boldsymbol{G}_{P1} = \boldsymbol{G}_P[A] = \int_{\Omega} \boldsymbol{g}_{\text{mon}}(\boldsymbol{P}) \mathrm{d}\boldsymbol{P} / m(\boldsymbol{\Omega}) \qquad (2.3.27)$$

It can be seen from formula (2.3.27) that $\boldsymbol{G}_{P1}$ is essentially the uniform weighted average of polarization projection set of an electromagnetic signal. Hence it can be called the uniform weighted cluster center of the electromagnetic signal. Now, the corresponding polarization divergence can be determined as

$$\boldsymbol{Div}_{(P1)}^{(k)}[A] = \int_{\Omega} \|\boldsymbol{g}_{\text{nom}}(\boldsymbol{P}) - \boldsymbol{G}_{P1}[A]\|^k \mathrm{d}\boldsymbol{P} / m(\boldsymbol{\Omega}) \qquad (2.3.28)$$

Let $a(\boldsymbol{P}) = g_0(\boldsymbol{P}) / \int_{\Omega} g_0(\boldsymbol{P}) \mathrm{d}P$, then

$$\boldsymbol{G}_{P2} \equiv \boldsymbol{G}_P[A] = \int_{\Omega} g(\boldsymbol{P}) \mathrm{d}P / \int_{\Omega} g_0(\boldsymbol{P}) \mathrm{d}P \qquad (2.3.29)$$

It can be seen from formula (2.3.29) that $\boldsymbol{G}_{P2}$ is essentially obtained by normalizing the total energy of electromagnetic wave in the spatial domain of interest after integral of full-polarization sub-vector of the spatial instantaneous Stokes vector of the electromagnetic wave. $\boldsymbol{G}_{P2}$ can also be deemed the weighted average energy of polarization projection set. Hence, it can be called the weighted energy polarization cluster center. Here, the corresponding polarization divergence can be determined using the formula:

$$\boldsymbol{Div}^{(k)}_{(P2)}[A] = \int\limits_{\Omega} g_0(\boldsymbol{P}) \|g_{\mathrm{nom}}(\boldsymbol{P}) - \boldsymbol{G}_{P2}\|^k \mathrm{d}\boldsymbol{P} / \int\limits_{\Omega} g_0(\boldsymbol{P}) \mathrm{d}\boldsymbol{P} \qquad (2.3.30)$$

The above concept of polarization divergence is formulated based on the general set. Thus it applies to the description of distribution of any spatial point set.

IV.   Included angle of antenna spatial instantaneous polarization

In discussing the effect of antenna parasitic polarization on polarization purity, we describe this effect using the angle between actual polarization and expected polarization Stokes sub-vector. This angle is recorded as "included angle of spatial instantaneous polarization $\beta(\boldsymbol{P})$". Suppose that expected polarization is $J_e(\boldsymbol{P})$ and actual polarization is $J(\boldsymbol{P})$, then the corresponding spatial polarization projection vectors are $\boldsymbol{g}_e(\boldsymbol{P})$ and $\boldsymbol{g}(\boldsymbol{P})$, respectively.

$$\cos\beta(\boldsymbol{P}) = \frac{\boldsymbol{g}_e^{\mathrm{T}}(\boldsymbol{P})\boldsymbol{g}(\boldsymbol{P})}{g_{0,e}(\boldsymbol{P})g_0(\boldsymbol{P})} \qquad (2.3.31)$$

We can see that, for a given spatial angular position $\boldsymbol{P}$, the smaller the absolute value of β is, the greater the value of $\cos\beta$ is, the closer of actual polarization is to the design value, and the higher the polarization purity of antenna is. To put it another way, the greater the absolute value of β is, the smaller the value of $\cos\beta$ is, the farther of actual polarization is to the design value, and the lower the polarization purity of antenna is. In fact, β is the spherical angle included by the point on the Poincare polarization sphere corresponding to the expected polarization $J_e(\boldsymbol{P})$ and the point on the Poincare polarization sphere corresponding to the actual polarization $J(\boldsymbol{P})$. Thus, it is possible to discuss the variation of antenna polarization characteristic during antenna scanning (i.e. $\boldsymbol{P}(\theta, \varphi)$ varying in a certain spatial domain) by analyzing the rule of $\beta(\boldsymbol{P})$ variation. For example, with $\boldsymbol{P}$ varying, finding the average $E[\beta(\boldsymbol{P})]$ and the variance $\mathrm{var}[\beta(\boldsymbol{P})]$ of $\beta(\boldsymbol{P})$ can obtain the distribution center and degree of dispersion of $\beta(\boldsymbol{P})$.

V.   Antenna spatial instantaneous polarization state variation rate

Polarization projection set is an ordered set, i.e. a spatial point set with spatial coordinates as order parameters. It describes the characteristic of instantaneous polarization of electromagnetic wave variation with spatial coordinates. This spatial variation characteristic is an inherent property of electromagnetic wave polarization projection set. If the electromagnetic wave has a uniform and continuous spatial energy spectrum, and its polarization projection set is continuous relative to the spatial variables, then the Stokes sub-vector must be continuous. From the definition of electromagnetic wave spatial instantaneous polarization projection set, it is known that the electromagnetic wave polarization projection set must be a continuous spatial curve. If the polarization cluster center and polarization divergence are described as the "static" spatial distribution characteristics of a polarization projection set, then, the "antenna spatial instantaneous polarization state variation rate" discussed below characterizes the "dynamic" evolution characteristic of the polarization projection set.

Suppose that the spatial instantaneous polarization projection set of electromagnetic wave is $\prod_P = \{g_{\text{norm}}(P), P \in \Omega\}$, then define the instantaneous polarization state variation rate vector as

$$V_P^{(n)}(P) = \frac{\mathrm{d}^n}{\mathrm{d}P^n} g_{\text{norm}}(P) \qquad (2.3.32)$$

where n is a positive integer, called the order of variation rate. In actuality, $n = 1$ is most commonly used. $V_P^{(1)}(P)$ is called the first-order instantaneous polarization state variation rate vector of electromagnetic wave. It can be seen from the definition that the vector by nature actually indicates the variation direction of electromagnetic wave instantaneous polarization in this spatial position. Modulus of the vector $\left\| V_P^{(1)}(P) \right\|$ describes the variation rate of electromagnetic wave instantaneous polarization in this spatial position. Hence, the instantaneous polarization state variation rate reflects the dynamic evolution characteristic of electromagnetic wave polarization.

VI. Description of dispersion of antenna spatial polarization characteristic

In actual application, generally it is necessary to perform dispersion sampling of electromagnetic signals for the sake of subsequent computerized/digital processing. Therefore, it is really necessary to define the method for describing dispersion of electromagnetic wave spatial instantaneous polarization. For the convenience of presentation, the IPPV of polarization sampling sequence is referred to as "IPPS" [6]. Since the antenna spatial polarization characteristic essentially is intended to describe the distribution situation of antenna instantaneous polarization states at different spatial points, including the overall distribution situation of polarization states, central position of polarization distribution, and density of distribution of spatial points, therefore, several typical polarization descriptors may be used to characterize the distribution situation.

On the horizontal and vertical polarization bases, for the plane electromagnetic wave propagating in space, $E(P)$, $P \in \Omega$, its dispersion sampling sequence (referred to as polarization sampling sequence) can be denoted by $E_{\text{HV}}(n)$, $n = 1, 2, \ldots, M$.

Then, the instantaneous Stokes vector and IPPV of the electromagnetic wave polarization sampling sequence are defined as follows

$$\dot{J}_{\text{HV}}(n) = \begin{bmatrix} g_{\text{HV0}}(n) \\ g_{\text{HV}}(n) \end{bmatrix} = RE_{\text{HV}}(n) \otimes E_{\text{HV}}^*(n) \qquad (2.3.33)$$

and

$$\tilde{g}_{HV}(n) = [\tilde{g}_{HV1}(n), \tilde{g}_{HV2}(n), \tilde{g}_{HV3}(n)]^T = \frac{g_{HV}(n)}{g_{HV0}(n)}, \quad n = 1, 2, \ldots, M \quad (2.3.34)$$

The dispersion forms of the spatial polarization cluster center and polarization divergence of IPPS can thus be given as:

$$\tilde{G}_{HV} = \frac{1}{M} \sum_{n=1}^{M} a(n) \tilde{g}_{HV}(n) \quad (2.3.35)$$

and

$$D_{HV}^{(k)} = \frac{1}{M} \sum_{n=1}^{M} a(n) \left\| \tilde{g}_{HV}(n) - \tilde{G}_{HV} \right\|^k \quad (2.3.36)$$

where, $k \in N$, called the order of polarization divergence, and $a(n)$ is the weighting factor sequence meeting the formula:

$$a(n) \geq 0, \quad \forall n \in [1, 2M] \ \& \ \sum_{(n=1)}^{M} a(n) = M \quad (2.3.37)$$

In particular, if

$$a(n) = 1, \quad \forall n \in [1, 2, \ldots, M] \quad (2.3.38)$$

Then

$$\tilde{G}_{HV} = \frac{1}{M} \sum_{n=1}^{M} \tilde{g}_{HV}(n) \quad (2.3.39)$$

is the uniform weighted cluster center.

Moreover, when the order of polarization divergence $k = 2$,

$$D_{HV}^{(2)} = \frac{1}{M} \sum_{n=1}^{M} \tilde{g}_{HV}(n) \tilde{g}_{HV}^T(n) - \frac{1}{M^2} \sum_{n=1}^{M} \sum_{j=1}^{M} \tilde{g}_{HV}(n) \tilde{g}_{HV}^T(j) \quad (2.3.40)$$

For the polarization descriptors such as antenna spatial instantaneous polarization ratio, antenna spatial instantaneous polarization purity angle, and antenna spatial instantaneous polarization state variation rate, dispersion forms may also be taken for characterization, which will not be detailed in this book.

The polarization trajectory of the electromagnetic wave radiated by an antenna is no longer an ellipse when the antenna is excited by non-time-harmonic signals. Therefore, it is difficult to characterize the polarization using the elliptical polarization descriptor. However, it is possible to extend such characterization methods as spatial polarization ratio and spatial polarization phase descriptor to describe the spatial dynamic variation information of the antenna radiated electromagnetic wave.

Chapter 3
Spatial Polarization Characteristics of Aperture Antenna

Antenna is the image and emblem of radar, which is closely related to the radar system. The design of new radar should first select the antenna system and scheme. There are many kinds of antenna, which can be roughly divided into two categories: wire antenna and aperture antenna. Wire antenna consists basically of metal wires. This category includes dipole antenna, monopole antenna, coil antenna, helical antenna, Yagi antenna, logarithm periodic antenna, and traveling-wave antenna. Aperture antenna also known as aperture antenna is usually an aperture in a plane or curved face. This category includes horn antenna, reflector antenna, slot antenna, and microstrip antenna.

Reflector antenna is notable for simple structure and low price. Being capable of realizing a good many kinds of beams (needle-shaped, sector-shaped, special-shaped, multi-beam), they can meet various requirements on tactical and technical performances. As the most important kind of aperture antenna, they have been widely used in active-duty radar equipment. Like an array antenna, a reflector antenna is capable of achieving high gain, low side-lobe, and wide coverage at elevation angle. Other benefits of reflector antenna are low cast, absence of grating lobe, wideband, as well as easy realization of circular or variable polarization. Reflector antenna is playing an important role in active-duty conical scanning and monopulse radars. With the users' demand for wideband and ultra wideband radars (to improve the capability of anti-electronic-jamming) and variable-polarization radars (to provide the target identification function), reflector antenna still have a great room for application and development.

Using the methods for characterization of antenna spatial polarization characteristics discussed previously, this chapter systematically studies the spatial polarization characteristics of typical antenna, with emphasis placed on the parabolic antenna. Spatial polarization characteristics of various wire antenna and aperture antenna, including short dipole antenna, orthogonal dipole antenna, coil antenna, helical antenna, typical waveguide radiators, horn antenna, and offset parabolic antenna, are theoretically deduced in this chapter with the computer simulation results provided. The spatial polarization characteristics of two kinds of antenna are

© National Defense Industry Press, Beijing and Springer Nature Singapore Pte Ltd. 2019   57
H. Dai et al., *Spatial Polarization Characteristics of Radar Antenna*,
https://doi.org/10.1007/978-981-10-8794-3_3

analyzed based on the data measured in anechoic chamber. Finally, four typical models of spatial polarization characteristics of antenna are established, bringing about a set of important conclusions. Due to the limitation on research time, this chapter only selects some typical antenna to discuss their spatial polarization characteristics. For subsequent use of antenna spatial polarization characteristics in target polarization characteristic measurement, anti-active-jamming, discrimination between true and false targets, and improvement on polarization information processing capability, the research results given in this chapter can provide a theoretical basis and strong evidence.

3.1 Spatial Polarization Characteristics of Typical Wire Antenna

Wire antenna is the most basic, earliest, and widely-used form of practical antenna. There are many types of wire antenna. In this section, short dipole antenna, orthogonal dipole antenna, coil antenna, and helical antenna are taken as examples for discussion.

3.1.1 Spatial Polarization Characteristics of Dipole Antenna

Dipole antennas are widely used in long and medium wave, short wave and ultra-short wave frequency bands. In microwave frequency bands, sometimes they are also used as the feeds for reflector antenna. Next, we first discuss the far field generated by the dipole pointing along the coordinate axis, and then obtain the spatial polarization characteristics of the dipole pointing in any direction.

In space spherical coordinates (r, θ, φ) direction, the far fields generated by the short dipole pointing along the X, Y and Z axes are respectively:

$$E_r = 0, E_\theta = -\frac{j\omega\mu Il}{4\pi r}\cos\theta\cos\varphi \cdot e^{-jkr}, E_\varphi = \frac{j\omega\mu Il}{4\pi r}\sin\varphi \cdot e^{-jkr} \qquad (3.1.1)$$

and

$$E_r = 0, E_\theta = -\frac{j\omega\mu Il}{4\pi r}\cos\theta\sin\varphi \cdot e^{-jkr}, E_\varphi = \frac{j\omega\mu Il}{4\pi r}\cos\varphi \cdot e^{-jkr} \qquad (3.1.2)$$

as well as

$$E_r = 0, \ E_\theta = \frac{j\omega\mu Il}{4\pi r}\sin\theta \cdot e^{-jkr}, \ E_\varphi = 0 \qquad (3.1.3)$$

where μ(H/m) is the permeability of the medium, and I is the current passing at all points of the dipole whose length is l.

For a short dipole pointing in any direction, initially place it horizontally on the X axis (Fig. 3.1a); in the horizontal plane rotate it counterclockwise to form an angle φ' with the X axis (Fig. 3.1b); and in the vertical plane, rotate it to form an angle θ' with the Z axis (Fig. 3.1c). Obvious, for the dipole placed on the X axis, $\theta' = \pi/2$, $\varphi' = 0$; for the dipole placed on the Y axis, $\theta' = \pi/2$, $\varphi' = \pi/2$; and for the dipole placed on the Z axis, $\theta' = 0$. Note: Fig. 3.1a shows the space angle coordinates of the point (θ, φ) to be measured.

Initially, the excitation current of the dipole in the X direction is I. For the dipole in any direction after rotation as shown in Fig. 3.1, its radiation field can be equivalent to the superposition of the radiation fields of the three dipoles placed at the point O and in the X, Y, and Z directions. The equivalent current components in the X, Y, and Z directions are:

$$I_X = I \cdot \sin\theta' \cdot \cos\varphi', \quad I_Y = I \cdot \sin\theta' \cdot \sin\varphi', \quad I_Z = I \cdot \cos\theta' \tag{3.1.4}$$

According to the electric field equation of short dipole [92], for the orientation (θ, φ) in the spatial to be measured, the synthetic electric field of the dipole at the orientation of (θ', φ') is:

$$E_\theta = -\frac{j\omega\mu Il}{4\pi r}e^{-jkr} \cdot \{\cos\theta \sin\theta' \cos(\varphi - \varphi') - \sin\theta \cos\theta'\} \tag{3.1.5}$$

$$E_\varphi = \frac{j\omega\mu Il}{4\pi r}e^{-jkr} \cdot \{\sin\theta' \sin(\varphi - \varphi')\} \tag{3.1.6}$$

The angles between the orientation (θ', φ') of the dipole and the orientation (θ, φ) of the spatial to be measured in the azimuth and elevation directions are recorded as:

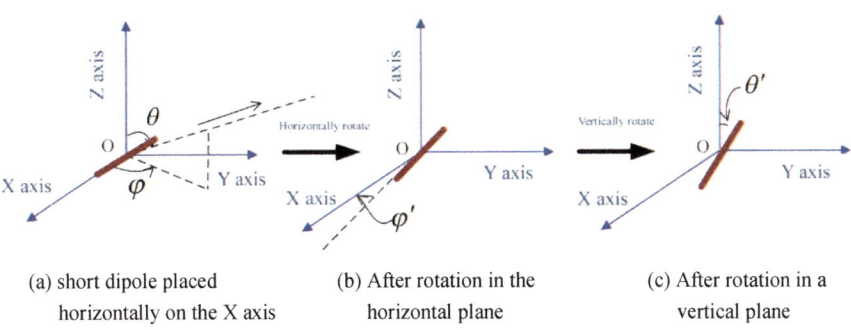

(a) short dipole placed   (b) After rotation in the   (c) After rotation in a
  horizontally on the X axis   horizontal plane   vertical plane

Fig. 3.1 Rotation of short dipole antenna

$$\Delta\theta = \theta - \theta', \quad \Delta\varphi = \varphi - \varphi' \tag{3.1.7}$$

Formula (3.1.5) can be written as

$$E_\theta = -\frac{j\omega\mu Il}{4\pi r} e^{-jkr} \cdot \{\cos(\theta' + \Delta\theta)\sin\theta'\cos\Delta\varphi - \sin(\theta' + \Delta\theta)\cos\theta'\} \tag{3.1.8}$$

$$E_\varphi = \frac{j\omega\mu Il}{4\pi r} e^{-jkr} \cdot \{\sin\theta'\sin\Delta\varphi\} \tag{3.1.9}$$

Using formulae (3.1.1)–(3.1.9), we can find the spatial polarization character-istics of the dipoles at different orientations:

(1) Polarization ratio of the dipole in the X direction

$$\rho_x = \frac{-E_\theta}{E_\varphi} = \cos\theta\,\mathrm{ctg}\varphi \tag{3.1.10}$$

It can be seen from formula (3.1.10) that the polarization ratio of the dipole in the X direction is the function of the orientation of antenna in spatial domain. For the dipole in the X direction, the orientation of the antenna is $\theta' = \pi/2$, $\varphi' = 0$, then $\Delta\theta = \theta - \pi/2$, $\Delta\varphi = \varphi$; and the above formula can be written as

$$\rho_x = -\sin\Delta\theta\,\mathrm{ctg}\Delta\varphi \tag{3.1.11}$$

Variation of the polarization ratio ρ_x with the space azimuth included angle $\Delta\varphi$ and elevation included angle $\Delta\theta$ is plotted in Fig. 3.2. Figure 3.2a is the three-dimensional map for the space distribution of polarization ratios. Figure 3.2b shows curves for variation of ρ_x with elevation included angle, with each curve

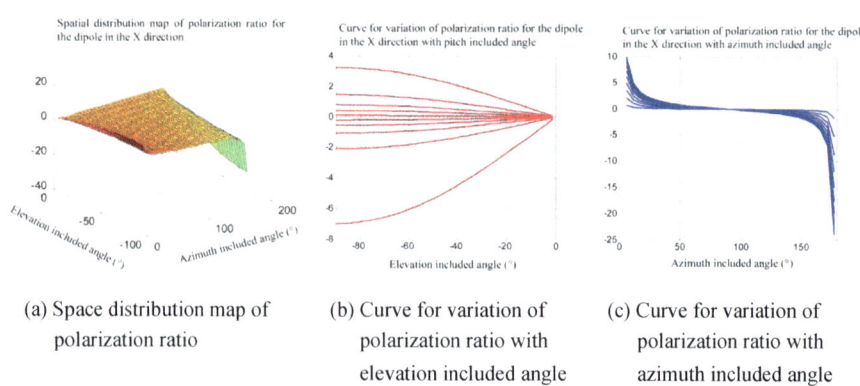

(a) Space distribution map of polarization ratio

(b) Curve for variation of polarization ratio with elevation included angle

(c) Curve for variation of polarization ratio with azimuth included angle

Fig. 3.2 Spatial distribution of polarization ratio for dipole antenna in the X direction

representing a certain azimuth included angle. Figure 3.2c shows curves for variation of ρ_x with azimuth included angle, with each curve representing a certain elevation included angle.

It can be seen from the above figure that, with the angle between the orientation of the spatial to be measured and that of the antenna varying, the polarization ratio of the dipole antenna in the X direction varies, experiencing the states of horizontal polarization, vertical polarization, and multiple linear polarization.

(2) Polarization ratio of the dipole in the Y direction

$$\rho_y = -E_\theta/E_\varphi = -\cos\theta \mathrm{tg}\varphi \tag{3.1.12}$$

Now, the orientation of the antenna is $\theta' = \pi/2$, $\varphi' = \pi/2$, then $\Delta\theta = \theta - \pi/2$, $\Delta\varphi = \varphi - \pi/2$; the above formula can be written as:

$$\rho_y = -\sin\Delta\theta \mathrm{ctg}\Delta\varphi \tag{3.1.13}$$

By comparing formula (3.1.11) with formula (3.1.13), it can be seen that the polarization ratio ρ_x of the dipole in the X direction and the polarization ratio ρ_y of the dipole in the Y direction have the same expression.

(3) Polarization ratio of the dipole in the Z orientation

$$\rho_z = -E_\theta/E_\varphi = \infty \tag{3.1.14}$$

It can be seen from the above formula that the dipole in the Z direction is vertical linear polarization.

(4) Polarization ratio of the dipole in any orientation

Record the orientation of the dipole antenna as $P'(\theta', \varphi')$, the orientation of the spatial to be measured as $P(\theta, \varphi)$, and the angle between them as $\Delta P(\Delta\theta, \Delta\varphi)$, the find the spatial polarization ratio:

$$\rho(P', \Delta P) = \frac{-E_\theta}{E_\varphi} = \frac{\cos(\theta' + \Delta\theta)\sin\theta'\cos\Delta\varphi - \sin(\theta' + \Delta\theta)\cos\theta'}{\sin\theta'\sin\Delta\varphi} \tag{3.1.15}$$

It can be seen from the above formula that the spatial polarization ratio of the dipole in any orientation is associated with spatial polarization ratio, elevation orientation θ', elevation included angle $\Delta\theta$ and azimuth included angle $\Delta\varphi$ between position to be measured and antenna beam. In order to more directly describe the variation of polarization ratio of short the dipole antenna in space, Fig. 3.3 shows the curves for variation of polarization ratio with azimuth included angle $\Delta\varphi$ and elevation included angle $\Delta\theta$ with the antenna beam pointing in different elevation

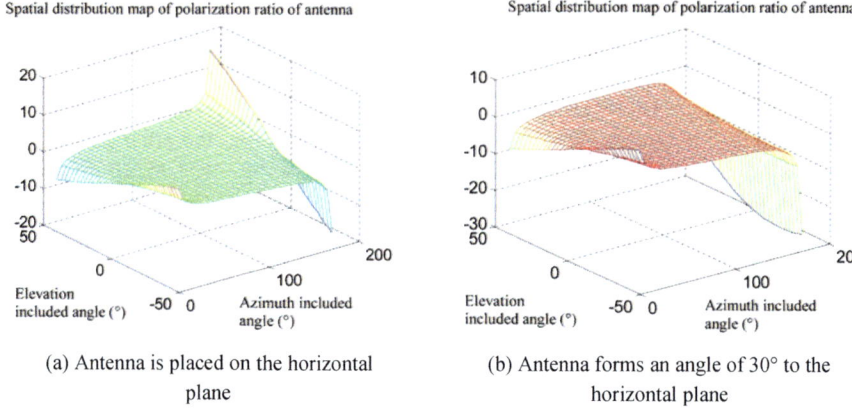

(a) Antenna is placed on the horizontal plane

(b) Antenna forms an angle of 30° to the horizontal plane

Fig. 3.3 Spatial distribution of polarization ratio of dipole antenna in any orientation

directions. Figure 3.3a, b show the situations $\theta' = 90°$ (i.e., the horizontal plane) and $\theta' = 60°$ (i.e., the plane forming an angle of 30° to the horizontal plane), respectively.

It can be seen from the above analysis that the spatial polarization characteristic of a short dipole antenna is not invariable, but varies according to a certain rule experiencing horizontal polarization, vertical polarization, and multiple intermediate polarization states.

3.1.2   Spatial Polarization Characteristics of Orthogonal Dipole Antenna

The orthogonal dipole antenna can be used to generate circularly-polarized waves. If the same dipoles are placed in the X and Y directions, and currents are fed with the same amplitude and phase difference of $\pi/2$, the waves radiated in the Z direction are circularly-polarized waves. If the current or voltage fed to the dipole in the X direction is taken as the reference, the current fed to the dipole in the Y direction is advanced by $\pi/2$, and the amplitude of the excitation current is I, then the electric field of the orthogonal dipole is the synthesis of the electric fields of the dipoles in the X and Y directions, i.e., the sum of the formula (3.1.1) and the formula (3.1.2) multiplied by j. and the expressions are as follows:

$$E_\theta = -\frac{j\omega\mu Il}{4\pi r} e^{-jkr}(\cos\theta\cos\varphi + j\cos\theta\sin\varphi) \tag{3.16}$$

$$E_\varphi = \frac{j\omega\mu Il}{4\pi r} e^{-jkr}(\sin\varphi - j\cos\varphi) \tag{3.1.17}$$

Polarization ratio

$$\rho = \frac{\cos\theta\cos\varphi + j\cos\theta\sin\varphi}{\sin\varphi - j\cos\varphi} = j\cos\theta \tag{3.1.18}$$

On the Z axis, $\theta = 0$, $\rho = j$, corresponding to the left-hand circularly-polarized wave propagating along the Z direction. On the XOY plane, $\theta = \pi/2$, $\rho = 0$, corresponding to the radiated horizontal linearly-polarized wave; different elliptically-polarized waves are radiated in other elevation directions. If the phase fed to the dipole in the Y direction lags by $\pi/2$ compared to that fed to the dipole in the X direction, the wave along the Z axis will be the right-handed circularly polarized wave.

Variation of spatial polarization characteristic of the orthogonal dipole rotating in the azimuth and elevation directions will be further discussed below. Set the initial state as shown in Fig. 3.4a. In the X and Y directions, the initial orientations of the dipole in the spherical coordinate system are $(\pi/2, 0)$ and $(\pi/2, 0)$, respectively. With horizontal rotation by φ', the orientation of the dipole in the X direction becomes $(\pi/2, \varphi')$, and that in the Y direction becomes $(\pi/2, \varphi' + \pi/2)$, as shown in Fig. 3.4b. With further rotation in the elevation direction, the orientation of the dipole in the X direction becomes (θ', φ'), and that in the Y direction becomes $(\theta', \varphi' + \pi/2)$, as shown in Fig. 3.4c.

The radiation field of the dipole after rotation can be equivalent to the superposition of the radiation fields of the three dipoles placed on the coordinate origin and in the X, Y, and Z directions. After rotation: the three equivalent current components of the dipole in the X, Y, and Z directions are respectively:

$$I_{Xx} = I \cdot \sin\theta' \cdot \cos\varphi, \quad I_{Xy} = I \cdot \sin\theta' \cdot \sin\varphi', \quad I_{Xz} = I \cdot \cos\theta' \tag{3.1.19}$$

$$I_{Yx} = I \cdot \sin\theta' \cdot \sin\varphi', \quad I_{Yy} = I \cdot \sin\theta' \cdot \cos\varphi', \quad I_{Yz} = I \cdot \cos\theta' \tag{3.1.20}$$

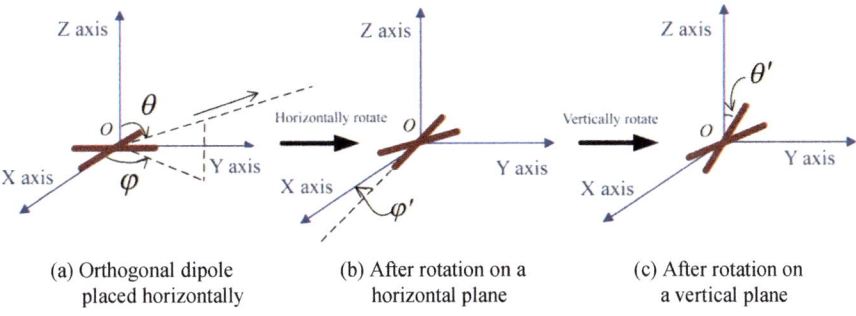

(a) Orthogonal dipole placed horizontally

(b) After rotation on a horizontal plane

(c) After rotation on a vertical plane

Fig. 3.4 Schematic of rotation of orthogonal dipole antenna

According to the electric field equation of the short dipole, the synthetic electric field of the orthogonal dipole in the spatial orientation (θ, φ) to be measured is:

$$E_\theta = -\frac{j\omega\mu Il}{4\pi r} e^{-jkr} \cdot \left\{ \begin{array}{l} [\cos\theta \sin\theta' \cos\Delta\varphi - \sin\theta \cos\theta'] \\ +j \cdot [\cos\theta \sin\theta' \sin\Delta\varphi - \sin\theta \cos\theta'] \end{array} \right\} \qquad (3.1.21)$$

$$E_\varphi = \frac{j\omega\mu Il}{4\pi r} e^{-jkr} \cdot \{\sin\theta' \sin\Delta\varphi - j \cdot \sin\theta' \cos\Delta\varphi\} \qquad (3.1.22)$$

where, $(\Delta\theta, \Delta\varphi)$ represents the included elevation and azimuth angles between antenna beam orientation and the spatial orientation to be measured.

Record the beam orientation of the dipole in the X direction as $P'(\theta', \varphi')$, the spatial orientation to be measured as $P(\theta, \varphi)$, and the corresponding included angle as $\Delta P(\Delta\theta, \Delta\varphi)$, then we can find the spatial polarization ratio:

$$\begin{aligned} \rho(P', \Delta P) &= j \cos\theta + (1-j) \sin\theta \operatorname{ctg}\theta' \cdot e^{-j\Delta\varphi} \\ &= j \cos(\theta' + \Delta\theta) + (1-j) \sin(\theta' + \Delta\theta) \operatorname{ctg}\theta' \cdot e^{-j\Delta\varphi} \end{aligned} \qquad (3.1.23)$$

It can be seen that the spatial polarization ratio of the orthogonal dipole antenna is the function of antenna beam orientation and the position of the antenna beam relative to the space orientation to be measured. Antenna spatial polarization characteristic variation with azimuth included angle and elevation included angle is discussed below for typical cases.

(1) If $\theta' = \pi/2$, the formula (3.1.23) can be simplified as $\rho = -j\sin\Delta\theta$. In addition, $\Delta\theta = \theta - \theta'$. Then, the expression of antenna polarization ratio is the same as the formula (3.1.18). Thus, if the orthogonal dipole antenna is placed on the XOY plane, the polarization state of the antenna in the horizontal direction remains unchanged; but in different elevation directions, the antenna polarization ratio varies sinusoidally with the elevation included angle $\Delta\theta$ experiencing multiple polarization states such as linear polarization, elliptical polarization, and circular polarization. As shown in Fig. 3.5, the shaded area indicates different elliptical polarization state of the antenna.

(2) With $\theta' = 0$, the dipoles placed on the X and Y axes become the dipoles in the Z direction. Then, the polarization ratio of the synthetic electric field is $\rho = \infty$, the antenna being at the vertical polarization state.

(3) In order to directly characterize the rule for space variation of antenna polarization characteristic with beam orientation of orthogonal dipole, we consider two typical situations $\theta' = 75°$ (the antenna forms an angle of 15° to the horizontal plane) and $\theta' = 60°$ (the antenna forms an angle of 30° to the horizontal plane). The curves for variation of polarization ratio with the elevation included

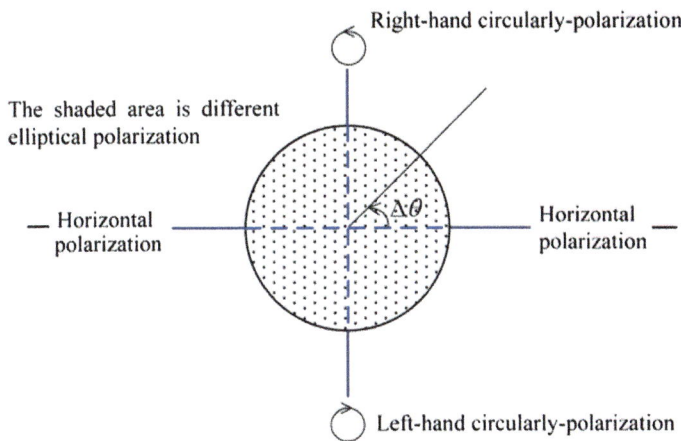

Fig. 3.5 Spatial distribution of polarization states of orthogonal dipole antenna

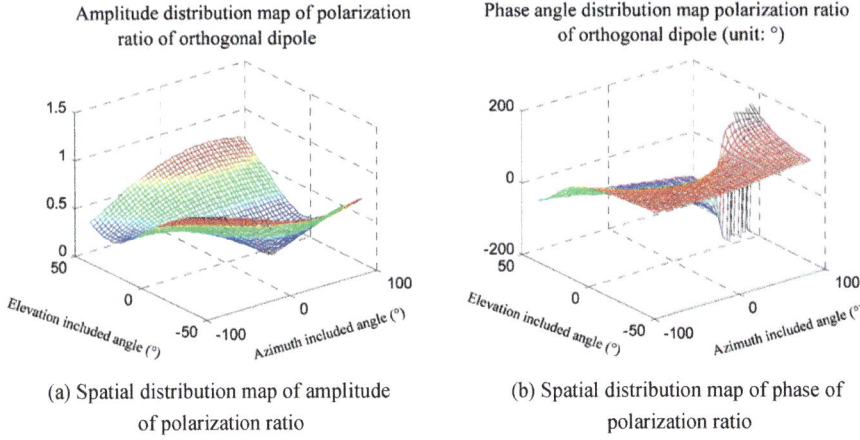

(a) Spatial distribution map of amplitude of polarization ratio

(b) Spatial distribution map of phase of polarization ratio

Fig. 3.6 Spatial distribution maps of polarization ratio of orthogonal dipole with an elevation angle of 15°

angle $\Delta\theta$ and the azimuth included angle $\Delta\varphi$ are plotted as shown in Figs. 3.6 and 3.7, respectively. Record $\rho = |\rho| \cdot e^{j\phi}$. Figures 3.6a and 3.7a are the spatial distribution maps of amplitude $|\rho|$ of polarization ratio of antenna; Figs. 3.6b and 3.7b are the spatial distribution maps of polarization phase descriptor ϕ.

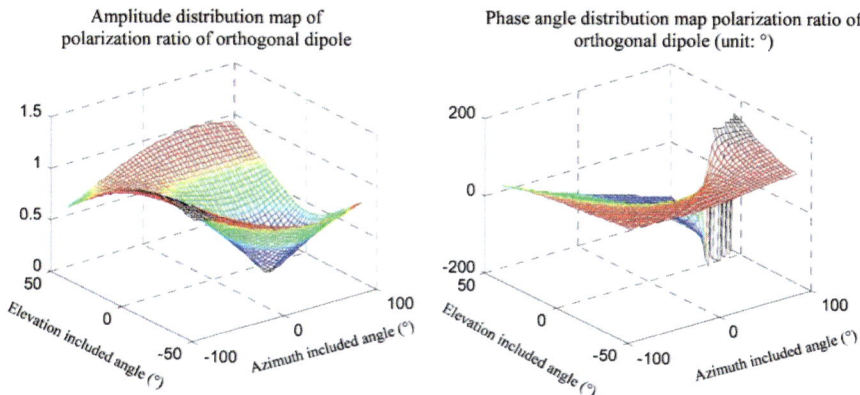

Fig. 3.7 Spatial distribution maps of polarization ratio of orthogonal dipole with an elevation angle of 30°

3.1.3   Spatial Polarization Characteristics of Coil Antenna

Coil antenna may be in the shape of a rectangle, a triangle, a circle or any other closed geometric shape. The characteristics of a coil antenna depends more on the dimension and current distribution of the coil, compared to the shape of the coil. As shown in Fig. 3.8, in this section we first take a small coil antenna with uniform current distribution to discuss its far-field distribution. On this basis, we then discuss the spatial polarization characteristic of a combination antenna that consists of a coil antenna and a dipole.

1. Circular antenna

The far-field of a small circular antenna with uniform in-phase current on the XOY plane is [7, 91]:

Fig. 3.8 Schematic of coil antenna and coordinate system

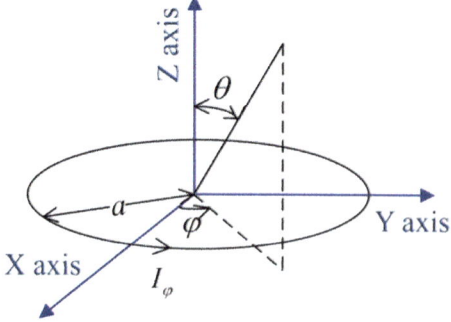

$$E_r = E_\theta = 0, \quad E_\varphi = \frac{\omega\mu k a^2 I_\varphi \sin\theta}{4r} \times e^{-jkr} \qquad (3.1.24)$$

where, μ(H/m) is the permeability of medium; a is the radius of the small circle; I_φ is the amplitude of current; $k = 2\pi/\lambda$ is the propagation constant in free space, λ being the operating wavelength.

The above formula is valid only when $a \ll \lambda$, i.e. in case the circle can be considered very small. In other cases, the coil antenna field can be expressed using the following general formula:

$$E_r = E_\theta = 0, \quad E_\varphi = \frac{\omega\mu a I_\varphi J_1(ka\sin\theta)}{2r} \times e^{-jkr} \qquad (3.1.25)$$

where, J_1 is the first kind of first-order Bessel function.

It can be seen from the formulae (3.1.24) and (3.1.25) that, the radiation field is in the horizontal polarization state, whether it is a small circular antenna or a large circular antenna.

2. Small circle and short dipole in the Z direction

It can be seen from the above analysis that the field of the small circle with uniform current distribution varies with $\sin\theta$ on the azimuth plane. It is shown by the formula (3.1.3) that, the field of the short dipole in the Z direction varies with $\sin\theta$ on the vertical plane (Note: the so-called "small" or "short" refers to the dimension is not greater than $\lambda/10$), and Fig. 3.9 is a schematic of the small circle and the short dipole.

Compare the field E_θ of the short dipole and the field E_φ of the small circle:

$$E_\theta = \frac{j\omega\mu I_d l}{4\pi r}\sin\theta \cdot e^{-jkr}, E_\varphi = \frac{\omega\mu k a^2 I_L \sin\theta}{4r} \times e^{-jkr} \qquad (3.1.26)$$

where, l is the length of the dipole, I_d is the current in the dipole, and I_L is the current in the circle.

Fig. 3.9 Combination of the small circle and the dipole in the Z direction

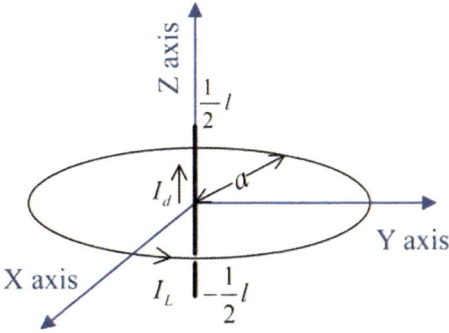

E_θ and E_φ are vertical to each other with a phase difference of 90°. Thus the synthetic electric field is elliptically polarized. The polarization ratio of the antenna is:

$$\rho = -j \cdot \frac{I_d l}{\pi k a^2 I_L} \tag{3.1.27}$$

Particularly, if the currents meet the following formula:

$$I_d / I_L = \pi k a^2 / l \tag{3.1.28}$$

A right-handed polarized wave is obtained. If any of the above currents reverses direction, then a left-handed polarized wave will be obtained.

3. A relatively large circle and a relatively long dipole in the Z direction

If a relatively large circle and a relatively long dipole are used in Fig. 3.1.9, the situation will change greatly. This combination antenna has remarkable spatial polarization characteristic.

The field of the dipole is

$$E_\theta = \frac{j Z_0 I_m}{2\pi r} \cdot \frac{\cos[(kl/2)\cos\theta] - \cos(kl/2)}{\sin\theta} \cdot e^{-jkr} \tag{3.1.29}$$

where, $Z_0 = \sqrt{\mu/\varepsilon}$ is the characteristic impedance in the medium, I_m is the current in the dipole, and l is the length of the dipole.

The field produced by the relatively large circle is

$$E_\varphi = \frac{\omega \mu a I_L J_1(ka \sin\theta)}{2r} \cdot e^{-jkr} \tag{3.1.30}$$

The polarization ratio can be obtained from the above two formulae:

$$\rho = -\frac{E_\theta}{E_\varphi} = C \cdot \frac{\cos[(kl/2)\cos\theta] - \cos(kl/2)}{\sin\theta \cdot J_1(ka\sin\theta)} \tag{3.1.31}$$

where, C is a constant, and $C = -j Z_0 I_m / \pi \omega \mu a I_L$.

It can be seen from the above formula that, the polarization ratio of antenna is an imaginary number. E_θ and E_φ are vertical to each other with a phase difference of 90°. Thus the synthetic electric field is elliptically polarized. The amplitude of antenna polarization ratio is a function of the spatial pitch θ. Obviously, by properly selecting the relative currents fed into the circle and the dipole, the antenna possibly radiates a circularly-polarized wave in the θ direction. If the wave is right-handed circularly polarized in the $\theta = \pi/2$ direction, then

$$C = \frac{-jJ_1(ka)}{1 - \cos(kl/2)} \tag{3.1.32}$$

Substituting this formula into the formula (3.1.31) can obtain the spatial polarization ratio of antenna at this moment:

$$\rho = \frac{-jJ_1(ka)}{1 - \cos(kl/2)} \cdot \frac{\cos[(kl/2)\cos\theta] - \cos(kl/2)}{\sin\theta \cdot J_1(ka\sin\theta)} \tag{3.1.33}$$

With the radius a of the coil antenna equal to the length l of the dipole (i.e. $a = l$), the normalized pattern on the elevation plane, the normalized pattern on the horizontal plane, and the spatial distribution map of amplitude of polarization ratio are shown in Fig. 3.10a–c, respectively, in the case of $l = 0.5\lambda$, and in Fig. 3.10d–f, respectively, in the case of $l = \lambda$.

It can be seen from the above figures that, with the increase of antenna dimension, the number of side-lobes of antenna pattern increases and the rule of antenna polarization ratio variation also changes.

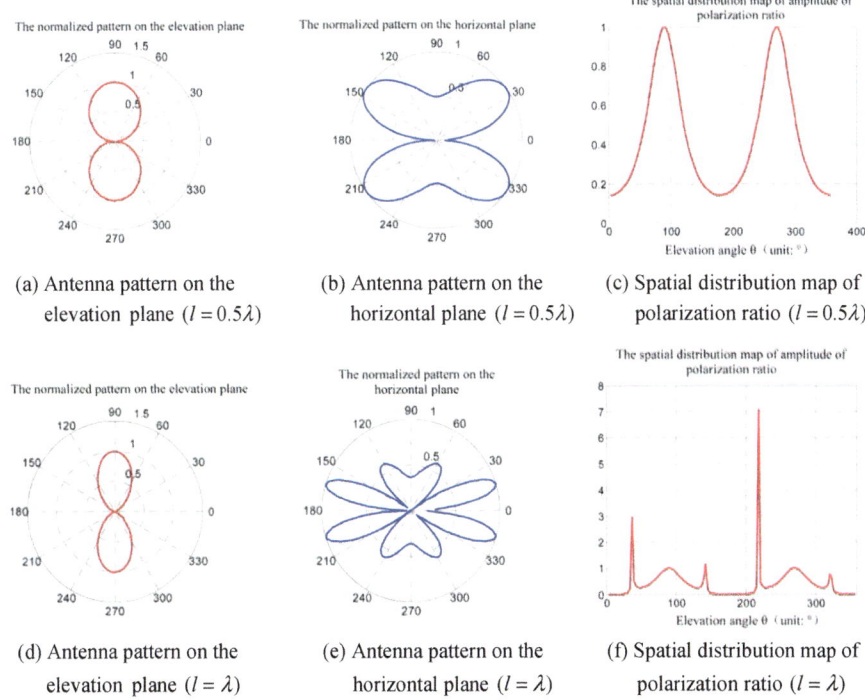

(a) Antenna pattern on the elevation plane ($l = 0.5\lambda$)

(b) Antenna pattern on the horizontal plane ($l = 0.5\lambda$)

(c) Spatial distribution map of polarization ratio ($l = 0.5\lambda$)

(d) Antenna pattern on the elevation plane ($l = \lambda$)

(e) Antenna pattern on the horizontal plane ($l = \lambda$)

(f) Spatial distribution map of polarization ratio ($l = \lambda$)

Fig. 3.10 Characteristic of the synthetic field of the large circle and the long dipole in the Z direction

3.1.4 Spatial Polarization Characteristics of Helical Antenna

Helical antenna is a kind of widely used wire antenna. In terms of value space for the ratio of helix diameter to wavelength, helical antenna can be divided into two types: normal-mode and axial-mode. The characteristic of normal-mode helical antenna is similar to that of monopole antenna. This section mainly discusses axial-mode helical antenna. The structure of axial-mode helical antenna is shown in Fig. 3.11, where a is the radius of the helix, h is the helical pitch, and L is the circumference of a turn.

For a single-turn planar circular antenna with traveling-wave current distribution, suppose that the current distribution on the circle is $J_l = Ie^{-jka\varphi'}\hat{\varphi}$, indicating the traveling wave along the direction of $+\varphi$. If the circumference of the circle is equal to the wavelength, $2\pi a = \lambda$, $ka = 1$. The radiation field of the current circle can be calculated, using the formula for calculating the far-field region vector potential of any current distribution, and the transformation between electric field and vector potential. For the convenience of calculation, let $\phi = \varphi' - \varphi$, and perform the transformation, to find the radiation field of the current circle as [91]:

$$E_\theta = \frac{\omega\mu}{4\pi r}e^{-jkr}e^{-j\varphi}Ia\cos\theta\int_0^{2\pi}\sin^2\phi\cos(\sin\theta\cos\phi)d\phi \tag{3.1.34}$$

$$E_\varphi = -j\frac{\omega\mu}{4\pi r}e^{-jkr}e^{-j\varphi}Ia\int_0^{2\pi}\cos^2\phi\cos(\sin\theta\cos\phi)d\phi \tag{3.1.35}$$

where, θ and φ represent the elevation and azimuth in the spherical coordinate system, respectively.

Thus, the polarization ratio of the current circle with the traveling wave distribution can be found as:

Fig. 3.11 Structure of a helical antenna

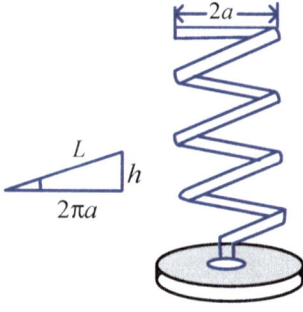

$$\rho = -\frac{E_\theta}{E_\varphi} = -j\cos\theta \cdot \frac{\int_0^{2\pi} \sin^2\phi \cos(\sin\theta\cos\phi)d\phi}{\int_0^{2\pi} \cos^2\phi \cos(\sin\theta\cos\phi)d\phi} \qquad (3.1.36)$$

Further,

$$\rho = j\cos\theta \cdot \frac{J_1(\sin\theta)}{J_2(\sin\theta) \cdot \sin\theta - J_1(\sin\theta)} \qquad (3.1.37)$$

where, J_1 is the first-kind first-order Bessel function, and J_2 is the first-kind second-order Bessel function.

It can be seen from the above two formulae that, in the direction of $\theta = 0$, the amplitude of E_θ is equal to that of E_φ, the phase of E_θ is in advance of that of E_φ by $\pi/2$, the antenna radiates right-handed circularly-polarized waves, and the polarization direction is consistent with the current direction. In the direction of $\theta = \pi/2$, $E_\theta = 0$, the radiation field has only the E_φ component, and the antenna radiates linearly-polarized waves. In other directions, the phases of E_θ and E_φ differ from each other by $\pi/2$, the amplitudes of E_θ and E_φ differ from each other, and the radiation field is elliptically polarized. Figure 3.12 shows the spatial distribution of amplitude and phase of polarization ratio.

For a multi-turn helical antenna, it has been demonstrated that, if the length of a turn is approximately one wave length, the current in the helical conductor is mainly the traveling wave along it. Compared with the single-turn plane circular antenna, the current in the helix has a small z component. In the formula (3.1.35), the integral has an item—a small phase difference caused by the displacement of the helix along the z axis. If the helical pitches h is sufficiently small, their effects are negligible. The radiation characteristic of single-turn helical current still exists, with the main radiation in the axial direction of the helix.

In this section, theoretical deduction and corresponding simulation of spatial polarization characteristic are performed for several typical wire antenna, including

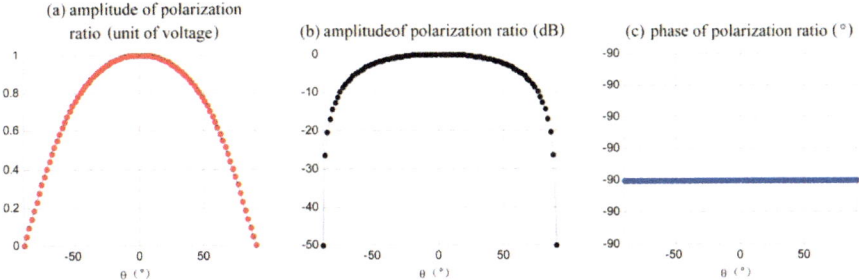

Fig. 3.12 Spatial distribution of amplitude and phase of polarization ratio for the single-turn planar circular antenna

Table 3.1 Spatial polarization characteristics of typical wire antenna

Wire antenna type		Spatial polarization ratio ρ
Short dipole and orthogonal dipole antenna	Dipole in the X direction	$\rho_x = -\sin\Delta\theta \operatorname{ctg}\Delta\varphi$
	Dipole in the Y direction	$\rho_y = -\sin\Delta\theta \operatorname{ctg}\Delta\varphi$
	Dipole in the Z direction	$\rho_z = \infty$
	Dipole in any direction	$\rho = \frac{\cos(\theta' + \Delta\theta)\sin\theta'\cos\Delta\varphi - \sin(\theta' + \Delta\theta)\cos\theta'}{\sin\theta'\sin\Delta\varphi}$
	Orthogonal dipole antenna	$\rho = \mathrm{j}\cos(\theta' + \Delta\theta) + (1 - \mathrm{j})\sin(\theta' + \Delta\theta)\operatorname{ctg}\theta' \cdot \mathrm{e}^{-\mathrm{j}\Delta\varphi}$
	Where, (θ', φ') represents the orientation of the dipole antenna in the spherical coordinate system, (θ, φ) represents the spatial orientation to be measured, and $\Delta\theta = \theta - \theta'$ and $\Delta\varphi = \varphi - \varphi'$ are the elevation included angle and the azimuth included angle between the spatial orientation to be measured and the orientation of the antenna, respectively	
Coil antenna, and combination of circle and dipole	Coil antenna	Horizontal linear polarization
	Small circle and short dipole in the Z direction	$\rho = -\mathrm{j} \cdot \frac{I_d l}{\pi k a^2 I_L}$
	Relatively large circle and relatively long dipole in the Z direction	$\rho = \frac{-\mathrm{j}J_1(ka)}{1-\cos(kl/2)} \cdot \frac{\cos[(kl/2)\cos\theta] - \cos(kl/2)}{\sin\theta \cdot J_1(ka\sin\theta)}$
	Where, θ represents the elevation angle defined in the spherical coordinate system as shown in Fig. 3.8	
Helical antenna	$\rho = \mathrm{j}\cos\theta \cdot \frac{J_1(\sin\theta)}{J_2(\sin\theta)\cdot\sin\theta - J_1(\sin\theta)}$, where, θ is the elevation angle in the spherical coordinate system	

short dipole antenna, orthogonal dipole antenna, coil antenna and helical antenna. It is shown by the analysis that, in different spatial positions, for most of wire antenna, their polarization states are not always invariable but vary complying with a certain rule, and different types of antenna follow different rules of spatial polarization characteristic variation. The spatial polarization characteristics of various wire antenna are shown in Table 3.1 for the convenience of comparison.

3.2 Spatial Polarization Characteristics of Typical Aperture Antenna

Common aperture antenna include horn antenna, parabolic antenna and lens antenna. Aperture antenna is also often known as aperture antenna. Aperture field radiation is the theoretical basis of aperture antenna. If the aperture field is known, the radiation of aperture field can be calculated by the equivalence principle. Waveguide opening can be considered as the simplest aperture antenna, but the aperture field distribution of waveguide opening is complex with poor radiation characteristic. Hence waveguide opening is seldom used directly as a radiator. In order to obtain better radiation characteristic, the waveguide opening is often gradually expanded into a horn. The horn antenna is structurally simple. Scattering factors have little impact on its radiation lobes. It is easy to separately control the radiation lobes on the two main planes. Horn antennas are usually used as the feed of parabolic antenna or the standard gain antenna. In this section, the polarization characteristics and their variation rules are firstly analyzed for the typical antenna of equivalent aperture field distribution. Then the spatial polarization characteristics of waveguide radiators and horn antenna are discussed. Spatial polarization characteristics of parabolic antenna, as a fairly important type of aperture antenna, will be discussed in Sect. 3.4.

3.2.1 Spatial Polarization Characteristics of Typical Equivalence Aperture Distribution Antenna

The planar aperture radiation can be calculated with the equivalent source method. For any antenna radiation source, an equivalent aperture can be found. According to the equivalent source theorem, the field of the source antenna can be approximately found by using the field distribution on the equivalent aperture, without having to care about the specific structure of the antenna and the specific generation method for the aperture distribution [120]. Since in many cases, it is impossible to accurately understand the type and parameter of a radar antenna, it is really necessary to investigate the spatial polarization characteristics of antenna with a common equivalent aperture (such as rectangular aperture, circular aperture, etc.). Figure 3.13 shows the aperture on the plane $z = 0$.

Applying the equivalence principle, set a conductive plane on the aperture. Performing a series of calculations can eventually find the radiation field of the aperture as [91]:

$$\begin{cases} E_\theta = jk_0 \frac{e^{-jk_0 r}}{2\pi r} (f_x \cos\varphi + f_y \sin\varphi) \\ E_\varphi = jk_0 \frac{e^{-jk_0 r}}{2\pi r} \cos\theta (f_y \cos\varphi - f_x \sin\varphi) \end{cases} \qquad (3.2.1)$$

Fig. 3.13 Schematic of
planar aperture

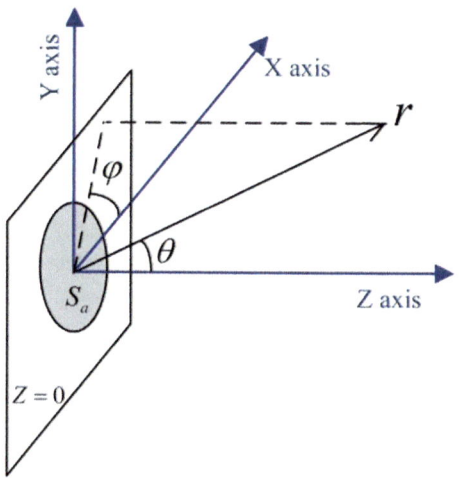

where, $f_x = \int_a E_{ax} e^{j(k_x x' + k_y y')} ds'$, $f_y = \int_a E_{ax} e^{j(k_x x' + k_y y')} ds'$, $k_x = k_0 \sin\theta \cos\varphi$, $k_0^2 = \omega^2 \varepsilon_0 \mu_0$ called wave number or propagation constant, ε_0(F/m) is the dielectric constant of free space, μ_0(H/m) is the permeability of free space and $k_y = k_0 \sin\theta \sin\varphi$.

Further, the polarization ratio can be found:

$$\mu = \frac{E_\theta}{E_\varphi} = \frac{f_x \cos\varphi + f_y \sin\varphi}{\cos\theta(f_x \sin\varphi - f_y \cos\varphi)} \tag{3.2.2}$$

For the aperture placed on the conductive ground, the aperture (other than the aperture itself) can be perfectly simulated into an infinite purely conductive plane. Then, the radiation field of the antenna can be calculated using the formula (3.2.1). Considering the boundary condition that the tangential electric field on the conductor is zero, the magnetic flux of the aperture electric field is zero outside the aperture. Assumption is usually made that the aperture field is a transverse electromagnetic (TEM) wave. For medium and high-gain antenna, this assumption is valid. It is usually successfully applied to the aperture dimensioned to only several wavelengths. Then the resulting aperture radiation field is [121]:

$$\begin{cases} E_\theta = jk_0 \dfrac{e^{-jk_0 r}}{2\pi r} \dfrac{1+\cos\theta}{2} (f_x \cos\varphi + f_y \sin\varphi) \\ E_\varphi = jk_0 \dfrac{e^{-jk_0 r}}{2\pi r} \dfrac{1+\cos\theta}{2} (f_y \cos\varphi - f_x \sin\varphi) \end{cases} \tag{3.2.3}$$

Further, the polarization ratio can be obtained:

$$\rho = -\frac{E_\theta}{E_\varphi} = \frac{f_x \cos\varphi + f_y \sin\varphi}{f_x \sin\varphi - f_y \cos\varphi} \tag{3.2.4}$$

Generally speaking, the formula (3.2.1) is often used for the aperture placed on the ground, while the formula (3.2.3) is used for the aperture antenna in free space. However, no matter the formula (3.2.2) or (3.2.4) is adopted to calculate antenna polarization ratio, we can reach the conclusion that, the polarization state of the aperture field radiated electromagnetic wave varies in the spatial domain, and the polarization ratio is a function of azimuth angle φ and elevation angle θ.

I. Radiation field of rectangular aperture

A rectangular aperture, the length and width of which are a and b, respectively, is placed on the plane $z = 0$. Suppose that the electric field of the aperture has only the y component:

$$E_\theta = jk_0 \frac{e^{-jk_0 r}}{2\pi r} f_y \sin\varphi, \quad E_\varphi = jk_0 \frac{e^{-jk_0 r}}{2\pi r} f_y \cos\theta \cos\varphi \tag{3.2.5}$$

and

$$f_y = \int_a E_{ay} e^{j(k_x x' + k_y y')} ds' \tag{3.2.6}$$

1. Aperture field of uniform distribution

Suppose that the field on the aperture is uniform, and:

$$E_a = \begin{cases} E_0 \hat{y} & |x| \le a/2, \quad |y| \le b/2 \\ 0 & \text{others} \end{cases} \tag{3.2.7}$$

The far-field radiation field of rectangular aperture of uniform distribution can be found to be:

$$\begin{cases} E_\theta = \frac{jk_0 abE_0}{2\pi r} e^{-jk_0 r} \frac{\sin u}{u} \frac{\sin v}{v} \sin\varphi \\ E_\varphi = \frac{jk_0 abE_0}{2\pi r} e^{-jk_0 r} \frac{\sin u}{u} \frac{\sin v}{v} \cos\varphi \cos\theta \end{cases} \tag{3.2.8}$$

where

$$u = \frac{k_x a}{2} = \frac{k_0 a}{2} \sin\theta \cos\varphi, \quad v = \frac{k_y b}{2} = \frac{k_0 b}{2} \sin\theta \sin\varphi \tag{3.2.9}$$

Further, the polarization ratio can be found:

$$\rho = -E_\theta / E_\varphi = -\text{tg}\varphi / \cos\theta \tag{3.2.10}$$

So, the polarization of this kind of aperture field is a function of spatial coordinates. On the azimuth plane, it varies with azimuth angle φ, showing the tangent relation. On the elevation plane, it varies with elevation angle θ, showing the secant relation.

2. Aperture field of non-uniform distribution

The truncated-cone distributions (i.e. small at both ends and large in the middle) are the most commonly used non-uniform distribution aperture fields. A typical example is cosine distribution. Taking the field distribution at the waveguide end with the TE_{10} mode propagating in the rectangular waveguide as an example, the aperture field is cosine distribution along the x direction and uniform distribution along the y direction:

$$E_{ay} = E_0 \cos \frac{\pi x}{a} \tag{3.2.11}$$

and

$$f_y = E_0 ab \cdot \frac{\cos u}{1 - (2u/\pi)^2} \cdot \frac{\sin v}{v} \tag{3.2.12}$$

Compared with the pattern of the aperture field of uniform distribution, the spatial polarization ratio of the aperture field of cosine distribution has the same expression $\rho = -E_\theta / E_\varphi = -\mathrm{tg}\varphi / \cos\theta$.

II. Radiation field of circular aperture
1. Aperture field of uniform distribution

Suppose that the circular aperture has a diameter of d and a uniform electric field distribution in the y direction. $\mathbf{E}_a = \hat{y} E_0$, and

$$f_y = \pi \left(\frac{d}{2}\right)^2 E_0 \cdot 2 J_1 \left(\frac{k_0 d}{2} \sin \theta\right) \bigg/ \left(\frac{k_0 d}{2} \sin \theta\right) \tag{3.2.13}$$

Radiation field of the antenna is:

$$\begin{cases} E_\theta = \frac{jk_0 E_0 d^2}{4r} e^{-jk_0 r} \cdot \sin \varphi \cdot J_1 \left(\frac{k_0 d}{2} \sin \theta\right) / \left(\frac{k_0 d}{2} \sin \theta\right) \\ E_\varphi = \frac{jk_0 E_0 d^2}{4r} e^{-jk_0 r} \cdot \cos \theta \cos \varphi \cdot J_1 \left(\frac{k_0 d}{2} \sin \theta\right) / \left(\frac{k_0 d}{2} \sin \theta\right) \end{cases} \tag{3.2.14}$$

Spatial polarization ratio is found to be:

$$\rho = -E_\theta / E_\varphi = -\mathrm{tg}\varphi / \cos\theta \tag{3.2.15}$$

We can see that this formula is the same as the formula (3.2.10).

2. Aperture field of non-uniform distribution

The approximate formula [123] for the aperture field distribution at the end of a circular waveguide excited by TE_{11} mode is:

$$\begin{cases} E_{ay} = E_0 \left[\frac{d}{2\chi p} J_1 \left(\frac{2\chi p}{d} \right) \sin^2 \varphi + J_1' \left(\frac{2\chi p}{d} \right) \cos^2 \varphi \right] \\ E_{ax} = E_0 \left[\frac{d}{2\chi p} J_1 \left(\frac{2\chi p}{d} \right) + J_1' \left(\frac{2\chi p}{d} \right) \right] \cos \varphi \sin \varphi \end{cases} \quad (3.2.16)$$

where, d is the diameter of the aperture, and χ is the first root of $J_1'(x)$.

Let $\frac{2\chi p}{d} = x$, then the above formula can be written as:

$$\begin{cases} E_{ay} = \frac{1}{2} E_0 [J_0(x) - J_2(x) \cos 2\varphi] \\ E_{ax} = \frac{1}{2} E_0 J_0(x) \sin 2\varphi \end{cases} \quad (3.2.17)$$

Let $y = k_0 p \sin \theta$, according to the Bessel formula [124], we can get:

$$\begin{cases} f_y = \int_0^{d/2} p \, \mathrm{d}p \int_0^{2\pi} E_{ay} e^{jk_0 p \sin \theta \cos \varphi} \mathrm{d}\varphi = 2SE_0 \frac{\chi^2 J_0(\chi) J_1'(u)}{\chi^2 - u^2} \\ f_x = \int_0^{d/2} p \, \mathrm{d}p \int_0^{2\pi} E_{ax} e^{jk_0 p \sin \theta \cos \varphi} \mathrm{d}\varphi = \frac{1}{2} E_0 \int_0^{d/2} p J_0(x) \mathrm{d}p \int_0^{2\pi} \sin 2\varphi e^{jy \cos \varphi} \mathrm{d}\varphi = 0 \end{cases} \quad (3.2.18)$$

where, $u = \frac{k_0 d}{2} \sin \theta$, and $S = \pi \left(\frac{d}{2} \right)^2$.

Using the formulae (3.2.1) and (3.2.18), the far-field distribution of the aperture is found to be:

$$\begin{cases} E_\theta = \frac{jk_0 SE_0}{\pi r} e^{-jk_0 r} \frac{\chi^2 J_0(\chi) J_1'(u)}{\chi^2 - u^2} \sin \varphi \\ E_\varphi = \frac{jk_0 SE_0}{\pi r} e^{-jk_0 r} \frac{\chi^2 J_0(\chi) J_1'(u)}{\chi^2 - u^2} \cos \theta \cos \varphi \end{cases} \quad (3.2.19)$$

The formula for the far-field radiation field of aperture antenna is very complex in expression, but the spatial polarization ratio formula $\rho = -\mathrm{tg}\varphi / \cos \theta$ is the same as the formula (3.2.10).

The above analysis deals with several most basic aperture field distributions. The far-field radiation fields may be different for different apertures, but they have the same formula for spatial polarization ratio $\rho = -\mathrm{tg}\varphi / \cos \theta$.

3.2.2 Spatial Polarization Characteristics of Typical Waveguide Radiator

The waveguide aperture can be considered as the simplest aperture antenna. The array antenna in the microwave frequency bands (especially C, X and above) usually use open-ended rectangular waveguides, circular waveguides, and slotted rectangular waveguides operating in the main mode as the array elements.

1. Open-ended waveguide on an infinite grounded plane

Take the floor as the XOY plane. The long edge of the waveguide transmits the TE_{10} mode in the x direction. On the infinite grounded plane, the far-field equation of the open rectangular waveguide is [7]:

$$\begin{cases} E_\theta = \dfrac{\omega ab E_0}{cr} \sin\varphi \dfrac{\cos[(\pi a/\lambda)\sin\theta\cos\varphi]}{\pi^2 - 4[(\pi a/\lambda)\sin\theta\cos\varphi]^2} \cdot \dfrac{\sin[(\pi b/\lambda)\sin\theta\sin\varphi]}{(\pi b/\lambda)\sin\theta\sin\varphi} \\[3mm] E_\varphi = \dfrac{\omega ab E_0}{cr} \cos\theta\cos\varphi \dfrac{\cos[(\pi a/\lambda)\sin\theta\cos\varphi]}{\pi^2 - 4[(\pi a/\lambda)\sin\theta\cos\varphi]^2} \cdot \dfrac{\sin[(\pi b/\lambda)\sin\theta\sin\varphi]}{(\pi b/\lambda)\sin\theta\sin\varphi} \end{cases} \quad (3.2.20)$$

where, a and b are the lengths of waveguide edges in the x and y directions, respectively, and c is the velocity of light. It can be seen from the above formula that, although the expression of the field component of antenna is very complex, the expression of its polarization ratio is rather simple

$$\rho = -\text{tg}\varphi/\cos\theta \quad (3.2.21)$$

It is noted that the radiated wave of the open-ended waveguide is linearly polarized everywhere. On the main E plane ($\varphi = \pi/2$), the polarization ratio of the open-ended waveguide radiated wave is $\rho = \infty$ being vertical polarization. On the main H plane ($\varphi = 0$), the open-ended waveguide radiates the horizontal polarization wave, with the polarization ratio $\rho = 0$. On the XOY plane ($\theta = \pi/2$), the open-ended waveguide radiates the vertical polarization wave.

2. Radiation field of the rectangular waveguide

The radiation field of the rectangular waveguide (H_{10} mode) with the opening size $a \times b$ and the reflection coefficient Γ is [120]

$$\begin{cases} E_\theta(\theta, \varphi) = \sin\varphi \left(1 + \dfrac{1-\Gamma}{1+\Gamma} \cdot \dfrac{\lambda}{\lambda_g} \cos\theta\right) \dfrac{\cos(\frac{1}{2}au)}{1 - \left(\frac{1}{\pi}au\right)^2} \cdot \dfrac{\sin(\frac{1}{2}bv)}{\frac{1}{2}bv} \\[3mm] E_\varphi(\theta, \varphi) = \cos\varphi \left(\cos\theta + \dfrac{1-\Gamma}{1+\Gamma} \cdot \dfrac{\lambda}{\lambda_g}\right) \dfrac{\cos(\frac{1}{2}au)}{1 - \left(\frac{1}{\pi}au\right)^2} \cdot \dfrac{\sin(\frac{1}{2}bv)}{\frac{1}{2}bv} \end{cases} \quad (3.2.22)$$

where, the propagation constant of the TE_{10} wave is $k_{10} = 2\pi/\lambda_g$, with $\lambda_g = \lambda/\sqrt{1 - (\lambda/2a)^2}$ as the wavelength, and u and v are generalized angular coordinates, $u = k\sin\theta\cos\varphi$, $v = k\sin\theta\sin\varphi$. Since it is very difficult to accurately calculate the reflection coefficient Γ at the waveguide opening, it is often determined with the experimental method, or approximately expressed as $|\Gamma| = (1 - \lambda/\lambda_g)/(1 + \lambda/\lambda_g)$.

The polarization ratio of the radiation field of the rectangular waveguide is calculated with the above formula:

$$\rho = -\frac{E_\theta}{E_\varphi} = -\text{tg }\varphi \cdot \left(1 + \frac{1-\Gamma}{1+\Gamma} \cdot \frac{\lambda}{\lambda_g} \cos\theta\right) \bigg/ \left(\cos\theta + \frac{1-\Gamma}{1+\Gamma} \cdot \frac{\lambda}{\lambda_g}\right) \qquad (3.2.23)$$

With the waveguide opening aperture used directly as antenna, the directivity is very weak due to small opening size and great lobe width. The matching between waveguide and free space is very bad. The reflection coefficient Γ of the waveguide can usually reach 0.25–0.3. In a special case where reflection is zero at the opening aperture, i.e. $\Gamma = 0$, the polarization ratio of radiation field of the waveguide is simplified as

$$\rho = -\text{tg}\varphi \qquad (3.2.24)$$

3. Radiation field of rectangular waveguide

The radiation field of a circular waveguide (H_{11} mode) with a diameter of $2a$ is:

$$\begin{cases} E_\theta(\theta, \varphi) = \left(1 + \frac{1-\Gamma}{1+\Gamma} \cdot \frac{\lambda}{\lambda_g} \cos\theta\right) \sin\varphi \frac{J_1(ka \sin\theta)}{ka \sin\theta} \\ E_\varphi(\theta, \varphi) = \left(\cos\theta + \frac{1-\Gamma}{1+\Gamma} \cdot \frac{\lambda}{\lambda_g}\right) \cos\varphi \cdot J_1'(ka \sin\theta) \bigg/ \left(1 - \left(\frac{ka}{1.841} \sin\theta\right)^2\right) \end{cases}$$
$$(3.2.25)$$

Using the formula, the polarization ratio of radiation field of the circular waveguide can be calculated to be:

$$\rho = -\frac{E_\theta}{E_\varphi} = -\text{tg }\varphi \cdot \frac{\left(1 + \frac{1-\Gamma}{1+\Gamma} \cdot \frac{\lambda}{\lambda_g} \cos\theta\right)}{\left(\cos\theta + \frac{1-\Gamma}{1+\Gamma} \cdot \frac{\lambda}{\lambda_g}\right)} \cdot \frac{\left(\frac{J_1(ka \sin\theta)}{ka \sin\theta}\right)}{\left(J_1'(ka \sin\theta) \bigg/ \left(1 - \left(\frac{ka}{1.841} \sin\theta\right)^2\right)\right)}$$
$$(3.2.26)$$

It can be seen from the formulae (3.2.21) (3.2.23) and (3.2.26) that, although various typical waveguides have different expressions for spatial polarization ratios, for the open-ended waveguides, rectangular waveguides, and circular waveguides on an infinite grounded plane, a common point among them regarding spatial polarization characteristics is that the polarization ratio ρ in the azimuth direction varies according to the tangent function $\text{tg}\varphi$ of azimuth.

3.2.3 Spatial Polarization Characteristics of Typical Horn Antenna

Horn antenna are widely used at microwave frequencies, above 1 GHz. Their advantages include high gain, low voltage standing wave ratio (VSWR), relatively great bandwidth, light weight, as well as easy set-up. Horns are usually used as the feed sources of parabolic antenna, or as standard gain antenna. In some cases, they are directly used as antenna. The horn antenna is basically formed in the gradual expansion of the rectangular waveguide and the circular waveguide. The horn formed with the rectangular waveguide wall expansion on only one plane is called sectorial horn. If the expansion is performed on the E plane, the horn is called E-plane sectorial horn. If the expansion is performed on the H plane, the horn is called H-plane sectorial horn. If the expansion in performed simultaneously on both the E plane and the H plane, the horn is called pyramid horn. The horn formed with the circular waveguide expansion is called conical horn.

Appendix A gives the formulae for far-field region radiation field calculation of H-plane sectorial horn, E-plane sectorial horn, pyramid horn, and circular horn. Spatial polarization ratios of various horn antenna can be got using the analysis results given in Appendix A. At the same time, we can arrive at the conclusion: the radiation field of horn antenna has a very complex expression, but the polarization ratio expression is fairly simple. For H-plane sectorial horns, E-plane sectorial horns, pyramid horns, and conical horns, the spatial polarization ratio can be expressed as follows:

$$\rho = -E_\theta/E_\varphi = -\text{tg}\,\varphi \qquad (3.2.27)$$

In this section, the spatial polarization characteristics of typical aperture antenna are theoretically deduced. For the sake of comparison, the spatial polarization characteristics of the above antenna are summarized as shown in Table 3.2.

Table 3.2 Spatial polarization characteristics of typical aperture antenna

Antenna type		Spatial polarization ratio ρ
Typical antenna of equivalent aperture distribution		$\rho = -\text{tg}\,\varphi/\cos\theta$
Waveguide radiator	Open-ended waveguide on infinite grounded plane	$\rho = -\dfrac{\text{tg}\,\varphi}{\cos\theta}$
	Rectangular waveguide	$\rho = -\text{tg}\,\varphi \cdot \left(1 + \frac{1-\Gamma}{1+\Gamma}\cdot\frac{\lambda}{\lambda_g}\cos\theta\right)/\left(\cos\theta + \frac{1-\Gamma}{1+\Gamma}\cdot\frac{\lambda}{\lambda_g}\right)$
	Circular waveguide	$\rho = -\text{tg}\,\varphi \cdot \dfrac{\left(1 + \frac{1-\Gamma}{1+\Gamma}\frac{\lambda}{\lambda_g}\cos\theta\right)}{\left(\cos\theta + \frac{1-\Gamma}{1+\Gamma}\frac{\lambda}{\lambda_g}\right)} \times \dfrac{\left(\frac{J_1(ka\sin\theta)}{ka\sin\theta}\right)}{\left(J_1'(ka\sin\theta)/1-\left(\frac{ka}{1.841}\sin\theta\right)^2\right)}$
Horn antenna		$\rho = -\text{tg}\,\varphi$

3.3 Spatial Polarization Characteristics of Offset Parabolic Antenna

For a single-reflection-aperture antenna, the reflection beam can be blocked by the feed source and the supporting rod. This blocking effect will not only reduce the antenna gain, but also increase the side lobe level. Cutting off a part of the circular paraboloid, with the focus situated outside the main beam of the reflection aperture, can form an offset parabolic antenna. In the case of a single-reflection-aperture offset parabolic antenna, only a part of the symmetric parabolic aperture is used, avoiding blockage to the reflector aperture by the feed source and the supporting rod, hence to reduce the near-axis side lobe level and input voltage standing wave ratio. Result is gain increase. A good many active-duty warning and guidance radars are single-reflection-aperture antenna. However, the offset structure destroys the symmetry of the antenna geometry, resulting in the rise of the cross polarization level and the beam tilt during application of circular polarization. Figure 3.14a, b are schematics of the rotational symmetric and offset paraboloids, respectively.

However, the so-called "shortcomings" are not always "shortcomings". In other words, clever use can turn them into "merits". For example, beam tilt is the shortcoming of the circular-polarization single-offset-paraboloid antenna, but it can be utilized cleverly to realize multi-beam [53, 125]. Similarly, high cross-polarization level is a serious shortcoming of linear-polarization single-offset-paraboloid antenna. Do not blindly spend big money in taking various measures to suppress cross polarization, but make reasonable and effective use of the shortcoming based on the existing situation. By designing advanced algorithm of signal processing, apply the antenna spatial polarization characteristics to such aspects as target polarization characteristics measurement, anti-active jamming, and discrimination between true and false targets. Capabilities of polarization information application and processing could be hence further improved realizing "turning shortcoming into merit". The analysis, design and application of

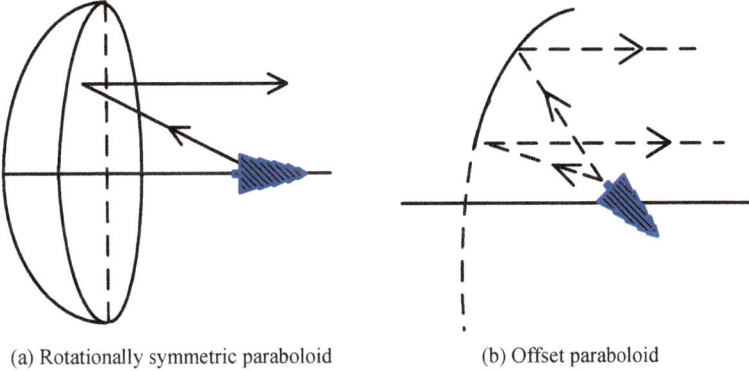

(a) Rotationally symmetric paraboloid   (b) Offset paraboloid

Fig. 3.14 Schematics of paraboloid antenna

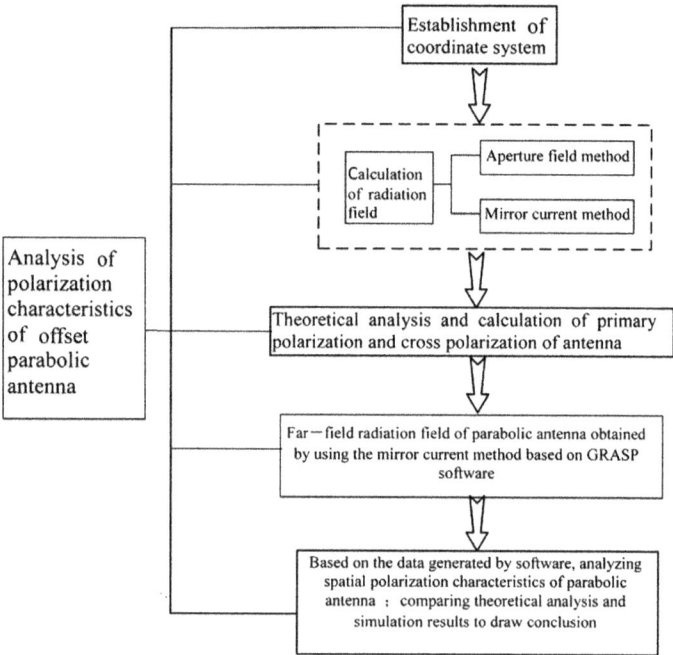

Fig. 3.15 Procedure for analysis of spatial polarization characteristics of offset parabolic antenna

paraboloid antenna are discussed in a good number of literatures. However, few of them discuss the polarization characteristics of parabolic antenna, let alone the spatial polarization characteristics. Therefore, in combination with theoretical analysis and software simulation [127, 128], this section studies the spatial polarization characteristics of single-offset-paraboloid antenna.

On the basis of the existing research results, we design a practical analysis procedure for the spatial polarization characteristics of single-offset-paraboloid antenna, as shown in Fig. 3.15.

3.3.1 Theoretical Calculation of Spatial Polarization Characteristics of Offset Parabolic Antenna

3.3.1.1 Establishment of Coordinate System

Nowadays, a variety of coordinate systems are adopted for the offset parabolic antenna. In the references, depending on respective emphasis, different methods are used for the selection and definition of coordinate systems. As shown in Figs. 3.16 and 3.17, five coordinate systems have been established for the discussion of spatial

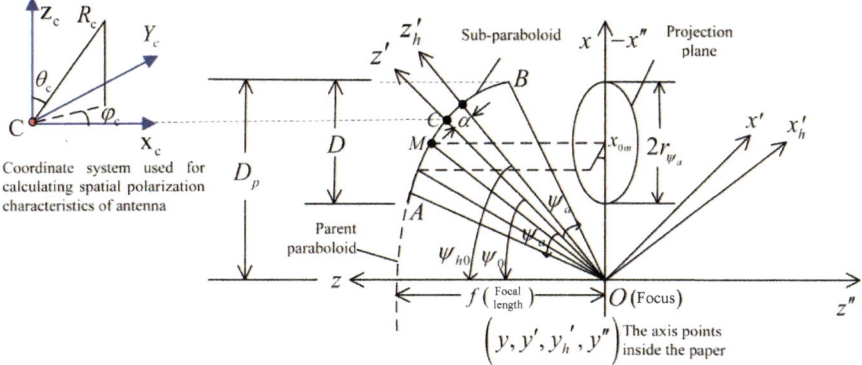

Fig. 3.16 Establishment of coordinate system of offset parabolic antenna

polarization characteristics of parabolic antenna, namely, the rectangular coordinate system (x, y, z) and the corresponding spherical coordinate system (r, θ, φ) used for indicating the parent paraboloid; the rectangular coordinate system (x', y', z') and the corresponding spherical coordinate system (r', ψ, ξ) used for calculating the parabolic aperture field; the rectangular coordinate system (x'_h, y'_h, z'_h) and the corresponding spherical coordinate system (r', ψ_h, ξ_h) used for indicating the feed source direction function; the rectangular coordinate system (x'', y'', z'') and the corresponding spherical coordinate system (R, Ψ, Φ) used for calculating the parabolic radiation field; and the rectangular coordinate system (x_c, y_c, z_c) and the corresponding spherical coordinate system $(R_c, \theta_c, \varphi_c)$ used for calculating the spatial polarization characteristics of antenna. In the figures, f is the focal length of the paraboloid, and $(\hat{x}, \hat{y}, \hat{z})$ and $(\hat{r}, \hat{\theta}, \hat{\varphi})$ are the unit vectors in the corresponding coordinate systems. Unit vectors in other coordinate systems may be determined similarly.

The parent paraboloid is a rotational symmetrical paraboloid of with focus being point O as focus, focal length being f, and Z axis as the symmetrical axis. Single offset paraboloid is a part of the parent paraboloid. Hence it is also known as sub-paraboloid. The edge of its profile is the line formed by the cone, the vertex of which is the point O, intersecting the parent paraboloid. The X' axis is not in the XOY plane. The Z' axis is the angular bisector of the sub-paraboloid and the Z axis intersects the paraboloid at the point C. Suppose that the angle between the OA axis (which is the line that connects the lower edge of the offset paraboloid to the focus O) and the OZ' axis is ψ_a, then the angle between the OB axis and the OA axis is $2\psi_a$. In discussion of the rule of polarization characteristic variation during parabolic antenna scanning in the azimuth and elevation directions, for the sake of convenience, the radar station rectangular coordinate system (x_c, y_c, z_c) and the corresponding coordinate system $(R_c, \theta_c, \varphi_c)$ used for calculating antenna spatial polarization characteristics are set up at the central point C of the paraboloid, with the elevation angle θ_c being the angle between the wave propagation direction and

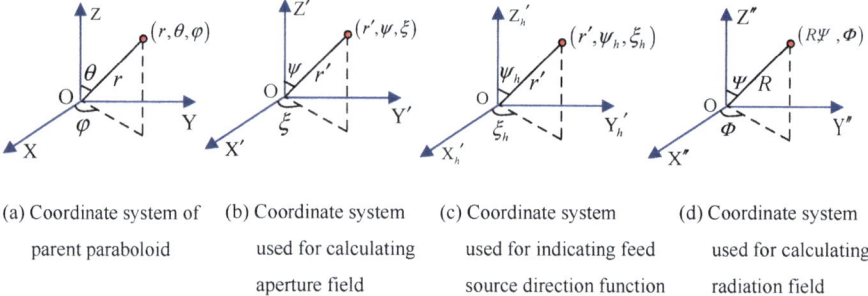

(a) Coordinate system of   (b) Coordinate system   (c) Coordinate system   (d) Coordinate system

parent paraboloid   used for calculating   used for indicating feed   used for calculating

                             aperture field   source direction function radiation field

Fig. 3.17 Correspondence between spherical coordinates and rectangular coordinates in coordinate systems of offset parabolic antenna

the axis Z_c, and the azimuth angle φ_c being the angle between the projection of the wave propagation direction on the X_cCY_c plane and the X_c axis. For the sake of clearness of expression, the radar station coordinate system (x_c, y_c, z_c) with point C as the center is given in the left of Fig. 3.17.

In Fig. 3.16, the offset angle of the offset paraboloid, recorded as ψ_0, is defined as the angle between the Z' axis and the Z axis. In particular, with the offset angle $\psi_0 = 0$, the offset paraboloid becomes a symmetric paraboloid (so-called normally-fed paraboloid). Z_h' is the symmetric axis of the feed source. X_h' is not in the XOY plane. Record the angle between the Z_h' axis and the Z' axis as α, and the angle between the Z_h' axis and the OZ axis as ψ_{h0}, then we have $\psi_{h0} = \psi_0 + \alpha$. Generally, the direction of feed source is such that the axis of feed source bisects the angle of the reflector ($\psi_{h0} = \psi_0$. In this case, the Z_h' axis coincides with the Z' axis, that is to say, the angle between them $\alpha = 0$) or the radial line along the feed source axis reaches the center of the projection aperture (i.e. the Z_h' axis coincides with the *OM* axis). For many designers of offset paraboloid, $\alpha = 0$ is the favorite structure [184]. In this case, however, the irradiation taper at the upper edge of the offset paraboloid differs greatly from that at the lower edge, resulting in uneven irradiation hence decrease in antenna gain. The conversions among coordinate systems are shown in Appendix B.

For the single-offset-paraboloid antenna, the actual radiation aperture is the aperture of the single-offset paraboloid projection on the XOY plane. So, in discussing the aperture field of a single-offset-paraboloid antenna, we can only discuss the XOY plane projection aperture field. Meanwhile, in calculating the far-field radiation field of antenna, we also calculate the integral of the projection on the XOY plane. Therefore, fully understanding the shape of paraboloid projection in the (x, y, z) coordinate system as well as the correspondence between the points on the paraboloid and the projective points on the XOY plane forms the basis for the discussion of the radiation field of paraboloid antenna and the analysis of antenna spatial polarization characteristics. For the shape of single-offset paraboloid projection on the XOY plane and the correspondence between the points on the paraboloid and the projective points on the XOY plane, refer to Appendix C.

Cross polarization is a serious shortcoming of linear-polarization single-offset-paraboloid antenna. But it is not a serious shortcoming for circular-polarization feed sources. Theoretically, circular-polarization balanced feed sources have no cross polarization. Their main shortcoming is beam tilt [121, 125]. Hence, in discussing spatial polarization characteristics of single-offset-paraboloid antenna, we mainly discuss those with linear-polarization feed sources.

3.3.1.2 Feed Model

Feed is one of the basic components of a parabolic antenna. With the geometrical structure of the paraboloid determined, performance of the antenna depends completely on the characteristics of the feed. The feed is called the primary antenna. The reflector is called the secondary antenna. The feed pattern is the primary pattern, while the pattern of entire antenna system is called the secondary pattern. The feed can be an oscillator, a horn, or a slot antenna. To achieve the best performance, the feed of a reflector antenna system must be fed with current properly. Therefore, for understanding spatial polarization characteristics of parabolic antenna, it is of immense importance to reasonably establish the model of electric field of the feed.

1. The establishment of typical feed model

The radiation field of a linear-polarization feed antenna can be expressed as:

$$E_f = \frac{e^{-jkr}}{r} [A_{\psi h}\hat{\boldsymbol{\psi}}_h + A_{\xi h}\hat{\boldsymbol{\xi}}] \tag{3.3.1}$$

where the radiation field of the feed antenna is defined as a pattern based on the Z'_h axis of the feed.

We can seldom know the $A_{\psi h}$ and $A_{\xi h}$ of all angles (ψ_h, ξ_h). Generally, only the pattern on the primary plane is known, i.e. $A_{\psi h}(\psi_h, \xi_h = 0) = A_E(\psi_h)$ on the E plane, and $A_{\xi h}(\psi_h, \xi_h = 90°) = A_H(\psi_h)$ on the H plane. The feed radiation field at any angle can be obtained using the method of interpolation approximation.

When the feed is linear polarization in the $\hat{x}'_h$ direction, the feed radiation field can be simulated based on the primary plane pattern:

$$E_f = \frac{e^{-jkr}}{r} [A_E(\psi_h) \cos \xi_h \hat{\boldsymbol{\psi}}_h - A_H(\psi_h)\sin\xi_h \hat{\boldsymbol{\xi}}_h] \tag{3.3.2}$$

we have

$$\begin{cases} A_{\psi h}(\psi_h, \xi_h) = A_E(\psi_h)\cos\xi_h \\ A_{\xi h}(\psi_h, \xi_h) = -A_H(\psi_h)\sin\xi_h \end{cases} \tag{3.3.3}$$

When the feed is linear polarization in the $\hat{y}'_h$ direction, the feed radiation field can be simulated based on the primary plane pattern:

$$\boldsymbol{E}_f = \frac{e^{-jkr}}{r}[A_E(\psi_h)\sin\xi_h\hat{\boldsymbol{\psi}}_h - A_H(\psi_h)\cos\xi_h\hat{\boldsymbol{\xi}}_h] \qquad (3.3.4)$$

we have

$$\begin{cases} A_{\psi h}(\psi_h, \xi_h) = A_E(\psi_h)\sin\xi_h \\ A_{\xi h}(\psi_h, \xi_h) = A_H(\psi_h)\cos\xi_h \end{cases} \qquad (3.3.5)$$

For example, for the short dipole polarized in the $\hat{x}'_h$ direction, its patterns on the E plane and H plane are respectively:

$$A_E(\psi_h) = \cos\psi_h, A_H(\psi_h) = 1 \qquad (3.3.6)$$

$\cos^q\psi_h(q = 1, 2, \ldots)$ is the most commonly used function form to simulate the feed pattern [89, 91]. Therefore, a universal feed model can be established

$$A_E(\psi_h) = \cos^{q_E}\psi_h, A_H(\psi_h) = \cos^{q_H}\psi_h, \psi_h < \pi/2 \qquad (3.3.7)$$

A feed with rotational symmetry is called a balanced feed. A balanced feed can be simplified as

$$A_E(\psi_h) = A_H(\psi_h) = A_0(\psi_h) = \cos^q\psi_h, \psi_h < \pi/2 \qquad (3.3.8)$$

Different feeds correspond to different values of q, which may be 1–4 and is often 2 in practice. It is possible to specify an edge illumination according to the expected performance, and then select the q value of the feed model $\cos^q\psi_h$ (or q_E and q_H) to somewhat match it with the pattern of the actual antenna except for the beam peak:

$$q = \log[A_0(\psi'_h)]/\log(\cos\psi'_h) \qquad (3.3.9)$$

where, ψ'_h is the matching point, such as -3 dB or -10 dB, or ψ_{hf} (ψ_{hf} is called edge illumination angle, i.e. the angle between the feed Z'_h axis and the reflector rim).

Using this feed model, the feed gain G_f and its edge illumination EI can be calculated conveniently:

$$G_f = \frac{2(2q_E + 1)(2q_H + 1)}{q_E + q_H + 1}, \quad EI = \frac{1 + \cos\psi_{hf}}{2}\cos^q\psi_{hf} \qquad (3.3.10)$$

Considering the formula (3.3.8) and (3.3.10), for the balanced feed, the pattern function can adopt the model below:

$$A_0(\psi_h) = \begin{cases} 2(n+1)\cos^q \psi_h & 0 \leqslant \psi_h \leqslant \pi/2 \\ 0 & \pi/2 \leqslant \psi_h \leqslant \pi \end{cases} \tag{3.3.11}$$

When the Z'_h axis coincides with the Z' axis, i.e. $\alpha = 0$, $\psi_{hf} = \psi_a$. Considering the formula (3.3.9) and (3.3.10), we can obtain:

$$q = \log \left[EI \left(\frac{2}{1 + \cos \psi_a} \right) \right] / \log(\cos \psi_a) \tag{3.3.12}$$

In order to obtain the best gain, the edge irradiation level of the aperture should be -11 dB or so [89, 93], i.e. $EI = 0.28$ (-11 dB).

2. Conversion between feed pattern coordinate systems

As discussed earlier, the actual feed pattern is based on the feed axis Z'_h. However, in the coordinate system (r', ψ_h, ξ_h), it is not convenient to calculate the aperture field and radiation field of the parabolic antenna. Therefore, we first convert the pattern of the feed into the patterns A_ψ and A_ξ based on the Z' axis, and then perform the calculation.

$$\begin{cases} A_\psi = (A_{\psi h}\hat{\psi}_h + A_{\xi h}\hat{\xi}_h) \cdot \hat{\psi} \\ A_\xi = (A_{\psi h}\hat{\psi}_h + A_{\xi h}\hat{\xi}_h) \cdot \hat{\xi} \end{cases} \tag{3.3.13}$$

Substituting the above formula into the formula (3.3.1) can obtain the feed patterns A_ψ and A_ξ based on the Z' axis. ψ_h, ξ_h is the spherical coordinate system where the feed phase center is the origin of coordinates. For describing the relation between ψ_h, ξ_h and ψ, ξ, Appendix B shows the deduction of $\hat{\psi}_h \cdot \hat{\psi}$, $\hat{\xi}_h \cdot \hat{\psi}$, $\hat{\psi}_h \cdot \hat{\xi}$, and $\hat{\xi}_h \cdot \hat{\xi}$.

For the linearly polarized feed that radiates rotational symmetrical beam:

$$\begin{cases} A_\psi = A_0(\psi_h) \left[\begin{smallmatrix} \cos \xi_h \\ \sin \xi_h \end{smallmatrix} \hat{\psi}_h + \begin{smallmatrix} -\sin \xi_h \\ \cos \xi_h \end{smallmatrix} \hat{\xi}_h \right] \cdot \hat{\psi} \\ A_\xi = A_0(\psi_h) \left[\begin{smallmatrix} \cos \xi_h \\ \sin \xi_h \end{smallmatrix} \hat{\psi}_h + \begin{smallmatrix} -\sin \xi_h \\ \cos \xi_h \end{smallmatrix} \hat{\xi}_h \right] \cdot \hat{\xi} \end{cases} \tag{3.3.14}$$

where, the upper formula is for the $\hat{x}'_h$ polarization feed, and the lower formula is for the $\hat{y}'_h$ polarization feed. In particular, when $\alpha = 0$, the Z'_h axis coincides with the Z' axis, $.\hat{\psi}_h = \hat{\psi}, \hat{\xi}_h = \hat{\xi}$

3.3.1.3 Aperture Field of Single Offset Parabolic Antenna

As is known to all, the reflection field of a single offset paraboloid aperture can be expressed:

$$E_r = 2(\hat{\boldsymbol{n}} \cdot \boldsymbol{E}_i)\hat{\boldsymbol{n}} - \boldsymbol{E}_i \tag{3.3.15}$$

where, $\hat{\boldsymbol{n}}$ is the unit normal vector of the single offset paraboloid, $\boldsymbol{E}_r$ is the reflection field of the single offset paraboloid aperture, and $\boldsymbol{E}_i$ is the incident field, i.e. the radiation field of the feed $\boldsymbol{E}_f$.

When the incident wave (i.e., the radiation wave of the feed) is linearly-polarized wave, it can be expressed in the spherical coordinate system (r', ψ, ξ)

$$\boldsymbol{E}_i = \frac{e^{-jkr}}{r}[A_\psi \hat{\boldsymbol{\psi}} + A_\xi \hat{\boldsymbol{\xi}}] \tag{3.3.16}$$

By deriving the projections of $\hat{\boldsymbol{n}}$ in the $\hat{\boldsymbol{\psi}}$, $\hat{\boldsymbol{\xi}}$ and $\boldsymbol{E}_i$ directions, we can obtain the tangential electric field distribution in the projection aperture of the single offset paraboloid. It can be expressed as a matrix [183]:

$$\begin{bmatrix} E_{ay}(\psi, \xi) \\ E_{ax}(\psi, \xi) \end{bmatrix} = \frac{e^{-j2kf}}{2f} \begin{bmatrix} -s_1 & c_1 \\ c_1 & s_1 \end{bmatrix} \begin{bmatrix} A_\psi(\psi, \xi) \\ A_\xi(\psi, \xi) \end{bmatrix} \tag{3.3.17}$$

where

$$\begin{cases} s_1 = (\cos \psi_0 + \cos \psi) \sin \xi \\ c_1 = \sin \psi_0 \sin \psi - (1 + \cos \psi_0 \cos \psi) \cos \xi \end{cases} \tag{3.3.18}$$

When the incident wave (i.e., the radiation wave of the feed) is circularly-polarized wave, we can obtain its matrix of projection aperture field [183]:

$$\begin{bmatrix} E_{aR} \\ E_{aL} \end{bmatrix} = \frac{e^{-j2kf}}{\sqrt{2}r} \begin{bmatrix} e^{jK} & -je^{jK} \\ e^{-jK} & je^{-jK} \end{bmatrix} \begin{bmatrix} A_\psi(\psi, \xi) \\ A_\xi(\psi, \xi) \end{bmatrix} \tag{3.3.19}$$

where, $K = \text{tg}^{-1}(s_1/c_1)$, and $\sqrt{c_1^2 + s_1^2}/2f = 1/r$ is used.

For incidence, the radiation field of the feed is propagated in the $+z$ direction. For reflection via the paraboloid, the propagation is in the $-z$ direction. Therefore, the rotation direction of circular polarization should also be reversed [53]. That is to say, when the incident wave is a left-hand circularly-polarized wave, the paraboloid reflects a right-hand circularly-polarized wave, due to the reverse of propagation direction; when the incident wave is a right-hand circularly-polarized wave, the paraboloid reflects a left-hand circularly-polarized wave.

Suppose that the spherical coordinates of field points are (R, Ψ, Φ). If the feed radiation wave is a linearly-polarized wave, E_Ψ and E_Φ, the radiation field components of Ψ and Φ respectively, can be expressed as

$$\begin{bmatrix} E_\Psi \\ E_\Phi \end{bmatrix} = \begin{bmatrix} -\cos\Phi & \sin\Phi \\ \cos\Psi\sin\Phi & \cos\Psi\cos\Phi \end{bmatrix} \begin{bmatrix} F_x \\ F_y \end{bmatrix} \tag{3.3.20}$$

where

$$F_x = \int_s E_{ax}\frac{e^{-jkR}}{R}\,dS, \; F_y = \int_s E_{ay}\frac{e^{-jkR}}{R}\,dS \tag{3.3.21}$$

representing the integrals of the transverse electric fields E_{ax} and E_{ay} on the aperture.

Considering the correspondence between rectangular coordinates and spherical coordinates (Appendix B), we have

$$\begin{bmatrix} \mathcal{E}_y \\ \mathcal{E}_x \end{bmatrix} = \begin{bmatrix} \sin\Phi & \cos\Phi \\ \cos\Phi & -\sin\Phi \end{bmatrix} \begin{bmatrix} E_\Psi \\ E_\Phi \end{bmatrix} \tag{3.3.22}$$

where, $\mathcal{E}_y$ and $\mathcal{E}_x$ represent the linear polarization components in the y and x directions of the parabolic antenna, respectively.

Considering also the formula (3.3.20) and (3.3.22), we can obtain:

$$\begin{bmatrix} \mathcal{E}_y \\ \mathcal{E}_x \end{bmatrix} = \begin{bmatrix} \sin\Phi & \cos\Phi \\ \cos\Phi & -\sin\Phi \end{bmatrix} \begin{bmatrix} -\cos\Phi & \sin\Phi \\ \cos\Psi\sin\Phi & \cos\Psi\cos\Phi \end{bmatrix} \begin{bmatrix} F_x \\ F_y \end{bmatrix} \tag{3.3.23}$$

Record $t = \mathrm{tg}\frac{\Psi}{2}$, the above formula can be written as:

$$\begin{bmatrix} \mathcal{E}_y(\Psi,\Phi) \\ \mathcal{E}_x(\Psi,\Phi) \end{bmatrix} = \frac{1+\cos\Psi}{2}\begin{bmatrix} 1-t^2\cos 2\Phi & -t^2\sin 2\Phi \\ \sin 2\Phi & -(1+t^2\cos 2\Phi) \end{bmatrix}\begin{bmatrix} F_y \\ F_x \end{bmatrix} \tag{3.3.24}$$

It can be seen from the above formula that, that even if the projection aperture field has only the primary polarization component and no cross polarization component, its radiation field also produces the cross polarization component.

3.3.1.4 Radiation Field of Single Offset Parabolic Antenna

Mirror current method and aperture field method are the two most commonly used methods for calculating the radiation field of reflector antenna. These two methods are based on geometrical optics approximation, so they are also called geometrical optics methods. For a large aperture antenna (for example, reflector with aperture more than 5 wavelengths), geometrical optics approximation is acceptable. However, if the radiation in the abaxial side lobe and back lobe is taken into consideration or the antenna aperture is not sufficiently large, neither of the two methods is satisfactory and correction must be performed according to the theory of geometrical diffraction. Generally, radars operate in microwave frequency bands. In the case of parabolic antenna, their aperture size is far greater than wavelength, and their radiation power is mainly concentrated axially in a narrow main lobe region. Therefore, both of the methods are used.

The aperture field method is essentially as follows: using the geometrical optics method, calculate the tangential electromagnetic field of the primary feed radiation field projected on the aperture after reflection by the reflection mirror aperture, and then, using the E-mode radiation field formula, or the Stratton-Chu radiation equation, calculate the radiation field. The mirror current method is essentially as follows: calculate the mirror current excited on the reflection mirror aperture by the primary feed, and then, using the Stratton-Chu radiation equation, calculate the radiation field of the mirror current. The difference between the two methods is that the first method uses the aperture current density on the reflection aperture as the initial data, and the second method uses the aperture field as the initial data. The aperture field method does not consider the field produced by the z direction component of the mirror current. For the axisymmetric reflector, generally the integral aperture is chosen to cover the reflection aperture. This choice is fairly natural. The integral aperture coincides with the physical aperture. It contains thus covers the reflector rim. In this case, the aperture field method and the mirror current method can get the same results. However, for the offset paraboloid, the projection aperture of the reflector is often used as the integral aperture. In this case, the aperture field method and the mirror current method will produce different results. People think that the mirror current method can produce more accurate results, especially for the cross polarization level. The z direction component of mirror current can cause increase in cross polarization level. By contrast, the aperture field method ignores the effect of longitudinal field component F_x. Therefore, the cross polarization level calculated with this method is a little smaller than the actual value. The accuracy of the aperture field method is slightly worse than that of the mirror current method. The mirror current method is more accurate than the aperture field method, in calculating the wide angle region of the side lobe and back lobe far away from the main lobe. Therefore, the radiation field of the offset parabolic antenna is calculated with the mirror current method below.

The Stratton-Chu radiation equation is [123, 125]

$$\varepsilon = \frac{jk}{4\pi} \int_s (\hat{\boldsymbol{n}} \times \boldsymbol{E}_s) \times \hat{\boldsymbol{R}} \frac{e^{-jkR_s}}{R_s} dS - \frac{j\omega\mu}{4\pi} \int_s \{\hat{\boldsymbol{R}}_s \times (\hat{\boldsymbol{n}} \times \boldsymbol{H}_s) \times \hat{\boldsymbol{R}}_s\} \frac{e^{-jkR_s}}{R_s} dS$$

$$(3.3.25)$$

where, $\hat{\boldsymbol{n}}$ is the normal unit vector of the mirror; $\boldsymbol{E}_s$ and $\boldsymbol{H}_s$ are the electric field and magnetic field of the mirror, respectively; R_s is the distance from the source point on the mirror to the point in the radiation field. Various geometrical quantities are related as shown in Fig. 3.18.

Fig. 3.18 Calculating the radiation field of the offset parabolic antenna using the mirror current method

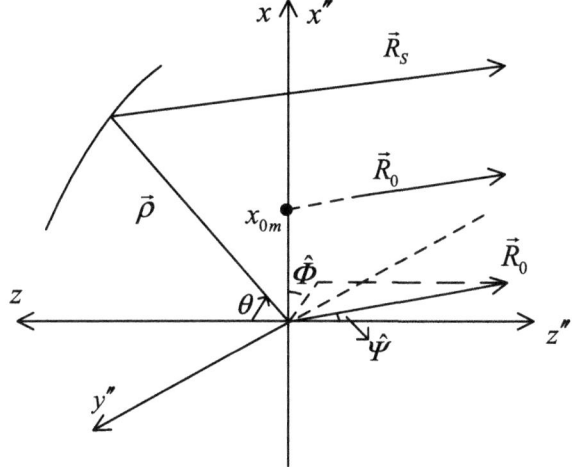

The electric field E_r and magnetic field H_r on the mirror are respectively

$$E_r = E_{rx}\hat{x} + E_{ry}\hat{y} = \frac{e^{-jkr}}{2f}\left\{[c_1A_\psi + s_1A_\xi]\hat{\xi} + [-s_1A_\psi + c_1A_\xi]\hat{y}\right\} \tag{3.3.26}$$

$$H_r = -\frac{1}{z_0}(\hat{n} \times E_r) \tag{3.3.27}$$

where, $z_0 = \sqrt{\mu_0/\varepsilon_0}$ is free space wave impedance.

Decompose the radiation field into two components ε_ψ and ε_ϕ, in the Ψ and Φ directions, respectively. The following formulae can hence be deduced:

$$\varepsilon_\psi = \varepsilon \cdot \hat{\Psi} = \frac{j}{\lambda}\int_s \left\{\begin{array}{l}\left(-\cos\Psi\cos\Phi - \text{tg}\frac{\theta}{2}\sin\Psi\cos\varphi\right)E_{rx} \\ + \left(\cos\Psi\sin\Phi - \text{tg}\frac{\theta}{2}\sin\Psi\sin\varphi\right)E_{ry}\end{array}\right\} \frac{e^{-jkR_s}}{R_s}\cos\frac{\theta}{2}dS \tag{3.3.28}$$

$$\varepsilon_\phi = \varepsilon \cdot \hat{\Phi} = \frac{j}{\lambda}\int_s (\sin\Phi E_{rx} + \cos\Phi E_{ry})\frac{e^{-jkR_s}}{R_s}\cos\frac{\theta}{2}dS \tag{3.3.29}$$

In calculating the integral, note that $dS_A = \cos\frac{\theta}{2}dS$, where dS_A is the projection of the cell area (where the source point on the offset paraboloid is located) on the projection aperture.

With the antenna operating with linear polarization, the linear polarization components of radiation field of the parabolic antenna in the y and x directions are:

$$
\begin{bmatrix} \varepsilon_y \\ \varepsilon_x \end{bmatrix} = \begin{bmatrix} \sin\Phi & \cos\Phi \\ \cos\Phi & -\sin\Phi \end{bmatrix} \begin{bmatrix} \varepsilon_\psi \\ \varepsilon_\phi \end{bmatrix}
$$

$$
= \frac{j}{\lambda}\frac{1+\cos\psi}{2}\left\{ \begin{bmatrix} t^2\sin 2\Phi & 1+t^2\cos 2\Phi \\ -(1-t^2\cos 2\Phi) & -t^2\sin 2\Phi \end{bmatrix}\begin{bmatrix} F_{rx} \\ F_{ry} \end{bmatrix} \right.
$$

$$
\left. -2\int_s \mathrm{tg}\frac{\Psi}{2}\begin{bmatrix} M\sin\Phi & N\sin\Phi \\ M\cos\Phi & N\cos\Phi \end{bmatrix}\begin{bmatrix} E_{rx} \\ E_{ry} \end{bmatrix}\frac{e^{-jkR_s}}{R_s}\mathrm{d}S_A \right\}
$$

$$
\tag{3.3.30}
$$

where

$$
\begin{cases} M = \mathrm{tg}\frac{\theta}{2}\cos\varphi = \dfrac{\sin\psi\,\cos\psi_0\cos\xi+\cos\psi\,\sin\psi_0}{1+\cos\psi_0\cos\psi-\sin\psi_0\sin\psi\,\cos\xi} \\[2mm] N = \mathrm{tg}\frac{\theta}{2}\sin\varphi = \dfrac{\sin\psi\,\sin\xi}{1+\cos\psi_0\cos\psi-\sin\psi_0\sin y\,\cos\xi} \end{cases}
\tag{3.3.31}
$$

With the antenna operating with circular polarization:

$$
\begin{bmatrix} \varepsilon_L \\ \varepsilon_R \end{bmatrix} = \frac{j}{\lambda}\frac{(1+\cos\psi)}{2}\left\{ \begin{bmatrix} 1 & -t^2e^{-j2\Phi} \\ -t^2e^{j2\Phi} & 1 \end{bmatrix}\begin{bmatrix} F_{rL} \\ F_{rR} \end{bmatrix} \right.
$$

$$
\left. +\frac{t(1+t^2)}{2}\int_s\begin{bmatrix} e^{-j\Phi}(M-jN) & e^{-j\Phi}(M+jN) \\ e^{j\Phi}(M-jN) & e^{j\Phi}(M+jN) \end{bmatrix}\begin{bmatrix} E_{rL} \\ E_{rR} \end{bmatrix}\frac{e^{-jkR_s}}{R_s}\mathrm{d}S_A \right\}
$$

$$
\tag{3.3.32}
$$

where F_{rx}, F_{ry}, F_{rL} F_{rL} and F_{rR} are defined as

$$
F_{ru} = \int_s E_{ru}\frac{e^{-jkR_s}}{R_s}\mathrm{d}S
\tag{3.3.33}
$$

where u can be x, y, R or L; dS is dS$_A$, the projection of the cell area (where the source point on the offset paraboloid is located) on the projection aperture. Thus the integral is still calculated on the projection aperture.

$R_S = R'_0 - \rho \cdot \hat{R}_S$ is the distance from the source point on the mirror to the field point, R'_0 is the distance from the origin O to the field point, ρ is the distance from the origin O to any point on the mirror; $\hat{\rho}$ and $\hat{R}_S$ are the unit vectors. Considering also

$$
\begin{cases} \hat{\boldsymbol{\rho}} = \sin\theta\,\cos\varphi\hat{\boldsymbol{x}} + \sin\theta\,\sin\varphi\hat{\boldsymbol{y}} + \cos\theta\hat{\boldsymbol{z}} \\ \hat{\boldsymbol{R}}_S = -\sin\Psi\,\cos\Phi\hat{\boldsymbol{x}} + \sin\Psi\,\sin\Phi\hat{\boldsymbol{y}} - \cos\Psi\hat{\boldsymbol{z}} \\ R'_0 = R_0 + x_{0m}\hat{\boldsymbol{x}}\cdot\hat{\boldsymbol{R}}_S = R_0 - x_{0m}\sin\Psi\,\cos\Phi \end{cases}
\tag{3.3.34}
$$

we can get

$$R_s = R_0 - \rho + 2f - (x_{0m} - x)\sin\Psi\cos\Phi - y\sin\Psi\sin\Phi - z(1 - \cos\Psi) \quad (3.3.35)$$

where, R_0 is the distance from the center of the projection aperture to the field point. For the calculation formula, see Appendix C.

Suppose that the coordinates on the XOY projection aperture are (x_1, y_1) corresponding to the point (r, ψ, ξ) in the spherical coordinate system of the offset paraboloid. We have:

$$x_1 = x - x_{0m}, \quad y_1 = y \quad (3.3.36)$$

After a series of operations, we obtain

$$F_u = \frac{e^{-jk(R_0 + 2f)}}{R_0} \int_0^{x_{1m}} \int_0^{y_{1m}} (E_{ru}e^{jk\rho}) e^{-jkx_1\sin\Psi\cos\Phi + jky_1\sin\Psi\sin\Phi + jkz(1-\cos\Psi)} dx_1 dy_1 \quad (3.3.37)$$

where, u may be x, y, R or L; $x_{1m} = y_{1m} = \frac{2f\sin\psi_a}{\cos\psi_0 + \cos\psi_a}$.

Note that, since E_{au} (u may be x, y, R, or L) is calculated in the coordinate system (r, ψ, ξ), E_{ax}, E_{ay}, E_{aR} and E_{aL} are all calculated based on the coordinate system (r, ψ, ξ), and the integral in the formula (3.3.37) is calculated on the projection aperture, it is necessary to perform coordinate conversion in the calculation. For details on the definition of x_{0m} and coordinate conversion, see Appendix C.

3.3.1.5   Qualitative Analysis of Spatial Polarization Characteristics of Single Offset Parabolic Antenna

In the previous analysis, the far-field radiation field of the parabolic antenna is calculated in the spherical coordinate system (R, Ψ, Φ). But use of the coordinate system $(R_c, \theta_c, \varphi_c)$ enhances the convenience in discussing the spatial polarization characteristics of parabolic antenna. Conversion between the coordinate systems $(R_c, \theta_c, \varphi_c)$ and (R, Ψ, Φ) is shown in Appendix B.

Since generally polarization is expected to occur in only one direction, the expected polarization is often selected as reference polarization (also known as "primary polarization". Polarization orthogonal to the primary polarization is called cross polarization. The definition of the primary polarization or reference polarization depends flexibly on the selected polarization basis. For example, if $\hat{\theta}_c$ and $-\hat{\varphi}_c$ are chosen as the polarization basis, the polarization in the $\hat{\theta}_c$ direction is set to be the primary polarization E_p, and the polarization in the $-\hat{\varphi}_c$ direction is set to be the cross polarization E_q, then the far-field radiation field vector of the parabolic antenna can be expressed as $E = E_{\theta_c}(\theta_c, \varphi_c)\hat{e}_{\theta_c} + E_{\varphi_c}(\theta_c, \varphi_c)\hat{e}_{\varphi_c}$, and its primary polarization and cross polarization are expressed as

$$\begin{cases} E_p(\theta_c, \varphi_c) = E_{\theta_c}(\theta_c, \varphi_c)\hat{e}_{\theta_c} \\ E_q(\theta_c, \varphi_c) = -E_{\varphi_c}(\theta_c, \varphi_c)\hat{e}_{\varphi_c} \end{cases} \tag{3.3.38}$$

This definition agrees with the second definition of cross polarization Ludwig formulated.

If the primary antenna is a linearly-polarized feed, another more direct definition method is selecting linear polarization bases $\hat{e}_x$ and $\hat{e}_y$. If the primary polarization is defined as the linear polarization in the x direction, then

$$\begin{cases} E_p(\theta_c, \varphi_c) = E_{\theta_c}(\theta_c, \varphi_c)\cos\varphi_c\,\hat{e}_{\theta_c} - E_{\varphi_c}(\theta_c, \varphi_c)\sin\varphi_c\,\hat{e}_{\varphi_c} \\ E_q(\theta_c, \varphi_c) = E_{\theta_c}(\theta_c, \varphi_c)\sin\varphi_c\hat{e}_{\theta_c} + E_{\varphi_c}(\theta_c, \varphi_c)\cos\varphi_c\,\hat{e}_{\varphi_c} \end{cases} \tag{3.3.39}$$

If the primary polarization is defined as the linear polarization in the y direction, then

$$\begin{cases} E_p(\theta_c, \varphi_c) = E_{\theta_c}(\theta_c, \varphi_c)\sin\varphi_c\,\hat{e}_{\theta_c} + E_{\varphi_c}(\theta_c, \varphi_c)\cos\varphi_c\,\hat{e}_{\varphi_c} \\ E_q(\theta_c, \varphi_c) = E_{\theta_c}(\theta_c, \varphi_c)\cos\varphi_c\hat{e}_{\theta_c} - E_{\varphi_c}(\theta_c, \varphi_c)\sin\varphi_c\,\hat{e}_{\varphi_c} \end{cases} \tag{3.3.40}$$

Formula (3.3.40) agrees with the third definition of cross polarization Ludwig formulated. It is also the most widely used one of the definitions of cross polarization Ludwig formulated. Similarly, for the circular polarization feed, we can choose the circular polarization bases $\hat{v}_R$ and $\hat{e}_L$. The previous analysis shows that the circular polarization direction of the reflector radiation is opposite to the circular polarization direction of the feed. Therefore, for the left-hand circular-polarization feed, we can choose the right-hand circular-polarization component ε_R of the parabolic antenna radiation field as the primary component E_p, and choose the left-hand circular-polarization component ε_L as the cross-polarization component E_q; similarly, for the right-hand circular-polarization feed, we can choose the left-hand circular-polarization component ε_L of the parabolic antenna radiation field as the primary component E_p, and choose the right-hand circular-polarization component ε_R as the cross-polarization component.

The theoretical calculation methods of spatial polarization characteristics of aperture antenna have been discussed in the above sections. Next, the spatial polarization characteristics of offset parabolic antenna will be theoretically analyzed qualitatively, and then simulated and discussed quantitatively in Sect. 3.3.2.

It is assumed that the primary antenna is the feed for radiating rotational symmetric beam, and the angle between the Z'_h axis and the Z' axis $\alpha = 0$. When the feed is linear polarization in the $\hat{x}'$ direction:

$$\begin{cases} A_\psi(\psi, \xi) = A_0(\psi)\cos\xi \\ A_\xi(\psi, \xi) = -A_0(\psi)\sin\xi \end{cases} \tag{3.3.41}$$

When the feed is linear polarization in the $\hat{y}'$ direction:

$$\begin{cases} A_\psi(\psi, \xi) = A_0(\psi)\sin\xi \\ A_\xi(\psi, \xi) = A_0(\psi)\cos\xi \end{cases} \tag{3.3.42}$$

Take the feed polarization in the $\hat{y}'$ direction as an example. Using the formulae (3.3.17) and (3.3.42), we obtain

$$\begin{cases} E_{ay}(\psi, \xi) = \frac{e^{-j2kf}}{2f} A_0(\psi) \left[\sin\psi_0 \sin\psi \cos\xi - \frac{1}{2}(1 + \cos\psi_0)(1 + \cos\psi) \right] \\ \qquad\qquad\qquad\quad \left[-\frac{1}{2}(1 - \cos\psi_0)(1 - \cos\psi)\cos\xi \right] \\ E_{ax}(\psi, \xi) = \frac{e^{-j2kf}}{2f} A_0(\psi) \left[\sin\psi_0\sin\psi \sin\xi - \frac{1}{2}\sin2\xi(1 - \cos\psi_0)(1 - \cos\psi) \right] \end{cases} \tag{3.3.43}$$

If the incident wave is x polarized, the y polarization component in the aperture field is defined as cross polarization. If the incident wave is y polarized, the x polarization in the aperture field is defined as cross polarization. It can be seen from the above formula that, with the incident wave being the y polarized wave, the principal value of the primary polarization component $E_{ay}(\psi, \xi)$ in the projection aperture field is $-1/2(1 + \cos\psi_0)(1 + \cos\psi)$, while the principal value of the cross polarization component $E_{ax}(\psi, \xi)$ is $\sin\psi_0 \sin\psi \sin\xi$. It is not difficult to see that the principal value of the cross polarization component increases with the increase of ψ_0, and that when $\xi = \pi/2$, the cross polarization component in the aperture field reaches the maximum $E_{ax}(\psi, \xi) = (e^{-j2kf}/2f) \cdot A_0(\psi) \cdot \sin\psi_0 \sin\psi$. In particular, when $\psi_0 = 0$, i.e. in the case of an axisymmetric paraboloid, $E_{ax}(\psi, \xi) = 0$, the cross polarization is zero in the aperture field. That is to say, for a balanced feed that has a rotational symmetric pattern and whose ideal phase center is at the focus, the cross polarization is zero in the aperture field. It can be seen from the formula (3.3.24) that, even if the projection aperture field has only the primary polarization component and no cross polarization component, its radiation field also produces cross polarization component. For an unbalanced feed, such as an oscillator, the cross polarization caused by the reflector is zero in the primary plane, but not zero on the other planes. In fact, the feed plays a major role in the generation of cross polarization of an axisymmetric reflector.

Offset degrades the cross polarization performance of a parabolic antenna. The cross polarization performance of the parabolic antenna worsens with the increase of offset angle ψ_0. Moreover, the F/D of the offset reflector is greater than the F/D_p of the parent reflector. The cross polarization level of the offset reflector increases with the decrease of F/D. A low cross polarization level is usually desired by antenna designers. For the sake of reducing cross polarization level, we should select a small offset angle ψ_0. However, it is known from the projection aperture radius formula that, with the radius fixed, to meet the specific design gain requirement, the focal length f of the paraboloid has to be increased. Consequently, the single offset paraboloid is longitudinally too large, increasing difficulty in structure realization. Therefore, in the current design of single offset paraboloid, the offset angle selected is not quite small. Generally $\psi_0 = 40°-44°$ or greater. Moreover, the antenna designers who wish a low cross polarization level are more willing to choose a large focal length-to-diameter ratio F/D. Benefits are straightening the current line on the mirror, and reducing d the cross polarization component. However, many factors must be taken into account when choosing the focal length-to-diameter ratio. A greater focal length can improve antenna performance, but increase its longitudinal dimensions. Thus, the F/D value chosen is generally not very great. Usually it is in the range of 0.3–0.8.

The polarization of the reflector antenna (secondary) pattern is affected jointly by the cross polarization of the feed (primary) pattern ($XPOL_F$) and the cross polarization introduced by the reflector ($XPOL_R$). In the previous analysis, we have ignored the cross polarization of any feed, i.e. $XPOL_F = 0$. However, the feed of a parabolic antenna generally is incapable of reaching the pure polarization state. That is to say, under normal conditions, $XPOL_F \neq 0$.

Usually, in order to provide a shaped beam, the offset paraboloid edge must be deformed accordingly (i.e., the edge is no longer in a plane, and is no longer an ellipse. The aperture projection is no longer a circle). The edge deformation can result in additional phase error. This phase error will cause the spatial redistribution of cross polarization lobes. In general, the spatial redistribution of cross polarization lobes can cause polarization purity deterioration. To achieve the best gain of the offset paraboloid, its Z' axis generally does not coincide with the axis Z'_h of its feed. This will lead to additional cross polarization component. Moreover, in actuality, restriction and influence by such factors as geometrical, dimensional, and machining errors and wave diffraction will also cause the deterioration of cross polarization. Therefore, when calculating the cross polarization of reflector antenna radiation field, consideration must be given to not only the cross polarization of the feed $XPOL_F$ and the cross polarization introduced by the reflector $XPOL_R$, but also the additional cross polarization caused by various actual factors $XPOL_0$.

Total cross polarization of the reflector system may be estimated using the formula below:

$$XPOL_S = XPOL_F + XPOL_R + XPOL_0 \tag{3.3.44}$$

For example, for a feed with cross polarization of -30 dB ($XPOL_F = 0.0316$) used in a reflector with cross polarization of -20 dB ($XPOL_R = 0.1000$), if various actual factors cause an additional cross polarization of -25 dB ($XPOL_0 = 0.0562$), then the total cross polarization of the reflector system can be calculated using the formula to be $XPOL_S = -14.5$ dB.

3.3.2 Simulation Analysis of Spatial Polarization Characteristics of Offset Parabolic Antenna

Because the integral calculation of parabolic antenna radiation field is very complicated, and the direct Matlab simulation is notorious for relatively low accuracy and efficiency, we use the powerful GRASP9 software developed by the Danish TICRA company specifically for the calculation of performance of reflector antenna to calculate the far field of single offset parabolic antenna, and based on the calculation analyze the spatial polarization characteristics of offset parabolic antenna.

GRASP9 is a commercial reflector antenna software using physical optics and mirror current integration. As a kind of advanced professional antenna design simulation software, GRASP9 has been widely recognized and used by many international well-known scholars and research centers. The software has also got the support from the European Space Agency. The calculation results can be directly used for the design and manufacturing of reflector antenna [126, 127].

3.3.2.1 Computer Simulation of Spatial Polarization Characteristics of Offset Parabolic Antenna with Different Feeds

First of all, based on the coordinate system $(R_c, \theta_c, \varphi_c)$ shown in Fig. 3.3.3 for the calculation of antenna spatial polarization, and based on numerical simulation with GRASP9 on the radiation field of the antenna; on this basis, to discuss the polarization characteristics of antenna changes with elevation angle θ_c and azimuth angle in space; then, to investigate the effect of the key factors such as feed polarization, and the focal diameter ratio of the paraboloid F/D and offset angle ψ_0 on the spatial polarization characteristics of parabolic antenna.

The schematic diagram of the offset parabolic antenna in the vertical section of the GRASP software as shown in Fig. 3.19. There are three main coordinate systems in the figure, the origin of the coordinate system XYZ is at the vertex of the paraboloid, the origin of the coordinate system $X_cY_cZ_c$ is at the center of the reflecting aperture, the coordinate system $X_fY_fZ_f$ is the feed coordinate system, and the corresponding Y axis of each coordinate system are pointing to the paper, and there is no indication in the figure. The polarization characteristics of the antenna are analyzed and calculated in the coordinate system $X_cY_cZ_c$. The angular coordinates (θ_c, φ_c) of the wave propagation direction of the antenna are in the coordinate system $X_cY_cZ_c$, where, θ_c is the angle between the direction of the radio wave propagation and the Z_c axis, and φ_c is the angle between the projection of the electric wave propagation direction and the X_c axis in the plane $X_cO_cY_c$. The coordinate system $X_cY_cZ_c$ is the same as that $X_cY_cZ_c$ defined in Fig. 3.16. For mechanical scanning radar, the antenna is scanned on the azimuth aperture, that is, scanning in the vicinity of $\varphi_c = 0°$; scanning on the elevation plane, that is, scanning in the vicinity of $\theta_c = 90°$. In order to facilitate the discussion of the

Fig. 3.19 the schematic
diagram of the offset
paraboloid coordinate system
in the GRASP software

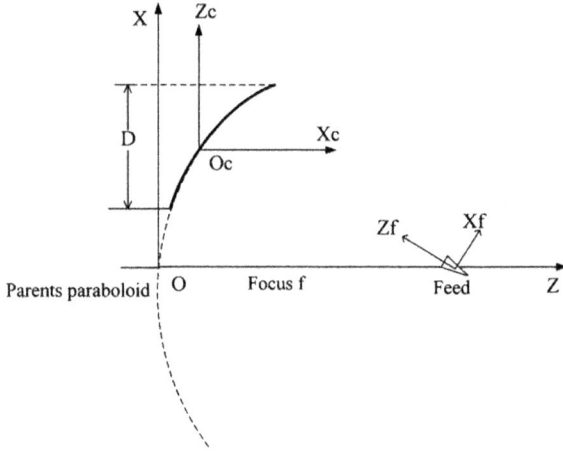

polarization characteristics of the antenna, the angle $\Delta\theta_c = 90° - \theta_c$ between the propagation direction and the plane $X_cO_cY_c$ of the radio wave is recorded, and the $\Delta\theta_c$ is called "elevation angle" in the subsequent analysis.

The working frequency is 5 GHz, diameter wavelength ratio $D/\lambda = 50$, focal diameter ratio F/D = 0.5 and offset angle $\psi_0 = 44°$, radar of C band (a domestic air surveillance radar guidance as the prototype) as an example, suppose that antenna feed is four typical polarization form as X direction linear polarization, Y direction linear polarization, left-hand circular polarization and right-hand circular polarization, respectively, and investigate the spatial polarization characteristics of parabolic antenna.

1. X direction linearly polarized feed

According to the definition of the spatial polarization ratio of the antenna in Sect. 2.4, the spatial polarization ratio of the antenna is obtained in the polarization basis $(\hat{h}, \hat{v})$

$$\rho = E_V/E_H = -E_\theta/E_\varphi \qquad (3.3.45)$$

First, according to set parameters obtain far radiation field E_θ and E_φ of parabolic antenna by GRASP software, and then discusses the variation of typical descriptor such as antenna spatial polarization ratio, polarization phase descriptor and spatial phase descriptors IPPV with azimuth angle φ_c and elevation angle $\Delta\theta_c$. The typical results are shown in Fig. 3.20. Among them, Fig. 3.20a, b are amplitude distribution map for horizontal polarization component E_H and vertical polarization component E_V, respectively; Fig. 3.20c shows the curves for variation of horizontal and vertical polarization component when antenna is scanned in the horizontal plane (elevation angle $\Delta\theta_c = 0$) with azimuth angle; Fig. 3.20d shows curves for variation of the antenna polarization ratio amplitude with the azimuth angle, and which selects several typical cases, as elevation angle $\Delta\theta_c = 0$, $\Delta\theta_c =$

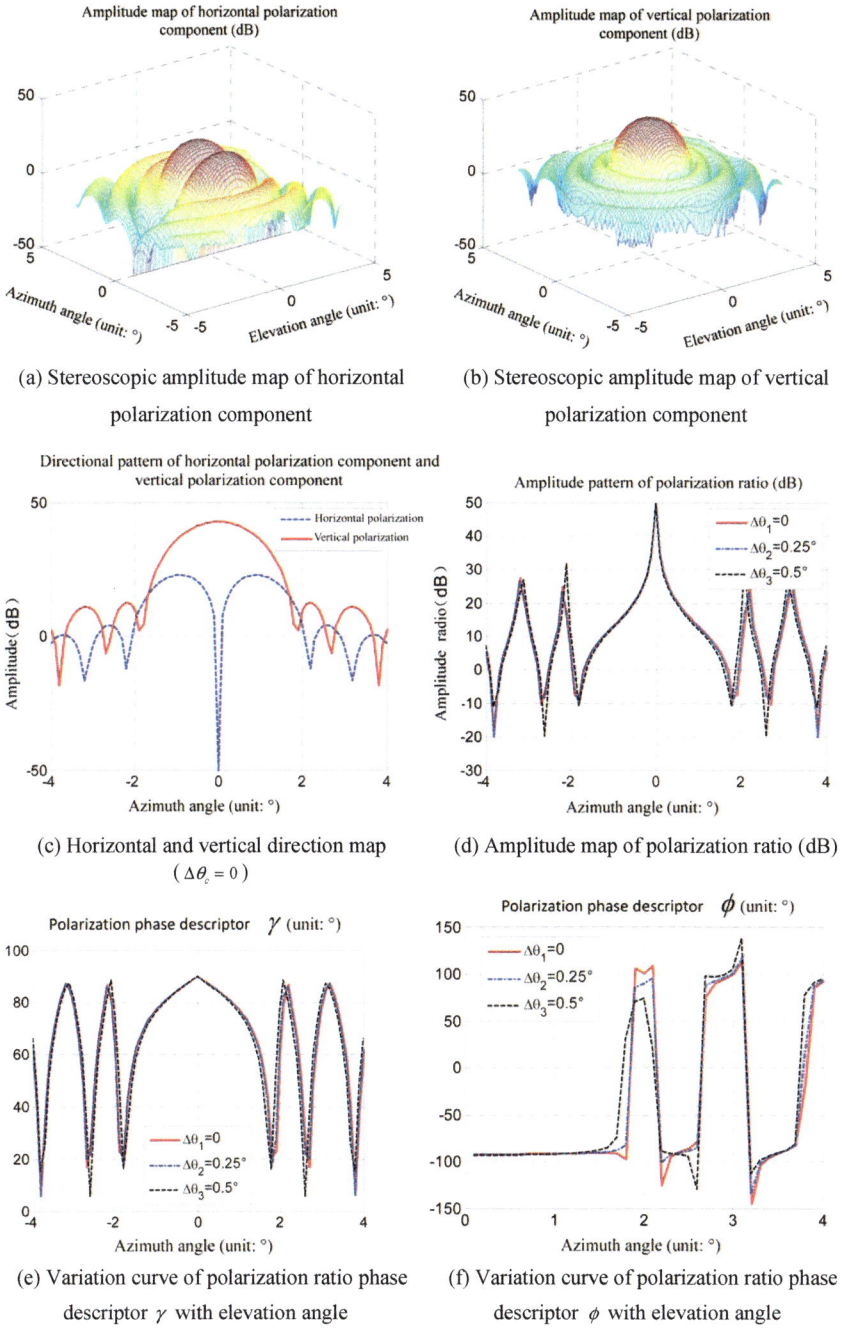

(a) Stereoscopic amplitude map of horizontal
polarization component

(b) Stereoscopic amplitude map of vertical
polarization component

(c) Horizontal and vertical direction map
($\Delta\theta_c = 0$)

(d) Amplitude map of polarization ratio (dB)

(e) Variation curve of polarization ratio phase
descriptor γ with elevation angle

(f) Variation curve of polarization ratio phase
descriptor ϕ with elevation angle

Fig. 3.20 Spatial polarization characteristics pattern of the paraboloid antenna with X linearly polarized feed

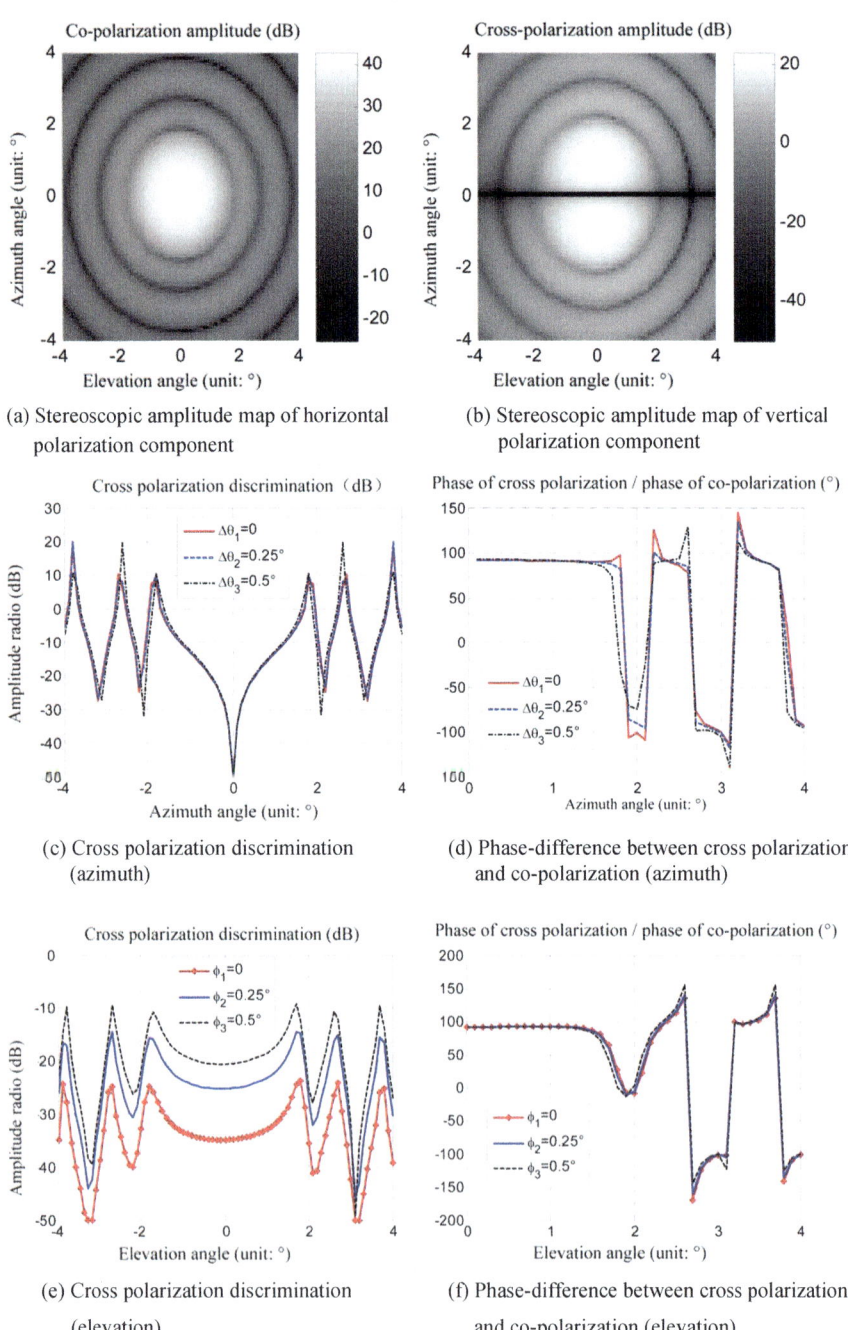

(a) Stereoscopic amplitude map of horizontal polarization component

(b) Stereoscopic amplitude map of vertical polarization component

(c) Cross polarization discrimination (azimuth)

(d) Phase-difference between cross polarization and co-polarization (azimuth)

(e) Cross polarization discrimination (elevation)

(f) Phase-difference between cross polarization and co-polarization (elevation)

Fig. 3.21 Cross polarization characteristics pattern of the paraboloid antenna with X linearly polarized feed

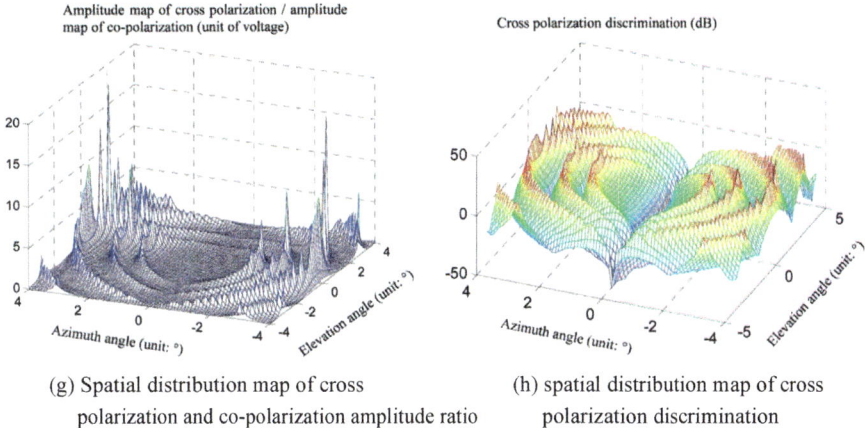

(g) Spatial distribution map of cross
polarization and co-polarization amplitude ratio

(h) spatial distribution map of cross
polarization discrimination

Fig. 3.21 (continued)

0.25° and $\Delta\theta_c = 0.5°$, and curves for variation of the spatial polarization phase descriptor γ and ϕ with azimuth angle as shown in Fig. 3.20e, f. Each curve represents a different elevation angle [129].

It can be seen from Fig. 3.20 that when the paraboloid antenna is scanned in the spatial, its polarization characteristics are not invariable, but constantly changing. From Fig. 3.20a, b, the co-polarization component of the antenna is vertical polarization, cross polarization component is horizontal polarization; from Fig. 3.20c the half power beam width of parabolic antenna in the azimuth plane is 1.4°, main lobe width is 3.6°. It can be seen in Fig. 3.20d, f, when the antenna is scanned in the azimuth direction, the polarization characteristics changed significantly, especially in main lobe range, and change basically monotonously. For example, when φ_c change in the range of $[0°, 1.8°]$, amplitude of the antenna polarization ratio decreases from $-\infty \rightarrow +10$ dB monotonically (Note: in order to easy to show, the value of $|\rho|$ which be greater than 50 dB was replaced by 50 dB), polarization phase descriptors γ decreases from $90° \rightarrow 20°$ monotonically, the polarization phase descriptors ϕ remained unchanged; at the same time, also can be seen when the change of elevation angle is small (for example, the elevation angle $\Delta\theta_c$ is 0°, 0.25° and 0.5°, respectively are shown in figure), the polarization characteristics of the antenna are almost unchanged.

According to the definition of the feed coordinate system in Fig. 3.19, it is shown that the co-polarization of the paraboloid antenna with X linearly polarized feed is vertical polarization, and the cross polarization component is horizontal polarization, and the conclusion is consistent with the simulation result of Fig. 3.20. Furthermore, the ratio of cross polarization to co-polarization and cross polarization discrimination are obtained, and the typical results are shown in Fig. 3.21.

Figure 3.21 shows that when the antenna scanning to the main lobe within range (i.e., φ_c changes in the range of $[0°, 1.8°]$), antenna cross polarization discrimination increases from $-\infty \rightarrow +10$ dB monotonically (Note: in order to easy to show, the value of XPD which is less than -50 dB was replaced by -50 dB); but in the elevation direction, the polarization state in the main lobe antenna range changes not obvious.

According to the analysis results in Figs. 3.20 and 3.21, it can be seen that in the usually concerned main lobe range, polarization state of antenna in azimuth direction changes obviously and monotonically; but the change in the elevation direction is not obvious, that is to say, even if the two coordinate radar for the elevation angle information is not accurate. For example, there is a certain measurement error δ_θ for elevation angle, but there is little effect on the variation of spatial polarization characteristics of parabolic antenna in azimuth.

2. Y direction linearly polarized feed

Figure 3.22 shows the typical analysis results of the spatial polarization characteristics of the paraboloid antenna when the antenna has Y direction linearly polarized feed.

From Fig. 3.22a, c, the main polarization of the paraboloid antenna with Y direction linearly polarized feed is horizontal polarization, and the half power beam width of the horizontal upward antenna is $1.4°$, the width of the main lobe is $3.6°$. From Fig. 3.22d, f, in the main lobe range, antenna polarization ratio amplitude and polarization phase descriptor γ increases monotonously, and the polarization phase descriptor ϕ remained unchanged, for example, when the azimuth angle φ_c increases gradually in the range of $[0°, 1.8°]$, the antenna polarization ratio amplitude $|\rho|$ is in range of $-\infty \rightarrow +10$ dB and polarization phase descriptor γ is in range of $0° \rightarrow 70°$.

According to the definition of the feed coordinate system in Fig. 3.19, it is shown that the main polarization of the paraboloid antenna with Y linearly polarized feed is horizontal polarization, and the cross polarization component is vertical polarization, and the conclusion is consistent with the simulation result. Moreover, further analysis shows that when the parabolic antenna is scanned in azimuth and elevation direction, the co-polarization component decreases and the cross polarization component increases gradually. In the azimuth of the main lobe, the polarization characteristics of the antenna change obviously and regularity, and a little elevation measurement error δ_θ has little influence on the spatial polarization characteristics in azimuth direction.

3. left-hand circular polarized feed

Figure 3.22 shows the typical analysis results of the spatial polarization characteristics of the paraboloid antenna when the antenna has left-hand circular polarized feed. Figure 3.23a, b shows the variation curves of polarization phase descriptor γ and ϕ with azimuth angle (each curve represents a different elevation angle). Figure 3.23c shows that when the azimuth angle φ_c and elevation angle $\Delta\theta_c$ change in the range of $[-4°, +4°]$, the distribution of all polarization antenna experience on the Poincare sphere, for the convenience of display, rotate the original figure appropriate, and rotate counter clockwise with $45°$ in azimuth, rotate upward

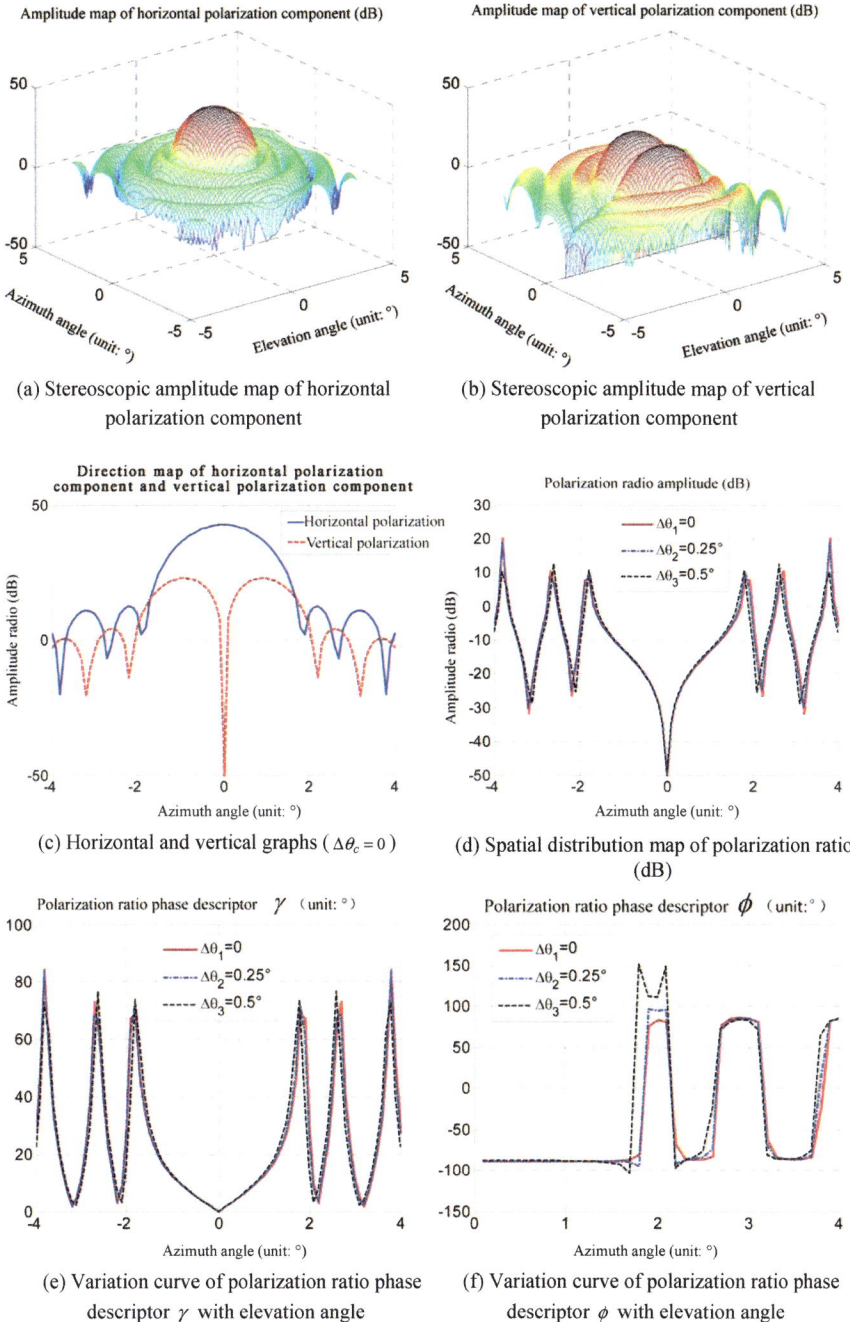

(a) Stereoscopic amplitude map of horizontal
polarization component

(b) Stereoscopic amplitude map of vertical
polarization component

(c) Horizontal and vertical graphs ($\Delta\theta_c = 0$)

(d) Spatial distribution map of polarization ratio
(dB)

(e) Variation curve of polarization ratio phase
descriptor γ with elevation angle

(f) Variation curve of polarization ratio phase
descriptor ϕ with elevation angle

Fig. 3.22 Spatial polarization characteristics pattern of the paraboloid antenna with Y linearly
polarized feed

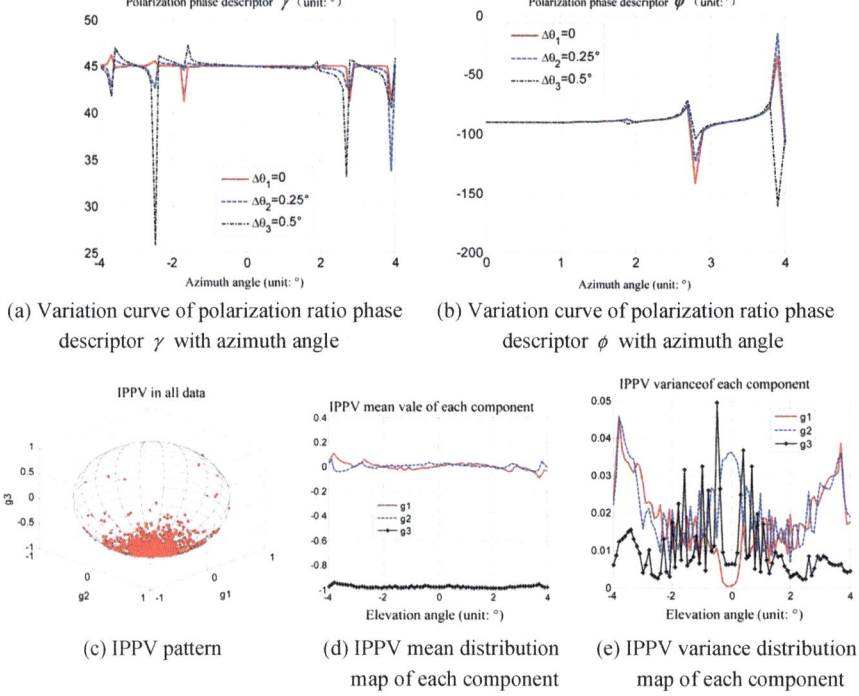

(a) Variation curve of polarization ratio phase
     descriptor γ with azimuth angle

(b) Variation curve of polarization ratio phase
     descriptor ϕ with azimuth angle

(c) IPPV pattern

(d) IPPV mean distribution
     map of each component

(e) IPPV variance distribution
     map of each component

Fig. 3.23 Spatial polarization characteristics pattern of the paraboloid antenna with left-handed circular polarized feed

clockwise with $30°$ in the elevation (here refers to the clockwise from left to right); as for each elevation angle, when the azimuth angle in the range of $[-4°, +4°]$ change, calculate corresponding Stokes vector to each polarization antenna experience, and then obtain the mean value and variance of g_1, g_2 and g_3 component, it's variation curve with the elevation angle as shown in Fig. 3.23d, e.

It can be seen from Fig. 3.23 that when the antenna is scanned in azimuth and elevation directions, the polarization states are distributed around the south pole of the Poincare sphere, that is, the main polarization of the antenna is right-handed circular polarization, which is consistent with the discussion of Sect. 3.3.1, i.e., the circularly polarized feed polarization will be reversed. At the same time, Fig. 3.23 shows that when the parabolic antenna has a left-handed circular polarization feed, the spatial polarization characteristics is not obvious, this is also consistent with the results discussed in Sect. 3.3.1, the cross polarization is the main problem of linear polarization feed parabolic antenna, but not circularly polarized parabolic line the main problem day.

Furthermore, the ratio of cross polarization to co-polarization and cross polarization discrimination are obtained, and the typical results are shown in Fig. 3.24.

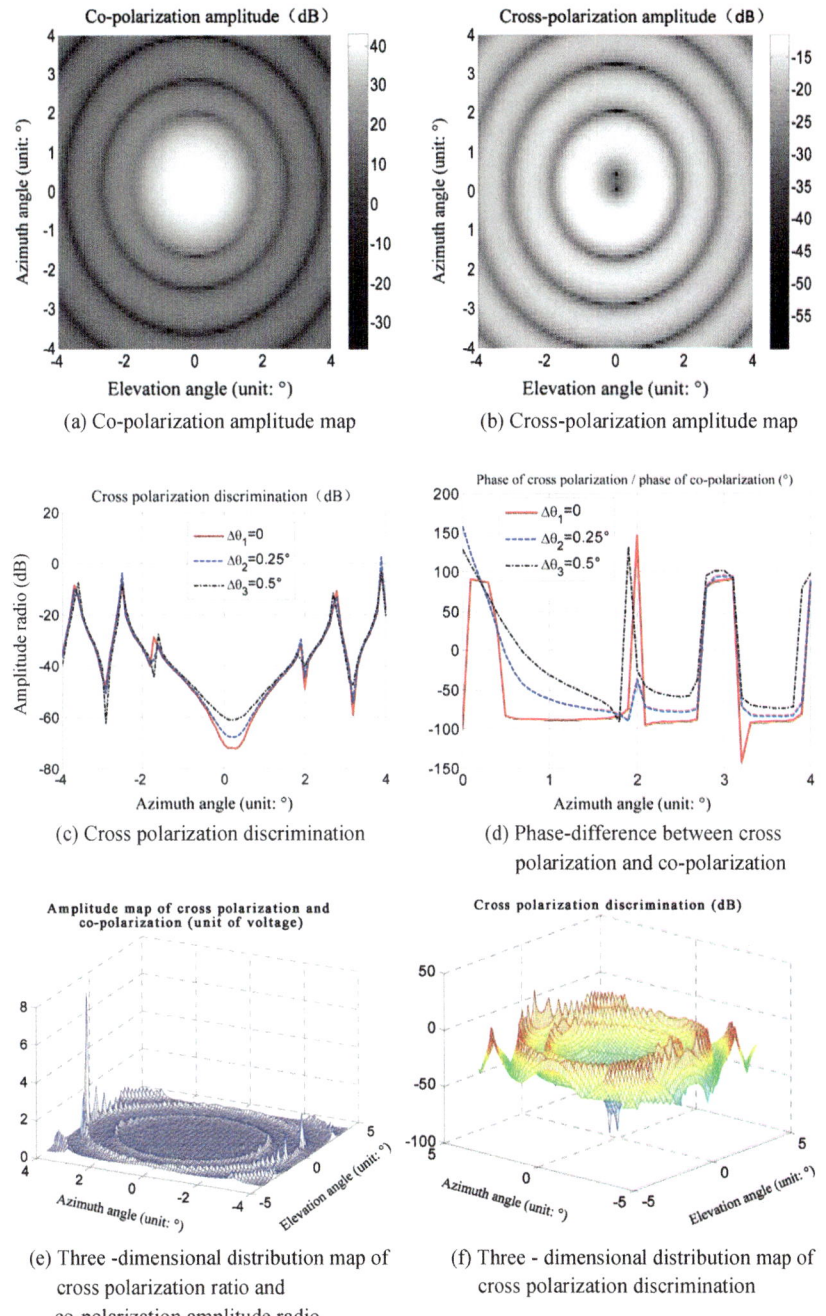

(a) Co-polarization amplitude map

(b) Cross-polarization amplitude map

(c) Cross polarization discrimination

(d) Phase-difference between cross
polarization and co-polarization

(e) Three -dimensional distribution map of
cross polarization ratio and
co-polarization amplitude radio

(f) Three - dimensional distribution map of
cross polarization discrimination

Fig. 3.24 Cross polarization characteristics pattern of the paraboloid antenna with left-handed
circular polarized feed

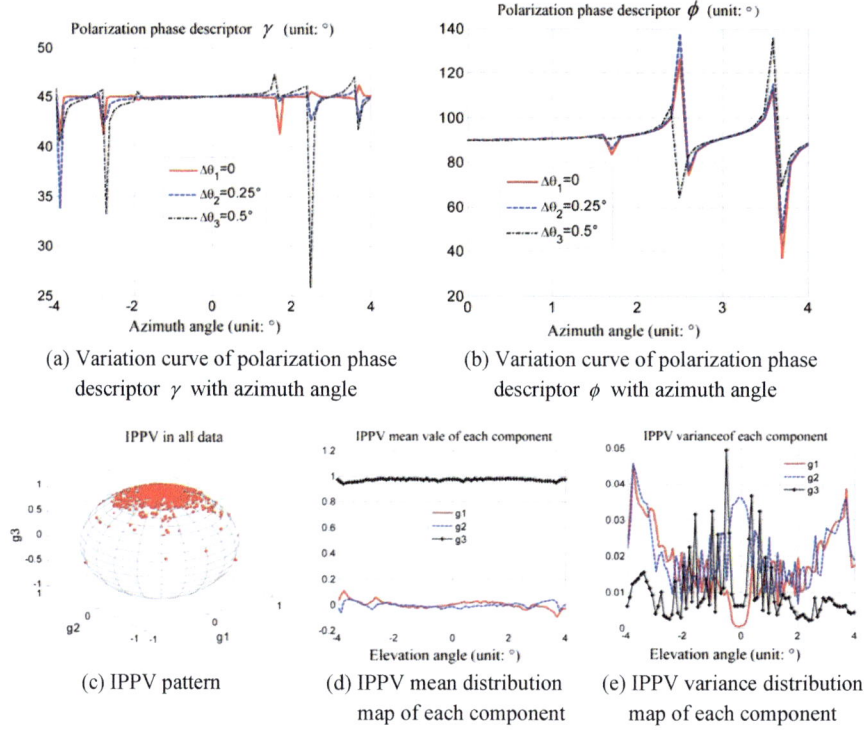

(a) Variation curve of polarization phase
descriptor γ with azimuth angle

(b) Variation curve of polarization phase
descriptor ϕ with azimuth angle

(c) IPPV pattern

(d) IPPV mean distribution
map of each component

(e) IPPV variance distribution
map of each component

Fig. 3.25 Cross polarization characteristics pattern of the paraboloid antenna with right-handed
circular polarized feed

As shown in Fig. 3.24 that, when the parabolic antenna is scanned in azimuth
and elevation direction, the co-polarization component decreases and the cross
polarization component increases gradually. For example, in azimuth direction,
when the antenna is scanned in the main lobe range, the cross polarization discrimination increases from -70 to -30 dB.

4. Right-handed circular polarized feed

Figure 3.25 shows the typical analysis results of the spatial polarization characteristics of the paraboloid antenna with right-handed circular polarized feed.

It can be seen from Fig. 3.25 that when the antenna is scanned in azimuth and
elevation directions, the polarization states are distributed around the north pole of
the Poincare sphere, that is, the main polarization of the antenna is left-handed
circular polarization, which is consistent with the discussion of Sect. 3.3.1.3, i.e.,
the circularly polarized feed polarization will be reversed, that is, co-polarization of
paraboloid antenna with right-handed circular polarized feed is right-handed circular polarization, cross polarization component is right-handed circular
polarization.

With Figs. 3.23, 3.24 and 3.25 and a large number of simulation results can be concluded: offset parabolic antenna with circular polarization feed position in different space, polarization characteristics change, but the change was not obvious, generally, the cross polarization component is very small relative to the co-polarization component is very small.

3.3.2.2 The Analysis of Effect of the Key Parameters on Spatial Polarization Characteristics of Parabolic Antenna

According to the analysis of Sect. 3.3.1.5, the focal diameter ratio and offset angle of the parabolic antenna are the key parameters affecting the polarization characteristics. The typical difference between the focal diameter ratio F/D of 0.25–0.8 and the offset angle ψ_0 of $30^\circ - 70^\circ$ is discussed, and the influence of the key parameters on the spatial polarization characteristics of the antenna is discussed.

1. The influence of the focal diameter F/D on spatial polarization characteristics of antennas

Taking the parabolic antenna with X direction linearly polarized feed as an example, suppose the offset angle of the paraboloid is $\psi_0 = 44^\circ$, and the different of the polarization characteristics of the antenna are discussed when the focal diameter ratio of the antenna is $F/D = 0.25$, $F/D = 0.35$, $F/D = 0.5$ and $F/D = 0.75$. The simulation results are shown in Fig. 3.26 when antenna lays flat (i.e., $\Delta\theta_c = 0$), where, Fig. 3.26a, b are the spatial distribution maps of the polarization ratio amplitude of the antenna and cross polarization discrimination with azimuth angle; Fig. 3.26c, d are the variation cure of polarization phase descriptors γ and ϕ.

As shown in Fig. 3.26, the smaller the focal diameter ratio F/D, the larger the cross polarization component of the offset parabolic antenna, the faster the polarization state of the antenna changes in space, or the more obvious the spatial polarization characteristics of the antenna, which is consistent with the theoretical analysis of Sect. 3.3.1.5.

2. The effect of offset angle ψ_0 on spatial polarization characteristics of the antenna

Taking the parabolic antenna with X direction linearly polarized feed as an example, suppose that the focal diameter ratio the paraboloid is $F/D = 0.5$, and the difference of the spatial polarization characteristics of the paraboloid antenna are discussed when the offset angle of the antenna is $\psi_0 = 30^\circ$, $\psi_0 = 44^\circ$, $\psi_0 = 55^\circ$ and $\psi_0 = 70^\circ$. When the antenna is flat (i.e., $\Delta\theta_c = 0$), the typical simulation results are shown in Fig. 3.27.

As shown in Fig. 3.27, the smaller the offset angle ψ_0, the larger the cross polarization component of the antenna, the faster the polarization state of the antenna changes in space, which is consistent with the theoretical analysis of Sect. 3.3.1.

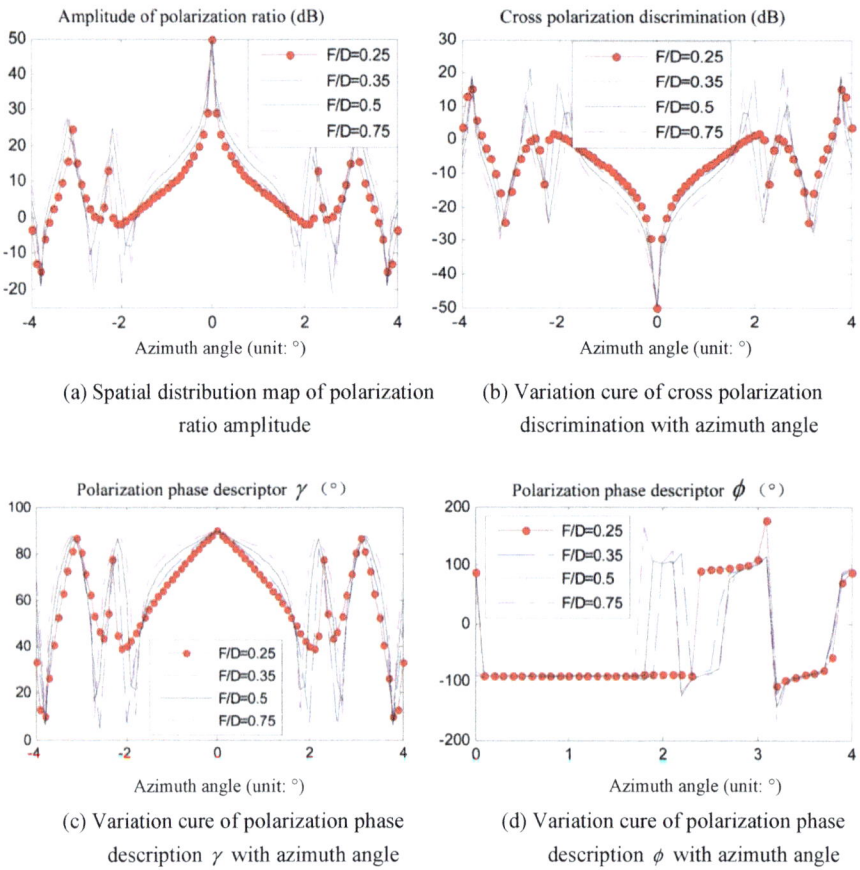

(a) Spatial distribution map of polarization
ratio amplitude

(b) Variation cure of cross polarization
discrimination with azimuth angle

(c) Variation cure of polarization phase
description γ with azimuth angle

(d) Variation cure of polarization phase
description ϕ with azimuth angle

Fig. 3.26 Spatial polarization characteristics comparison pattern of the paraboloid antenna with different focal diameter ratio *F/D*

3.3.2.3 The Simulation Analysis of Spatial Polarization Characteristics of Typical Normally Fed Antenna

In particular, the offset angle $\psi_0 = 0$, the offset paraboloid is defined as the symmetrical paraboloid, (i.e., the normally fed paraboloid. Suppose that paraboloid have Y direction linear polarization feed, working frequency of antenna is 10 GHz, section diameter is 30 cm (with the analysis in Sect. 3.5.2 of a parabolic antenna corresponding to the actual paraboloid antenna), parabolic focal diameter ratio *F/D* = 0.5, spatial polarization characteristics are fed parabolic antenna by computer simulation, the typical simulation results are given.

Figure 3.28a, b are the spatial distribution maps of the co-polarization component of the antenna and cross polarization component amplitude, respectively.

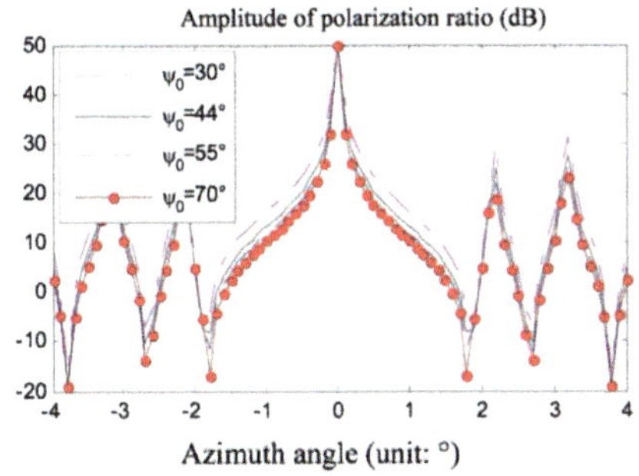

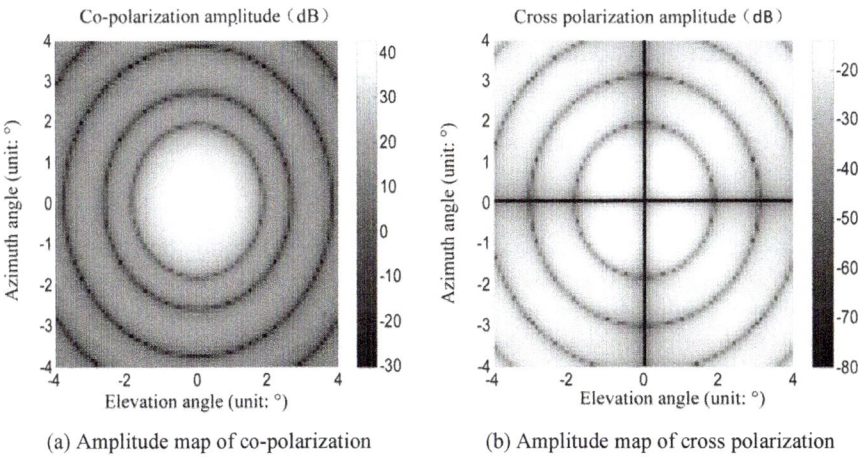

Fig. 3.27 Spatial polarization characteristics comparison pattern of the paraboloid antenna with different offset angle ψ_0

(a) Spatial distribution map of polarization ratio amplitude (dB)

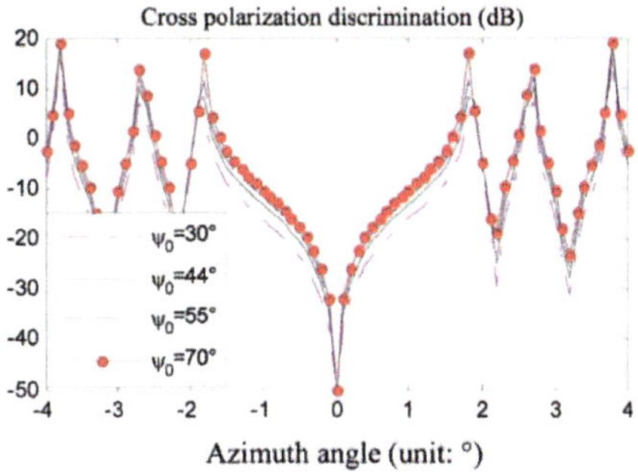

(b) Variation cure of cross polarization discrimination with azimuth angle

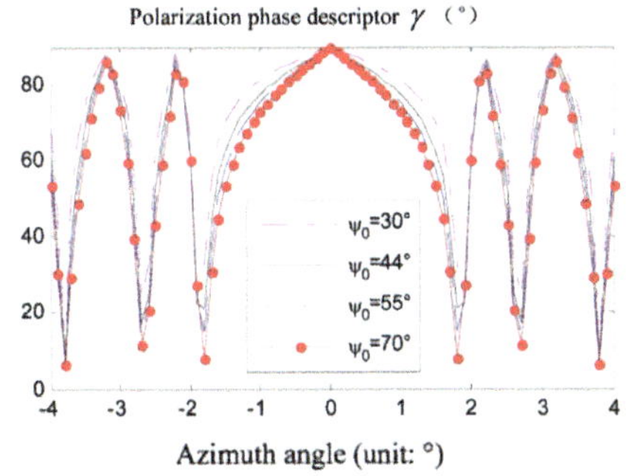

(c) Variation cure of polarization phase description γ with azimuth angle

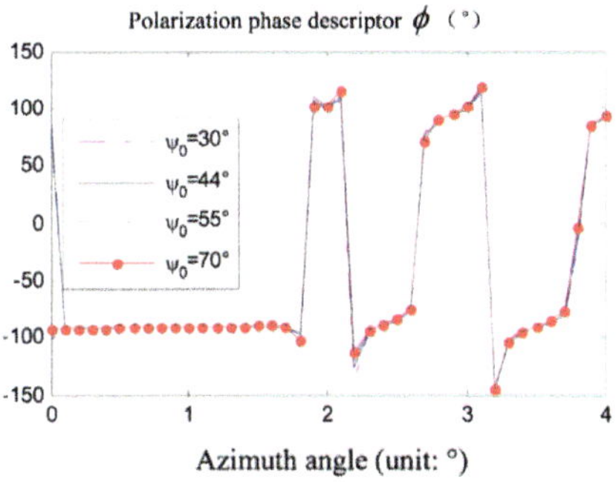

(d) Variation cure of polarization phase description ϕ with azimuth angle

(a) Amplitude map of co-polarization

(b) Amplitude map of cross polarization

Fig. 3.28 Amplitude spatial distribution pattern of normally fed paraboloid co-polarization and cross polarization

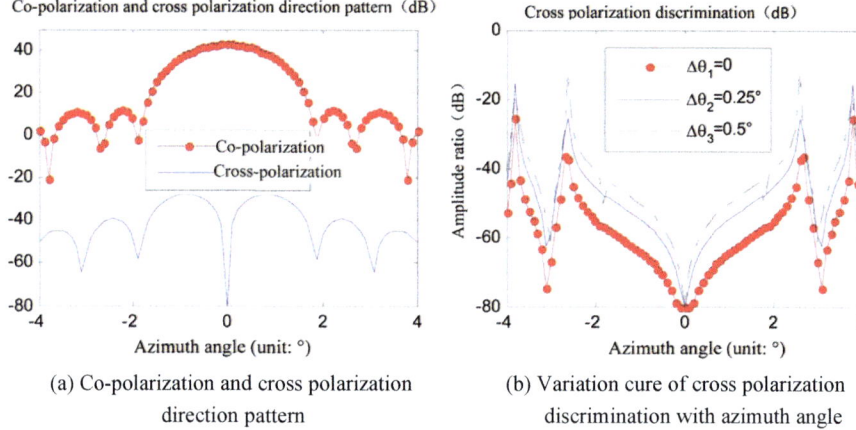

(a) Co-polarization and cross polarization   (b) Variation cure of cross polarization
direction pattern   discrimination with azimuth angle

Fig. 3.29 Spatial distribution pattern of cross polarization discrimination of normally fed paraboloid antenna

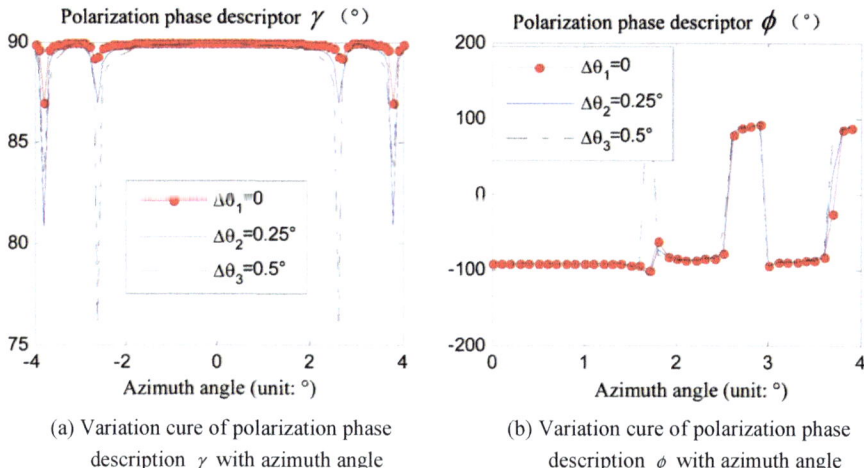

(a) Variation cure of polarization phase   (b) Variation cure of polarization phase
description γ with azimuth angle   description ϕ with azimuth angle

Fig. 3.30 Spatial distribution pattern of polarization phase descriptor of normally fed paraboloid antenna

Figure 3.29a shows the co-polarization and cross polarization direction patterns when the antenna is flat (i.e., elevation angle of antenna $\Delta\theta_c = 0$); Fig. 3.29b shows the cross polarization discrimination of the antenna varies with the azimuth angle, in which each curve represents a certain elevation angle.

Figure 3.30a shows the polarization phase descriptors γ and ϕ of the antenna varies with the azimuth angle, in which each curve represents a certain elevation angle.

It can be seen from Figs. 3.28, 3.29 and 3.30 that when the antenna is scanned in azimuth direction to the main lobe range, the cross polarization discrimination is only increased from $-\infty$ to -50 dB, and the polarization phase descriptor γ is only reduced from 90° to 89.7°, and the range of variation is very small. At the same time, the simulation results can be seen, the polarization purity of normally fed paraboloid antenna is far higher than the offset feed parabolic antenna, especially in main lobe range, polarization characteristics of antenna has some changes but is not obvious, the cross polarization component is very small.

3.3.2.4 The Analysis of the Influence of Working Frequency on Spatial Polarization Characteristics of Paraboloid Antennas

Finally, radar antenna at S wave band with the working frequency of 2 GHz and the wavelength-diameter ratio of $D/\lambda = 20$ (based on domestically produced remote surveillance guide radar); radar antenna at C wave band with the working frequency of 5 GHz and the wavelength-diameter ratio of $D/\lambda = 50$ (based on domestically produced air surveillance guide radar); radar antenna at X wave band with the frequency of 10 GHz and the wavelength-diameter ratio of $D/\lambda = 100$ are taken as examples for calculation to analyze the similarities and differences of the polarization characteristics of paraboloid antenna operating in different frequency bands.

Figure 3.31 shows the typical simulation results of the paraboloid antenna with X direction linearly polarized feed operating in different frequency bands when focal diameter ratio is F/D = 0.5 and the offset angle $\psi_0 = 44°$. Assume the antenna is placed flat (that is $\Delta\theta_c = 0$), Fig. 3.31a, c are the directional map of the co-polarization and cross-polarization when antenna works at different typical frequency bands; Fig. 3.31d, e are the spacial distribution curve of antenna polarization ration amplitude and cross-polarization discrimination quantity varying with the azimuth angles; Fig. 3.31f, g are the spacial distribution curve of polarization phase descriptor γ and ϕ varying with the azimuth angle.

From Fig. 3.31a–c we can see that the higher the frequency of the antenna is, the higher the gain is and the narrower the beam is. At the same time, from Fig. 3.31d– we can see that within the same scanning scope, the higher the operating frequency of the antenna is, the greater the change degree of the antenna polarization characteristics is, that is to say, the faster the antenna spatial polarization characteristics changes.

Table 3.3 shows the main performance parameters and the spatial polarization characteristics of the antenna at different working frequencies when scanning in azimuth angle, including the maximum gain, beam width, the monotonic variation range φ_Δ of the antenna polarization characteristic (that is to say, when the range of scanning in azimuth plane is $[-\varphi_\Delta/2, +\varphi_\Delta/2]$, the antenna polarization phase descriptor has linear change). The size of the antenna cross-polarization discrimination quantity is very small when at the half power beam width, 2 times the half power beam width, 2.5 times the half power beam width, the edge of antenna polarization phase descriptor in linear variation range.

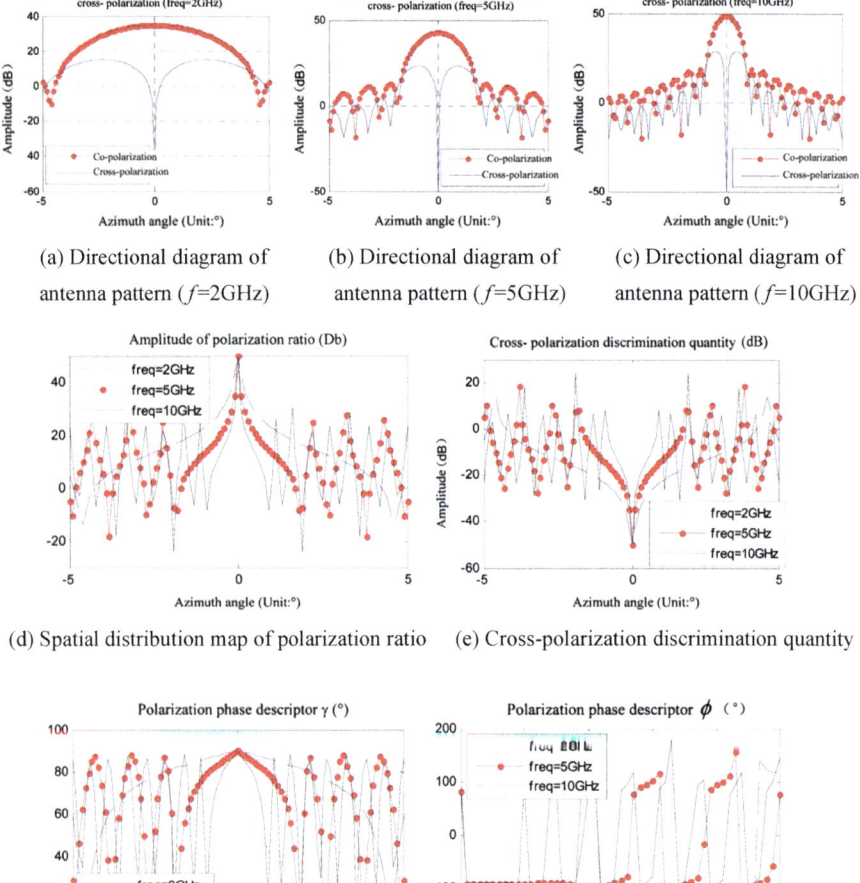

(a) Directional diagram of
antenna pattern (f=2GHz)

(b) Directional diagram of
antenna pattern (f=5GHz)

(c) Directional diagram of
antenna pattern (f=10GHz)

(d) Spatial distribution map of polarization ratio

(e) Cross-polarization discrimination quantity

(f) Variation curve of polarization phase
descriptor γ varying with elevation angle

(g) Variation curve of polarization phase
descriptor ϕ varying with elevation angle

Fig. 3.31 The comparison diagram of paraboloid antenna spatial polarization characteristics
operating at different frequency

Table 3.3 shows that the main performance parameters of the antenna like the
maximum gain, beam width are quite different at different working frequencies;
however, the change degree of antenna polarization characteristics are pretty much
the same when the antenna operates at different frequencies and in its typical
corresponding range of the half power beam width, 2 times the half power beam
width, 2.5 times the half power beam width, linear variation range of antenna
polarization phase descriptor. For example, when the azimuth angle φ_c is from

Table 3.3 Main performance parameters of single offset paraboloid antenna under different operating frequency bands

Working frequency band		f = 2 GHz	f = 5 GHz	f = 10 GHz
Main performance parameters				
Maximum gain (dB)		34.76	42.72	48.74
Half power beam width $\varphi_{3\,dB}$		3.44°	1.38°	0.68°
2 times the half power beam width $2\varphi_{3\,dB}$		6.88°	2.76°	1.36°
2.5 times the half power beam width $2.5\varphi_{3\,dB}$		8.6°	3.45°	1.7°
The monotonic variation range φ_Δ of antenna polarization characteristics		9.2°	3.6°	1.8°
Polarization characteristic at $\varphi = \frac{\varphi_{3dB}}{2}$	Cross-polarization discrimination quantity	−17.37 dB	−17.36 dB	−17.57 dB
	Polarization phase descriptor γ	82.28°	82.27°	82.36°
Polarization characteristic at $\varphi = \varphi_{3dB}$	Cross-polarization discrimination quantity	−7.98 dB	−7.91 dB	−8.19 dB
	Polarization phase descriptor γ	68.24°	68.05°	68.6°
Polarization characteristic at $\varphi = 1.25\varphi_{3dB}$	Cross-polarization discrimination quantity (dB)	1.26	2.26	2.45
	Polarization phase descriptor γ	40.85°	38.35°	38.30°
Polarization characteristic at $\varphi = \frac{\varphi_\Delta}{2}$	Cross-polarization discrimination quantity (dB)	13.92	8.27	8.34
	Polarization phase descriptor γ	11.4°	21.1°	21°

$0 \rightarrow \varphi_\Delta/2$, and the antenna works at three bands as f = 2 GHz, f = 5 GHz, f = 10 GHz, its cross-polarization discrimination quantity increases respectively from $-\infty \rightarrow -17.37\,dB$, $-\infty \rightarrow -17.36\,dB$, $-\infty \rightarrow -17.57\,dB$, and the polarization angle decreases respectively from $90° \rightarrow 11.4°$, $90° \rightarrow 21.1°$, $90° \rightarrow 21°$, while the polarization angle ϕ remains unchanged; at the same time, we can also see that the monotonic variation range of the antenna polarization characteristics is approximately equal to 2.5 times the half power beam width of the antenna, that is $\varphi_\Delta \approx 2.5 \times \varphi_{3\,dB}$.

From the analysis of Sect. 3.3, it can be seen that the symmetry of the antenna structure is destroyed due to the offset aperture, so the cross-polarization will increase when the single offset reflector antenna is used, and the cross-polarization component is larger, and the spatial polarization characteristics of the antenna are obvious and have strong regularity.

3.4   Actual Measurement Spatial Polarization Characteristics of Antenna

The polarization of the antenna is closely related to the antenna type, and the changing rules of polarization characteristics of different antennas in space are not the same. Considering the measured data at the microwave anechoic chamber when a certain jammer antenna operating in the C wave frequency band and a certain parabolic antenna operating in the X wave frequency band are scanning in the azimuth and elevation directions, analyze the spatial polarization characteristics of the two antenna to provide a theoretical basis for subsequent application research. Figure 3.32 is a picture of the microwave anechoic chamber in the experiment.

3.4.1   Spatial Polarization Characteristics of a Certain Jamming Antenna

In this section, as for the actual jammer antenna operating in the C wave frequency band, we analyzed the co-polarization pattern and cross-polarization pattern of the antenna by using the measured data at the anechoic chamber when the antenna is scanning in the azimuth and elevation directions within a certain region. Then we discussed the classical descriptors of antenna polarization ratio, polarization purity and other spatial polarization characteristics as well as the distribution of spatial instantaneous polarization characteristics descriptor such as the spacial IPPV, polarization cluster center and polarization divergence.

The measurement mode is as follows: ① The scan range in azimuth direction: −60° to +60°, the scanning interval is 0.5°; ② The scan range in elevation direction: −45° to +45°, the scanning interval is 5°; the working frequency range:

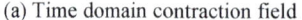

(a) Time domain contraction field   (b) Microwave anechoic chamber

Fig. 3.32 Picture of the microwave anechoic chamber in the experiment

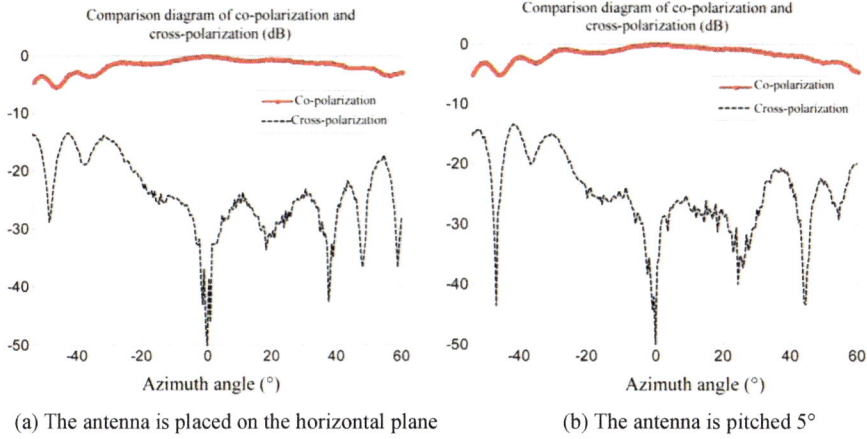

Fig. 3.33 Measured pattern of co-polarization and cross-polarization of the jammer antenna

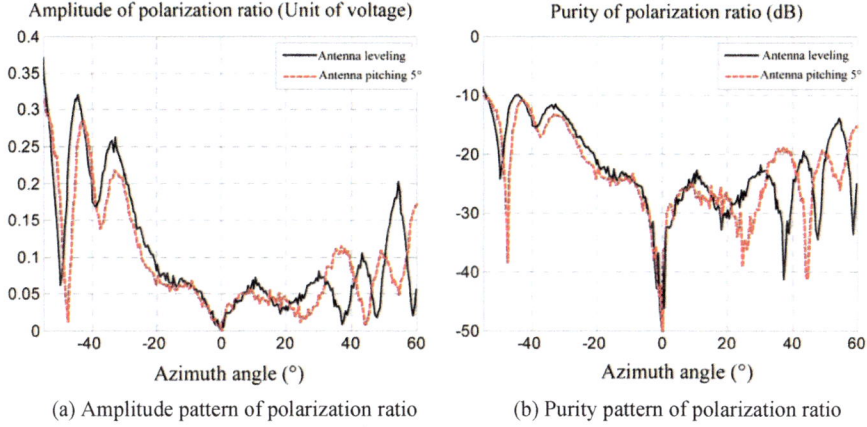

Fig. 3.34 Measured spacial distribution map of polarization ratio and polarization purity of jammer antenna

3.9–6.2 GHz; ④ the voltage data is received by using two channels of vertical and horizontal polarizations.

Figure 3.33 shows the normalized co-polarization pattern and cross-polarization pattern of the antenna when the antenna operates at a center frequency band of $f = 5.05$ GHz ± 12.2 MHz. In this figure, the horizontal axis is the azimuth angle, the vertical axis represents the voltage amplitude of the antenna, and the unit is dB; Fig. 3.33a shows when the antenna is placed on a horizontal aperture while Fig. 3.33b is when the antenna pitches 5°.

Figure 3.34a is the variation curve of the amplitude ratio of the cross-polarization and the co-polarization changing with the azimuth angle of the antenna; Fig. 3.34b gives the variation curve of the polarization purity of the antenna changing with the azimuth angle.

A large number of simulation results in various situations show that the jammer antenna is a wide wave beam antenna. When the antenna scans in azimuth and elevation direction, the polarization characteristics change obviously in certain rules. When the antenna is placed flat, the IPPV of the co- polarization component is $[-0.0524, 0.9986, 0.0034]^T$. It points at the wave beam center, with high antenna polarization purity and the cross-polarization discrimination quantity is as low as -50 dB. It's cross-polarization component gradually increases, and the polarization purity decreases, while the cross-polarization discrimination quantity gradually increases, reaching to -7 dB at the maximum as the pointing of the antenna wave beam deviates from the center position.

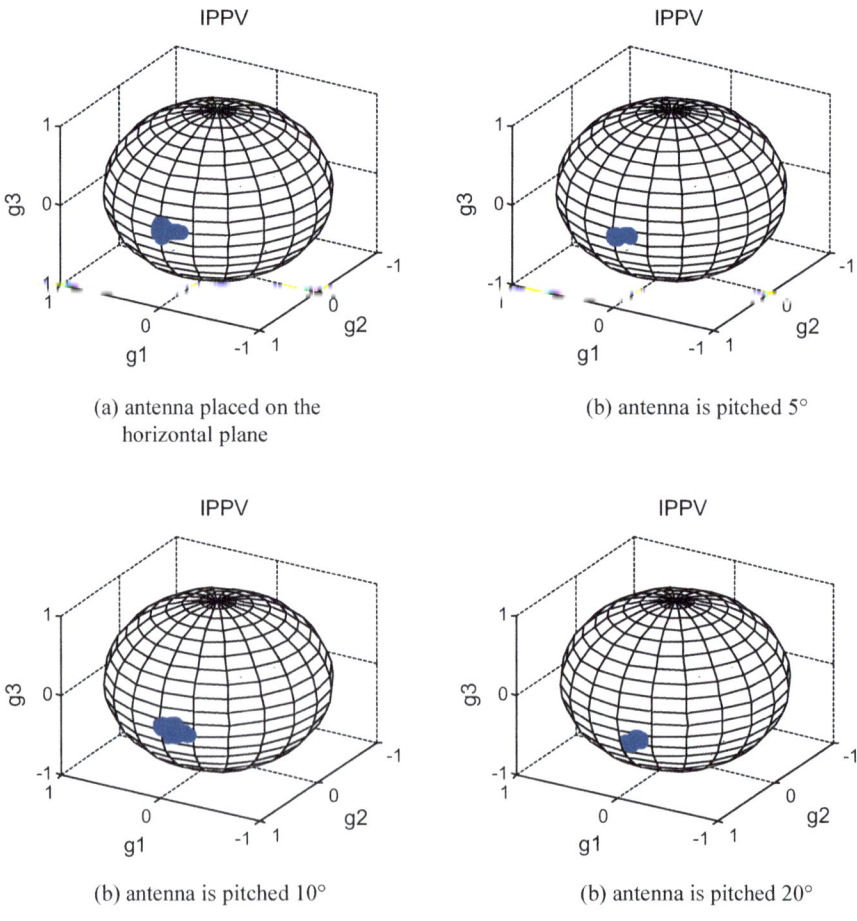

(a) antenna placed on the horizontal plane

(b) antenna is pitched 5°

(b) antenna is pitched 10°

(b) antenna is pitched 20°

Fig. 3.35 Measured spacial IPPV of jammer antenna

The spatial polarization projection set of the antenna fully describes the spatial polarization characteristics of the antenna, and the distribution of the instantaneous polarization projection vector (IPPV) on the unit sphere of the Poincare shows the distribution of the polarization state of the antenna at different spatial points. According to the definition of the spacial instantaneous polarization characteristics of the antenna in the second chapter, Fig. 3.35 shows the spatial IPPV diagram of the antenna at different elevation angles when the antenna works in the central frequency band and scans in the azimuth within the main lobe range. Figure 3.35a–d is the situation of antenna leveling, antenna pitching 5°, pitching 10° and pitching 20°. In order to facilitate the display, the original picture was rotated properly, and rotated 150° counterclockwise in the azimuth direction, and rotated clockwise 3° upward in the elevation direction (clockwise here refers to the cross section from left to right).

Assume the rectangular coordinates of the polarization state of the electromagnetic wave on the Poincare sphere are defined as (g_1, g_2, g_3), and the corresponding spherical coordinates can be defined as $(\theta_{polar}, \varphi_{polar})$. The relationship between the rectangular coordinates and the spherical coordinates of the Poincare sphere is as follows

$$\begin{cases} \mathrm{tg}\theta = \sqrt{g_1^2 + g_2^2}/g_3 & \theta_{polax} \in [0, \pi) \\ \mathrm{tg}\varphi = g_2/g_1 & \varphi_{polar} \in [0, \pi) \end{cases} \tag{3.4.1}$$

In order to better express the relative relationship between the co-polarization and the various polarization state experienced by the antenna when it is scanning in the space, the relative relationship diagram of co-polarization and the various polarization under the rectangular coordinates and the spherical coordinates of the Poincare sphere when the antenna is placed on the horizontal plane, operates in the center frequency and scans in azimuth direction in the main lobe range are given in Fig. 3.36a, b.

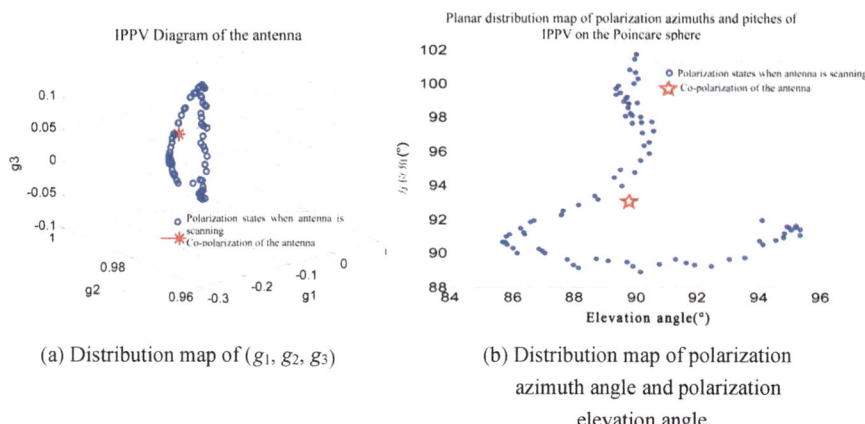

(a) Distribution map of (g_1, g_2, g_3)

(b) Distribution map of polarization azimuth angle and polarization elevation angle

Fig. 3.36 Measured distribution map of spacial IPPV in the rectangular coordinates and the spherical coordinates of jammer antenna

Table 3.4 Measured distribution range list of the polarization state of the jammer antenna in azimuth direction scanning in the main lobe

	g_1 component	g_2 component	g_3 component	Polarization elevation angle φ_{polar} (°)	Polarization azimuth angle φ_{polar} (°)
Co-polarization	−0.0524	0.9986	0.0034	93.0064	89.8052
Minimum value	−0.2036	0.9791	−0.0934	85.6597	88.8840
Maximum value	0.0195	0.9999	0.0757	95.3615	101.7447

It can be seen from Figs. 3.35 and 3.36 that when the antenna is scanning in azimuth and elevation direction, the polarization state of the antenna changes, and the polarization states are distributed around the co-polarization in certain rules. In the working scenario of Fig. 3.36 and Table 3.4 shows the co-polarization state of the antenna, the g_1, g_2, g_3 components of each polarization state experienced by the antenna, and the range of the θ_{polar} and φ_{polar} values.

In order to further study the size of the deviation of the various polarization state of the antenna from the co-polarization and its dispersion degree in the spatial distribution, the typical spatial instantaneous polarization polarization descriptors are analyzed. According to the discussion in 2.4.2, when a(n) = 1, the definition of the formula (2.4.39) and the formula (2.4.40) shows that the value of the center of the uniformly weighted clustering is the mean value of the components of the instantaneous Stokes vector of the antenna, and the second order polarization divergence is the variance of the components of the instantaneous Stokes vector of the antenna. When the antenna works in the center frequency band, as for each elevation angle, Eq. (3.4.2) is used to calculate the mean value of each polarization state deviating from the co-polarization when the antenna is scanning in azimuth direction within the main lobe range. Figure 3.37a shows the variation curve of the mean value of the deviation changing with the antenna elevation pointing; Eq. (3.4.3) is used to calculate the antenna polarization divergence when the antenna is scanning in azimuth direction within the main lobe range; Fig. 3.37b shows the variation curve of the antenna polarization divergence changing with the antenna elevation pointing.

$$\Delta \tilde{G}_{HV} = \left(\frac{1}{M} \sum_{n=1}^{M} \tilde{g}_{HV}(n) \right) - \tilde{g}_{main} \tag{3.4.2}$$

$$D_{HVm}^{(2)} = \frac{1}{M} \sum_{m=1}^{M} a(n) \left| \tilde{g}_{HVm}(n) - \tilde{G}_{HVm} \right|^2, \quad m = 1, 2, 3 \tag{3.4.3}$$

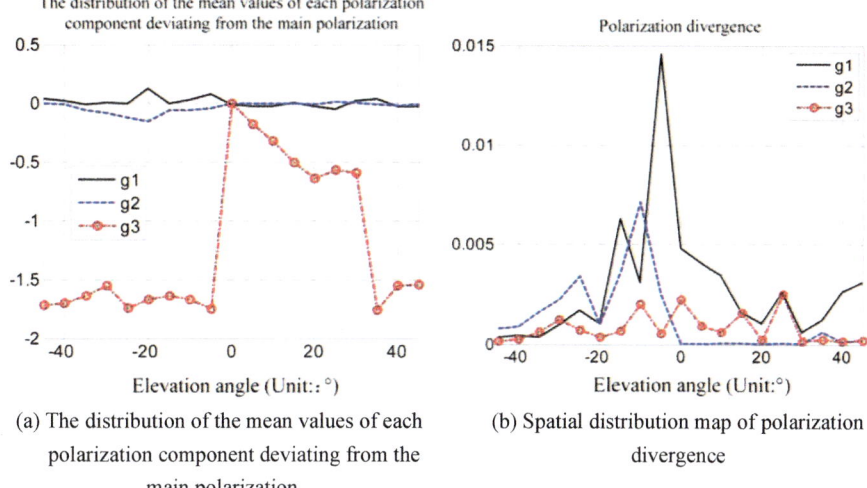

(a) The distribution of the mean values of each polarization component deviating from the main polarization

(b) Spatial distribution map of polarization divergence

Fig. 3.37 The measured variation curve of the statistical characteristics of the polarization descriptor of the jammer antenna changing with the elevation angle

among them,

$$\tilde{G}_{HVm} = \frac{1}{M} \sum_{m=1}^{M} \tilde{g}_{HVm}(n), \quad m = 1, 2, 3 \tag{3.4.4}$$

It can be seen from Fig. 3.37a that the g_3 component deviates from the corresponding co-polarization component farther than g_1 component and g_2 component; Fig. 3.37b shows that the g_2 and g_3 component is more concentrated in the spatial distribution, and the g_1 component is relatively sparse in the spatial distribution. At the same time, we can see that when the antenna is placed on the horizontal plane and scanning in azimuth direction, the various polarization state experienced by it are closely distributed around the co-polarization and its statistical mean value is basically the same as the co-polarization. When the antenna is scanning in the elevation direction, the various polarization of the antenna will gradually deviates from the co-polarization and its distribution will gradually be sparse.

In particular, for a better characterization of the relative relationship between each polarization state of the antenna and the co-polarization, according to the definition in Sect. 2.4.2 "Spatial polarization angle", assume that the co-polarization $\boldsymbol{h}_0$ of the antenna corresponds to the points $(\tilde{g}_{10}, \tilde{g}_{20}, \tilde{g}_{30})$ on the Poincare polarization sphere, a polarized state $\boldsymbol{h}_i$ experienced by the antenna in spacial scanning corresponds to $(\tilde{g}_{1i}, \tilde{g}_{2i}, \tilde{g}_{3i})$ and the sphere center angle β corresponding to the two points meet the following formula:

$$\cos\beta_i = \sum_{k=1}^{3} \tilde{g}_{k0}\tilde{g}_{ki}, \quad \cos^2\frac{\beta_i}{2} = \frac{|\boldsymbol{h}_0^T\boldsymbol{h}_i|^2}{\|\boldsymbol{h}_0\|^2\|\boldsymbol{h}_i\|^2} = |\boldsymbol{h}_0^T\boldsymbol{h}_i|^2 \qquad (3.4.5)$$

where, i = 1,2, …, N, N means the number of polarization states experienced when the antenna is scanning in a certain region of space.

The following can be got from the above equation

$$\beta_i = \arccos\left(\sum_{k=1}^{3} \tilde{g}_{k0}\tilde{g}_{ki}\right), \quad \beta_i \in (0,\pi) \qquad (3.4.6)$$

When the antenna is placed on the horizontal plane, operates at the center frequency band, and scans in the azimuth direction, Fig. 3.38a gives the statistical histogram of spatial polarization angles between co-polarization and the various polarization states the antenna has experienced. At the same time, as for the antenna points at different elevation direction, we calculated the standard deviation of various spatial polarization angles when the antenna is scanning in azimuth direction within the main lobe range and drew the curve as shown in Fig. 3.38b.

It can be seen from Fig. 3.38 that when the antenna is placed horizontally, its polarization angle is distributed around 4.6° in certain rules, and the range of variation is (0.974°, 8.7433°); moreover, the dispersion of the polarization angle distribution changes with the changing of scanning of the antenna in the elevation direction.

This section analyzes the spatial distribution characteristics and value range of the typical spatial instantaneous polarization descriptors of a jammer antenna. Due

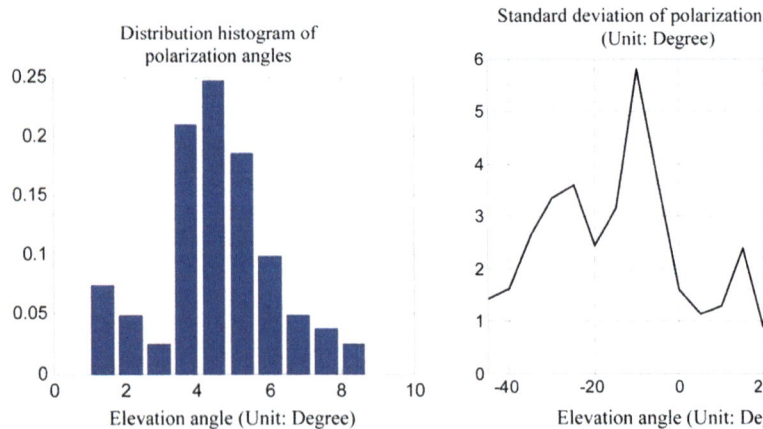

(a) Statistical histogram of polarization angle   (b) Mean value distribution map of polarization
             in antenna leveling   angle under different elevation angles

Fig. 3.38 Measured statistical spacial distribution map of polarization angle of jammer antenna

to limited data, we are unable to accurately derive the mathematical expressions of statistical distributions for such polarization descriptors. But the existing analysis results, from the actual angle of the antenna, shows that when scanning in azimuth and elevation direction, the polarization state of the antenna is changeable, and distributes around the co-polarization in certain rules. This not only fully proved the existence of instantaneous polarization characteristics of antenna, but showed its distribution characteristics and evolution rules.

3.4.2 Spatial Polarization Characteristics of a Certain Parabolic Antenna

A normal feed parabolic antenna operating in the X wave frequency band is shown in Fig. 3.39. The diameter of the paraboloid is 30 cm. The working center frequency is 10 GHz, and the bandwidth is about 10%. The expected polarization of the antenna is horizontal polarization. The antenna is placed on the turntable of the microwave anechoic chamber and scanned in azimuth direction. The voltage data are received by two channels of vertical polarization and horizontal polarization simultaneously.

The co-polarization component of the antenna is horizontal polarization E_H, and the cross-polarization component is vertical polarization E_V. Figure 3.40a–c shows the normalized co-polarization pattern and cross-polarization pattern of the antenna operating at the center frequency 10 GHz, with its adjacent frequency bands 9.8 and 10.2 GHz respectively.

It can be seen from Fig. 3.40 that the paraboloid antenna is a narrow beam antenna with a beam width of about 5°. According to the definition equation

Fig. 3.39 Measured photographs of paraboloid antenna

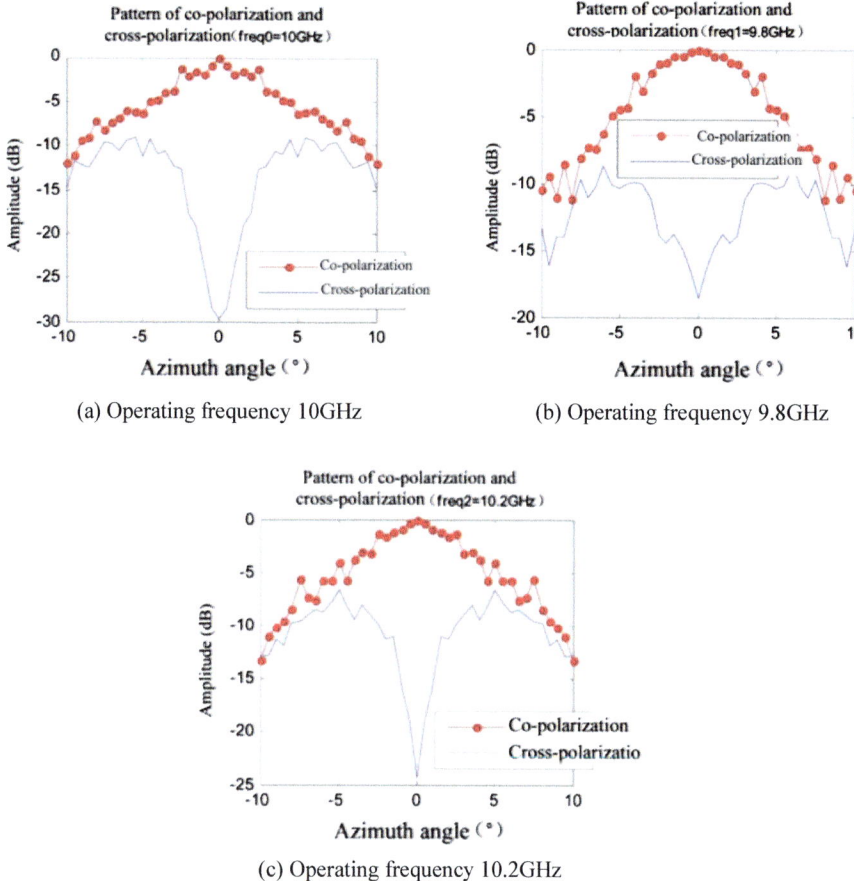

Fig. 3.40 Measured directional map of co-polarization and cross-polarization of paraboloid antenna

$\rho = E_V/E_H$ of polarization ratio in the second chapter and the definition equation $\text{XPD} = 20 \lg(E_{\text{cross}}/E_{\text{co}})$ of cross-polarization discrimination quantity (or polarization purity), it can be seen that for the horizontal polarization parabolic antenna, the latter is the decibel expression of the former. Figures 3.41 and 3.42 shows respectively the variation curve of polarization ratio amplitude and polarization phase descriptor of the antenna changing with the elevation angle, among which, each curve represents different central working frequency.

Figures 3.41 and 3.42 show that the polarization characteristics of antenna change significantly when it in spacial scanning. For example, when the antenna is scanning in azimuth direction in the half power beam width, antenna cross-polarization discrimination quantity increases from $-\infty$ to -7 dB, and the polarization phase descriptor γ increases from $0°$ to $25°$. When the antenna is scanning in azimuth direction within the main lobe, its cross-polarization

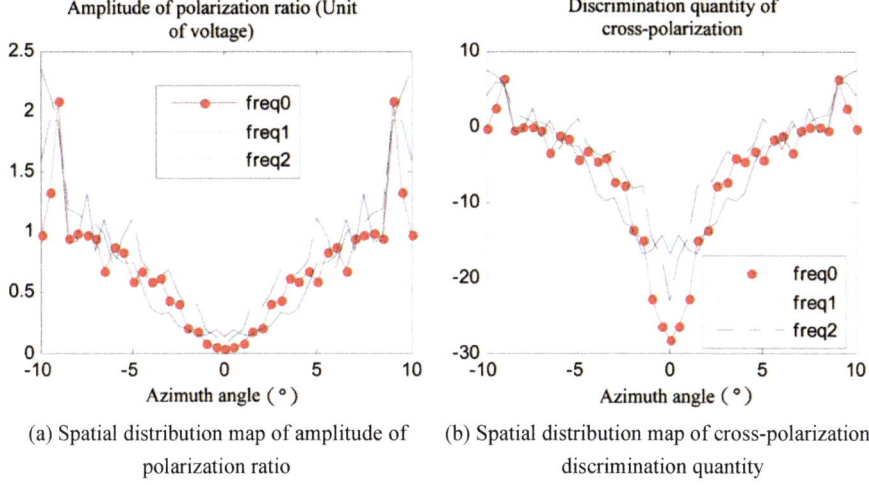

(a) Spatial distribution map of amplitude of polarization ratio

(b) Spatial distribution map of cross-polarization discrimination quantity

Fig. 3.41 Measured spacial distribution map of polarization ratio amplitude of paraboloid antenna

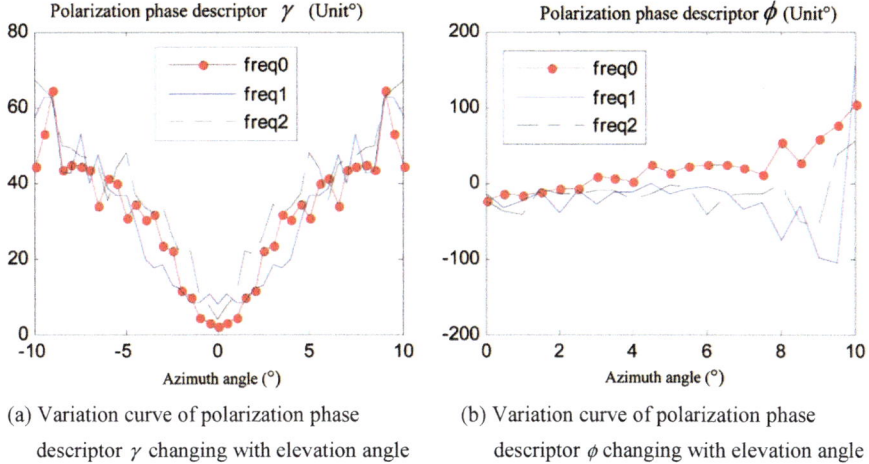

(a) Variation curve of polarization phase descriptor γ changing with elevation angle

(b) Variation curve of polarization phase descriptor ϕ changing with elevation angle

Fig. 3.42 Measured spatial distribution maps of the polarization phase descriptor of the paraboloid antenna

discrimination quantity is 7 dB at the maximum and polarization phase descriptor γ is 65° at the maximum.

Figure 3.43a–c shows the distribution of the polarization states experienced by the antenna on the Poincare sphere when the antenna scans in the azimuth direction within the main lobe range and operates at the center frequency 10 GHz, with its nearby frequency at 9.8 and 10.2 GHz.

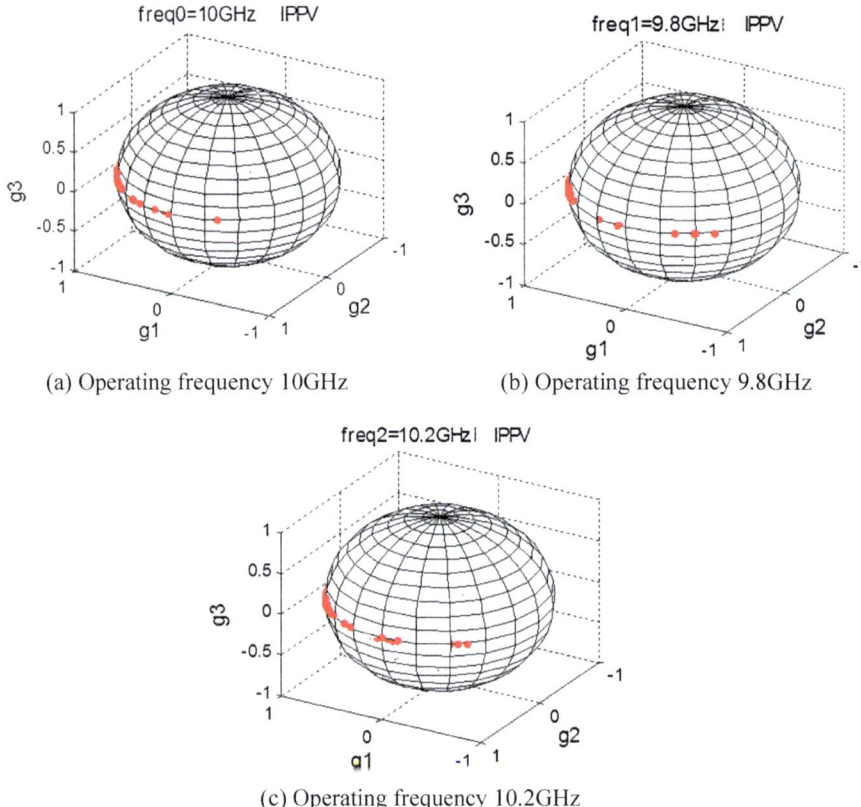

(a) Operating frequency 10GHz   (b) Operating frequency 9.8GHz

(c) Operating frequency 10.2GHz

Fig. 3.43 Measured spacial IPPV diagram of the paraboloid antenna

It can be seen from the above figure that the polarization states experienced the antenna are basically distributed on the equator of the Poincare sphere and scattered near the intersection point of the +x axis and the Poincare sphere when the antenna is scanning in azimuth direction.

We can see from the simulation results from Figs. 3.41, 3.42, 3.43, 3.44 and 3.45 that the polarization characteristics of the antenna change obviously when the normal feed paraboloid antenna is scanning in azimuth direction. Moreover, within the scope of the main lobe, the antenna polarization phase descriptor γ monotonically increases while the polarization phase descriptor ϕ stays in $0°$.

In Sect. 3.4.2.3, we conduct the computer simulation of the spacial polarization characteristics of typical normal feed paraboloid antenna by using GRASP9 software. The simulation results show that the polarization of the normal feed paraboloid antenna is highly pure, especially within the scope of main lobe. The antenna cross-polarization component is very small. By comparing the analysis results of Sect. 3.4.2.3 and this section, we can see that due to the impact of various realistic

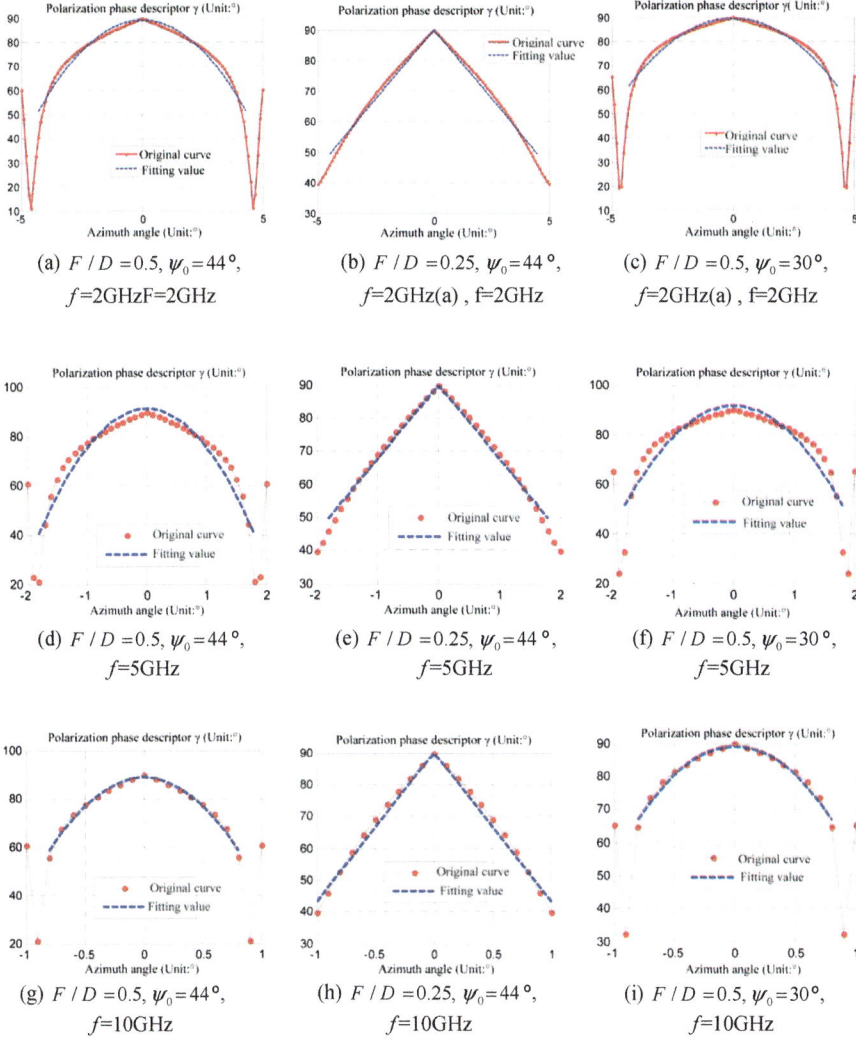

(a) $F/D=0.5$, $\psi_0=44°$,
$f=2$GHzF=2GHz

(b) $F/D=0.25$, $\psi_0=44°$,
$f=2$GHz(a) , f=2GHz

(c) $F/D=0.5$, $\psi_0=30°$,
$f=2$GHz(a) , f=2GHz

(d) $F/D=0.5$, $\psi_0=44°$,
$f=5$GHz

(e) $F/D=0.25$, $\psi_0=44°$,
$f=5$GHz

(f) $F/D=0.5$, $\psi_0=30°$,
$f=5$GHz

(g) $F/D=0.5$, $\psi_0=44°$,
$f=10$GHz

(h) $F/D=0.25$, $\psi_0=44°$,
$f=10$GHz

(i) $F/D=0.5$, $\psi_0=30°$,
$f=10$GHz

Fig. 3.44 Fitting chart of polarization phase descriptor γ of paraboloid antenna (vertical polarization as main polarization)

factors, the actual antenna polarization purity is much worse than the theoretical derivation and the simulation results, that is to say, the actual space variant characteristics of antenna polarization is much more significant than the derived theoretical results.

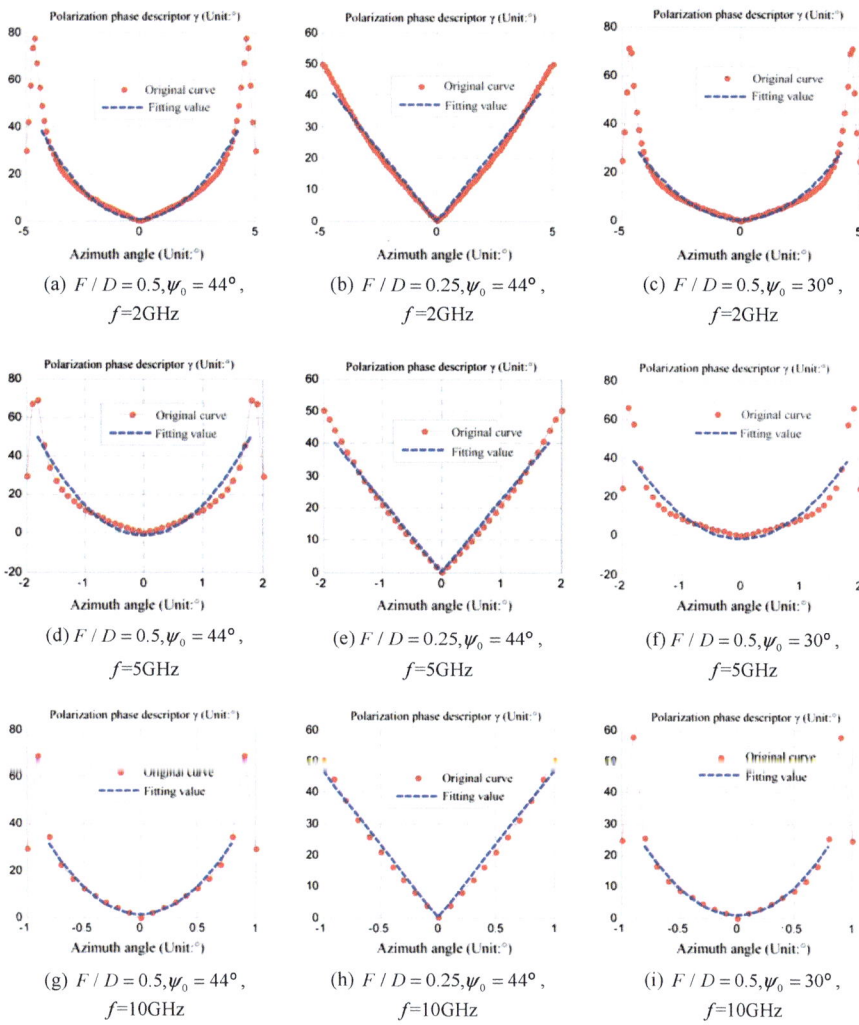

Fig. 3.45 Fitting chart of polarization phase descriptor γ of the paraboloid antenna (vertical polarization as main polarization)

3.5   Modelling and Simulation of Antenna Spatial Polarization Characteristics

Several wire antennas, several aperture antennas, offset paraboloid antennas and two kinds of practical antennas are respectively considered as the research objects, to discuss the spatial polarization characteristics of typical antennas in Sects. 3.1, 3.2, 3.3 and 3.4. Based on the previous analysis results, in this section, polynomial fitting is conducted on the typical spatial polarization characteristics descriptor of the offset paraboloid antenna firstly, and then the rules are summarized, and the spatial polarization characteristic model of the radar antenna is established.

As discussed in Sect. 2.4, a double coordinate mechanical scanning radar usually adopt the fan beam to scan azimuth plane more precisely to get accurate azimuth information; and on the elevation plane, the scanning is relatively sketchy, but at the same time, it works with a nod height measurement radar to get elevation information. In such case, the three-dimensional problem of "Searching the rule of spatial polarization characteristics of antenna varying with azimuth and elevation angle" can be simplified into two dimensions, that is: for a certain elevation angle, the changing rules of the polarization characteristics varying with the azimuth angle when the antenna is scanning on azimuth plane. In Sect. 3.4.2, it is concluded from the influence of different elevation measurement deviation on the variation rule of polarization characteristics of offset paraboloid antenna on the horizontal azimuth plane that even if elevation angle information obtained by the two-coordinate radar is not accurate (for example there are certain elevation measurement deviation), the influence on the variation rules of the spatial polarization characteristics of the offset paraboloid antenna on the azimuth direction is minimal. Based on the above considerations, this section will establish a typical model of polarization characteristics varying with azimuth angle when the antenna is scanning on the azimuth plane.

The descriptors such as "polarization ratio" and "polarization phase descriptor", "cross-polarization discrimination quantity" and "antenna spatial IPPV" can better characterize the variation of the antenna polarization state in the space. Theoretically, we can use any of these descriptors of the antenna polarization characteristic to model spatial polarization characteristics, but due to the magnitude of the polarization ratio and cross-polarization discrimination quantity often will appear similar infinitesimal or infinite values like ∞ and $-\infty$, modeling is not very convenient. Therefore, it is a very appropriate modeling method to model the spatial polarization characteristics of antenna with "polarization phase descriptor".

A large number of simulation results under different conditions show that whatever feed source of polarization form the paraboloid antenna has, or whatever the geometric structure parameters of paraboloid is, or however the working frequency range is set up, when the antenna is scanning in azimuth plane, polarization phase descriptors γ will change obviously, and in the main lobe region where people concern, with first order or second order polynomial can be well fitted for variation rules of γ in azimuth direction. At the same time, in the scope of the main lobe, the phase descriptors ϕ remained unchanged.

In Sect. 3.4.2.4, it is pointed out that the monotonic variation range of the polarization phase descriptor of the paraboloid antenna is approximately equal to 2.5 times the half power beam width. Therefore, the spatial variation law of γ is modeled for the antenna scanning in the range of $[-1.25\varphi_{3dB}, +1.25\varphi_{3dB}]$ on azimuth plane. The polarization phase descriptors γ are simulated by first order/second order polynomial, respectively, for the paraboloid antennas operating in with S and C, X band, with focus-diameter ratio F/D, and different typical values of offset angles ψ_0 and with vertical polarization as main polarization. It is given in Table 3.5 that under different circumstances, the fitting polynomial of the antenna polarization phase descriptor γ varying with the spacial distribution of azimuth angles, whereas the variable x representing the azimuth angle φ, and its unit is "degree". The unit for the polarization phase descriptor is also "degree", and $\varphi_{3\,dB}$ is the half power beam width. φ_Δ is the monotonic variation scope of antenna polarization phase descriptor change, namely in the range of $[-\varphi_\Delta/2, +\varphi_\Delta/2]$ near the antenna spindle, the antenna polarization phase descriptor varies monotonically.

Figure 3.44a–i are the comparison charts for variation curves and polynomial fitting curves of γ varying with azimuth angle, respectively for paraboloid antennas with the central working frequency of 2, 5 and 10 GHz, whereas the dotted line is a polynomial fitting curve and the solid line is the true curve. Note: The ranges of the horizontal axis in Fig. 3.44a–i are different.

It can be seen from Table 3.5 and Fig. 3.44 that the first order or second order polynomial can be used to better fit the antenna polarization phase descriptor γ varying with azimuth angle. Besides, the analysis results also show that the higher the working frequency of the antenna is, the narrower of the beam is, γ will vary faster with azimuth angle. But the variation degrees of antenna polarization characteristics are almost the same when the antenna is working in various frequency bands, and in the typical ranges of the half power beam width, 2 times the half power beam width, 2.5 times the half power beam width, the monotonic variation of γ near the primary axis of the antenna. For example, when the antenna parameters F/D = 0.5 and $\psi_0 = 44°$, and azimuth angle φ changes from $0 \rightarrow \varphi_{3dB}/2$, (Note: φ_{3dB} is not the same when the antenna works in different frequency band), as for the paraboloid antenna with the center frequency of f = 2 GHz, f = 5 GHz, f = 10 GHz, the polarization angle γ decreases from $90° \rightarrow 82.3°$, and $90° \rightarrow 82.2°$ respectively but the polarization angle ϕ remains almost unchanged at $-90°$. When azimuth angle φ changes from $0 \rightarrow \varphi_\Delta/2$, the polarization angle γ of the antenna declines respectively from $90° \rightarrow 11.4°$, $90° \rightarrow 21.1°$ and $90° \rightarrow 21°$, but the polarization angle ϕ remains unchanged. Polynomial fitting of polarization phase descriptor γ of the paraboloid antenna with vertical polarization as main polarization is given in Table 3.6 and Fig. 3.45.

It can be seen from the above analysis results that the first order or second order polynomial using the spacial scanning angle can better fit the variation rules of the

Table 3.5 Fitting polynomial table of polarization phase descriptor γ of paraboloid antenna (vertical polarization as main polarization)

Operating frequency (GHz)	$F/D = 0.5, \psi_0 = 44°$	$F/D = 0.25, \psi_0 = 44°$	$F/D = 0.5, \psi_0 = 30°$
2	$\gamma = -2.04x^2 + 89.2$ (°)	$\gamma = -9.01x^2 + 90.0$ (°)	$\gamma = -1.53x^2 + 89.7$ (°)
	$\varphi_{3dB} = 3.4°, \varphi_0 = 9.2°$	$\varphi_{3dB} = 3.6°, \varphi_0 = 10.4°$	$\varphi_{3dB} = 3.4°, \varphi_0 = 9.2°$
5	$\gamma = -15.8x^2 + 91.1$ (°)	$\gamma = -22.4x^2 + 90.0$ (°)	$\gamma = -12.5x^2 + 91.0$ (°)
	$\varphi_{3dB} = 1.4°, \varphi_0 = 3.6°$	$\varphi_{3dB} = 1.4°, \varphi_0 = 4.0°$	$\varphi_{3dB} = 1.4°, \varphi_0 = 3.6°$
10	$\gamma = -47.4x^2 + 88.7$ (°)	$\gamma = -46.8x + 90.0$ (°)	$\gamma = -34.9x^2 + 89.2$ (°)
	$\varphi_{3dB} = 0.68°, \varphi_0 = 1.8°$	$\varphi_{3dB} = 0.8°, \varphi_0 = 2.0°$	$\varphi_{3dB} = 0.68°, \varphi_0 = 1.8°$

Table 3.6 Fitting polynomial table of polarization phase descriptor γ of the paraboloid antenna (vertical polarization as main polarization)

Operating frequency (GHz)	$F/D = 0.5, \psi_0 = 44°$	$F/D = 0.25, \psi_0 = 44°$	$F/D = 0.5, \psi_0 = 30°$
2	$\gamma = 2.04x^2 + 0.80\ (°)$	$\gamma = 9.02x^2 + 0.0\ (°)$	$\gamma = 1.52x^2 + 0.35\ (°)$
	$\varphi_{3dB} = 3.4°, \varphi_0 = 9.2°$	$\varphi_{3dB} = 3.6°, \varphi_0 = 10.4°$	$\varphi_{3dB} = 3.4°, \varphi_0 = 9.2°$
5	$\gamma = 15.8x^2 - 1.10\ (°)$	$\gamma = 22.36x^2 + 0.0\ (°)$	$\gamma = 12.45x^2 - 1.50\ (°)$
	$\varphi_{3dB} = 1.4°, \varphi_0 = 3.6°$	$\varphi_{3dB} = 1.4°, \varphi_0 = 4.0°$	$\varphi_{3dB} = 1.4°, \varphi_0 = 3.6°$
10	$\gamma = 47.42x^2 - 1.3\ (°)$	$\gamma = 46.79x + 0.0\ (°)$	$\gamma = 34.90x^2 + 0.80\ (°)$
	$\varphi_{3dB} = 0.68°, \varphi_0 = 1.8°$	$\varphi_{3dB} = 0.8°, \varphi_0 = 2.0°$	$\varphi_{3dB} = 0.68°, \varphi_0 = 1.8°$

polarization phase descriptor γ in azimuth angle within the main lobe range while the polarization phase descriptor ϕ remains unchanged in main lobe range.

At the same time, the analysis results of Sects. 3.2 and 3.3 in combination with Tables 3.1 and 3.2 can show that, although there are different structures and properties of various antennas, but the spacial polarization ratio ρ of many wire antennas and aperture antennas is proportional to the tangent function $tg\varphi$ of its azimuth scanning angle φ. By combining the relation formula $\rho = tg\ \gamma e^{j\phi}$ between polarization ratio ρ and the phase polarity descriptors (γ, ϕ), we can learn that it is equivalent to the polarization phase descriptor γ and the spatial scanning angle φ are in the linear relationship and the polarization phase descriptor φ remains basically unchanged. It can be seen from the analysis results of Table 3.5 and Fig. 3.6 that the first order or second order polynomial using the spacial scanning angle can better fit the variation pattern of the polarization phase descriptor γ of antenna in azimuth angle but the phase descriptor ϕ remains almost unchanged. On this basis, typical models of four kinds of antenna spatial polarization characteristics are established.

1. The main polarization is "horizontal polarization", and the polarization angle γ is two times polynomial with the spatial scanning angle φ.

Assume that the expected polarization of the antenna is "horizontal polarization", then the initial polarization vector of the antenna is $\boldsymbol{h} = \begin{bmatrix} 1 & 0 \end{bmatrix}^{T}$. When the antenna is scanning in a certain space region, its polarization purity decreases gradually, and the cross-polarization component increases [55, 120]. Assume that the polarization angle ϕ of the antenna remains unchanged, for example, $\phi = -90°$, the polarization angle γ and the spatial azimuth scanning angle φ are in quadratic polynomial relationship, then we have

$$\gamma(\varphi) = K_{polar} \cdot \varphi^2, \quad \varphi \in [-\varphi_0/2, +\varphi_0/2] \tag{3.6.1}$$

In the upper equation, K_{polar} is the "spatial change rate of the antenna polarization angle", and there is $K_{polar} > 0$; φ_0 is the width of the spatial scanning. The larger the K_{polar} is, the faster the polarization characteristics of the antenna changes in the space, and the more obvious the spatial polarization characteristics of the antenna in the same scanning range.

The antenna polarization vector $\boldsymbol{h} = [\cos\gamma \, \sin\gamma \cdot e^{j\phi}]^T$, at this time, the polarization ratio of the antenna is

$$\rho(\varphi) = \mathrm{tg}\gamma(\varphi) \cdot e^{j\phi} = \mathrm{tg}(K_{polar} \cdot \varphi^2) \cdot e^{j\phi} \qquad (3.6.2)$$

For example, for wide beam antenna operating at 2 GHz with half power beam width $\varphi_{3dB} = 3.4°$, when the focal diameter ratio $F/D = 0.5$ and offset angle $\psi_0 = 44°$, spatial change rate of antenna polarization angle $K_{polar} \approx 2.04$; if $F/D = 0.5$ and $\psi_0 = 30°$, $K_{polar} \approx 1.52$. For example, for medium wide beam antenna operating at 5 GHz with half power beam width $\varphi_{3dB} = 1.4°$, when the focal diameter ratio is $F/D = 0.5$ and offset angle is $\psi_0 = 44°$, spatial change rate of antenna polarization angle $K_{polar} \approx 15.8$; when $F/D = 0.5$ and offset angle is $\psi_0 = 44°$, $K_{polar} \approx 12.45$. For example, for narrow beam antenna operating at 10 GHz with half power beam width $\varphi_{3dB} = 0.68°$, when the focal diameter ratio $F/D = 0.5$ and offset angle $\psi_0 = 44°$, spatial change rate of antenna polarization angle $K_{polar} \approx 47.4$; when $F/D = 0.5$ and $\psi_0 = 30°$, $K_{polar} \approx 34.9$. It is noteworthy that the units of spatial scanning angle φ and polarization phase descriptor γ are both "degrees", and if the following models are not specified, the units are "degrees".

2. The main polarization is "horizontal polarization", and the polarization angle γ is in linear relationship with the spatial scanning angle ψ.

If the expected polarization of an antenna is "horizontal polarization", the polarization angle ϕ remains unchanged, for example $\phi = -90°$, the polarization angle γ and the spatial azimuth scanning angle φ is in linear relationship, so we get

$$\gamma(\varphi) = K_{polar} \cdot |\varphi|, \quad \varphi \in [-\varphi_0/2, +\varphi_0/2] \qquad (3.6.3)$$

where, K_{polar} is the spatial change rate of the antenna polarization angle, and there is $K_{polar} > 0$; φ_0 is the spatial scanning width of the antenna.

At this time, the polarization ratio of the antenna is

$$\rho(\varphi) = \mathrm{tg}\left(K_{polar} \cdot |\varphi|\right) \cdot e^{j\phi} \qquad (3.6.4)$$

For example, when the paraboloid antenna parameter is $F/D = 0.25$ and $\psi_0 = 44°$, $K_{polar} \approx 9.02$ for the wide beam antenna operating at 2 GHz, $K_{polar} \approx 22.3$ for medium beam antenna operating at 5 GHz, $K_{polar} \approx 46.7$ for narrow beam antenna operating at 10 GHz.

3. The main polarization is "vertical polarization", and the polarization angle γ is in quadratic polynomial relationship with the spatial scanning angle φ

Assume the expected polarization of the antenna is "vertical polarization", the initial polarization vector of the antenna is $h = [1 \quad 0]^{\mathrm{T}}$. The polarization angle ϕ remains unchanged all the time, for example $\phi = -90°$ and the polarization angle γ is in quadratic polynomial relationship with the spatial scanning angle φ, then we get

$$\gamma(\varphi) = -K_{\mathrm{polar}} \cdot \varphi^2 + 90°, \quad \varphi \in [-\varphi_0/2, +\varphi_0/2] \qquad (3.6.5)$$

where, K_{polar} is the spatial change rate of the antenna polarization angle, and there is $K_{\mathrm{polar}} > 0$; φ_0 is the spatial scanning width of the antenna.

For example, for wide beam antenna operating at 2 GHz with half power beam width $\varphi_{3\mathrm{dB}} = 3.4°$, when the focal diameter ratio $F/D = 0.5$ and offset angle $\psi_0 = 44°$, antenna spatial change rate of polarization angle is $K_{\mathrm{polar}} \approx 2.05$; when $F/D = 0.5$ and $\varphi_{3\mathrm{dB}} = 3.4°$, $K_{\mathrm{polar}} \approx 1.53$. For example, for medium beam antenna operating at 5 GHz with half power beam width $\varphi_{3\mathrm{dB}} = 1.4°$, when the focal diameter ratio $F/D = 0.5$ and offset angle $\psi_0 = 44°$, spatial change rate of antenna polarization angle is $K_{\mathrm{polar}} \approx 15.8$; when $F/D = 0.5$ and $\psi_0 = 30°$, $K_{\mathrm{polar}} \approx 12.5$. For example, for narrow beam antenna operating at 10 GHz with half power beam width $\varphi_{3\mathrm{dB}} = 0.68°$, when the focal diameter ratio $F/D = 0.5$ and offset angle $\psi_0 = 44°$, spatial change rate of antenna polarization angle is $K_{\mathrm{polar}} \approx 47.4$; when $F/D = 0.5$ and $\psi_0 = 30°$, $K_{\mathrm{polar}} \approx 34.9$.

4. The main polarization is "horizontal polarization", and the polarization angle γ is in linear relationship with the spatial scanning angle φ

Assume the expected polarization of the antenna is "vertical polarization", the antenna polarization angle ϕ remains unchanged all the time, for example $\phi = -90°$ and the polarization angle γ and the spatial azimuth scanning angle φ is in linear relationship, then we have

$$\gamma(\varphi) = -K_{\mathrm{polar}} \cdot |\varphi| + 90°, \quad \varphi \in [-\varphi_0/2, +\varphi_0/2] \qquad (3.6.6)$$

where, K_{polar} is the spatial change rate of the antenna polarization angle, and there is $K_{\mathrm{polar}} > 0$; the φ_0 is the spatial scanning width of the antenna.

For example, when the paraboloid antenna parameter is $F/D = 0.25$ and $\psi_0 = 44°$, for the wide beam antenna operating at 2 GHz, $K_{\mathrm{polar}} \approx 9.01$; for the medium beam antenna operating at 5 GHz, $K_{\mathrm{polar}} \approx 22.4$; for the narrow beam antenna operating at 10 GHz, $K_{\mathrm{polar}} \approx 46.8$.

Chapter 4
Spatial Polarization Characteristics of Phased Array

With rapid development of polarization measurement technology, vector signal process and microwave technology, polarization information is sufficiently used to provide a potential technical approach for impairment of influence of adverse electromagnetic environment on the radar system, resistance to active jamming, suppression of environmental clutter, anti-stealth and target recognition and other aspects, being capable of effectively improving working performance of modern radars. As indicated above, antenna polarization is not always kept constant, but is closely related to measurement position; that is to say, polarization state of electromagnetic wave radiated by the antenna in different observation direction (equivalent beam scanning angle) shows a significant and regular variation, i.e. the antenna polarization is a "Slow spatial variable". This kind of characteristic is called "Spatial polarization characteristic". Meanwhile, due to restriction and influence of non-ideal process and element level, "Spatial polarization characteristic" of the antenna will become more significant. For the reason that existing radar antenna has this property, receive signal of the antenna can be considered linear combination of spatial polarization state and incoming wave signal polarization state in scanning period of the antenna, and receive signal at different scanning angle modulates antenna polarization characteristic at the same angle. We first propose to a method of using spatial polarization characteristic of a single-polarized mechanical scanning antenna to acquire full polarization scatter information of the target [116,117]. Although this method is a suggestive method, but it cannot be used to obtain full polarization scatter matrix of a moving target and multi-target tracing condition, whose application is restricted due to short residence time.

Due to its excellent performance, the phased array radar system is widely applied in long range early warning and short range missile defense system. The phased array antenna is an important component of the phased array radar, and is a key to differentiate the phased array radar from the ordinary mechanical scanning radar. The phased array radar overcomes imperfections of mechanical antenna's inertial scanning, which has beam agility and is able to jump to another wave position from one wave position in an inertialess manner. In existing Refs. [131–138], studies on

© National Defense Industry Press, Beijing and Springer Nature Singapore Pte Ltd. 2019   133
H. Dai et al., *Spatial Polarization Characteristics of Radar Antenna*,
https://doi.org/10.1007/978-981-10-8794-3_4

characteristic of the phased array antenna focus on reducing minor lobe level, increasing antenna gain, enhancing antenna polarization purity, giving spatial adaptive beam formation algorithm, developing conformal array and etc., and most studies on polarization characteristics are only restricted to characteristics shown when it operates at cardinal plane, but less studies on polarization characteristics in other spatial direction (orientation, tilt scanning angle).

In fact, when phased array antenna scans at individual wave positions in spatial, its beam center direction varies in relation to direction of the antenna aperture, and the direction pattern structure is also different. It has certain difference from mechanical scanning antenna. Study on spatial polarization characteristics of the mechanical scanning antenna only considers two degree of freedom, i.e. evolution law of polarization characteristics at different elevation angle and azimuth angle. Nevertheless, spatial scanning of the phased array antenna is achieved by scattered scanning wave position, which can be called "Quasi continuous scanning", and has four degree of freedom including beam azimuth scanning angle, beam elevation scanning angle, azimuth angle, and elevation angle. So, it is more complicated to analyze spatial polarization characteristics of the phased array antenna. Hence, study on polarization characteristics of the phased array antenna during beam spatial scanning does not only has certain theoretical research value, but also has important military meaning for making efficient use of this characteristic to improve target recognition and anti-jamming ability of the radar.

With respect to problems mentioned above, Sect. 4.1 hereof further elaborates connotation of antenna spatial polarization and concludes several description methods for spatial polarization characteristic of the antenna; Sect. 4.2 gives a modeling method for phased array antenna spatial polarization characteristic, and takes area array and wave-guided slot array composed of dipoles as example for simulation analysis to study variation law of polarization characteristic; Sect. 4.3 utilizes a high-frequency electromagnetic simulation platform XFDTD to design a phased array antenna with 625 elements, and calculate radiation electric field pointed by multiple beam to obtain approximate linear variation law of the polarization characteristic. Study in this Chapter give a crucial complementation to theoretical research on antenna polarization characteristic, so as to offer fundamentals and practical basis for further application study on the same characteristic. However, although present theoretical research analysis, simulation model and experimental results are not refined, and more complicated conditions are not considered, it is expected to describe "Polarization characteristic of phased array spatial scanning" by virtue of above-mentioned model and analysis results, so as to lay a solid basis for further research, measurement and application of the characteristic.

4.1   Antenna Spatial Polarization Characterization

4.1.1   Antenna Spatial Polarization Characteristic Connotation

The so-called polarization means polarization of specific electric field vector of the electromagnetic wave. At a fixed space point, polarization of a single-frequency electric field vector means shape, orientation and rotating direction of the vector end-point moving path [39]. Rotating direction of the electric field vector end-point path is specified as that observed in wave propagation direction. Accordingly, the wave can be illustrated as linear polarization wave, circular polarization wave or elliptical polarization wave.

The antenna is used to convert the guided wave on the transmission structure into free space wave. IEEE gives following official definition: "the part used to radiate or receive electromagnetic wave in the transmission or receive system". In all cases, the antenna has direction characteristic, i.e. the electromagnetic power density is radiated from the antenna, and its intensity is related to the antenna angle. The radiation field of the antenna can be approximately seen as a kind of interference field, i.e. if the charge is accelerated in round trip, a regular interference is built, while the radiation will continue to exist, and the antenna is a device used to support charge oscillation, which provide a means to radiate and receive radio wave. In other words, information can be transmitted between different places without any intermediate mechanism. In the example of accelerating moving charge, the radiation is directional. The maximum interference occurs in a direction perpendicular to the electric charge acceleration direction, that is to say, the maximum radiation will be produced in a direction perpendicular to the antenna direction.

According to principle of the antenna [123], it can be known that, in addition to co-polarization component in the electromagnetic wave radiated by the antenna, it includes some orthogonal polarization component, called "Cross polarization component". Percentage of co-polarization component power density occupied in all power density is called polarization purity of the antenna. For a well-designed antenna, its polarization purity at center frequency and center orientation remains high, but after the orientation is deviated from the center, the polarization purity decreases, and its variable quantity is a function about spatial angle. This characteristic can be called as "Spatial polarization characteristic" of the antenna.

In traditional concept, no difference between electromagnetic wave polarization and antenna polarization exists essentially, and polarization characteristic of the antenna can be seen as polarization characteristic of electromagnetic wave radiated by the antenna. According to linear polarization or circular polarization of the electromagnetic wave radiated by the antenna, corresponding antenna is called linear polarization antenna or circular polarization antenna. Hence, all discussions about wave polarization are applicable to antenna polarization.

However, it is noted that polarization of antenna's radiated wave is varied by direction, and a factor is traditionally implied in the definition of the antenna polarization, i.e. polarization of the beam center position. Usually, polarization characteristic of the antenna remains relatively constant in center direction of the main lobe. In this sense, polarization of peak position of the main lobe is used to describe antenna polarization, while polarization of minor lobe may be different from that of the main lobe.

When measuring radiation of the antenna, it is required to measure E_θ and E_ϕ for the reason of completeness. When pointing direction of a probe antenna is to respond to E_θ, main plane direction pattern of a pure linear polarization antenna (such as line source at z axis) can be measured precisely (Fig. 4.1).

To sum up, antenna polarization, as it is often called, means polarization in maximum direction or maximum receive direction. For mechanical scanning antenna, it can be defined by use of normal direction of the antenna aperture; however, for the phased array system antenna, polarization of the phased array antenna is not only related to space direction discussed, but also to scanning direction of the relative front main lobe. When beam scanning is in progress, the maximum radiation direction of the beam is not coincided with normal direction of the antenna array, but angle between both directions increases as scanning angle increases.

To completely describe polarization characteristic of the phased array antenna, scope of the antenna polarization characteristic should be expanded to different spatial pointing direction, i.e. taking orientation, elevation observation angle or beam scanning angle as variables to analyze evolution law of antenna polarization state. Although definition and characterization about antenna spatial polarization characteristic in Ref. [3] are for mechanical scanning system, but their physical nature is same for polarization characteristic analysis in center orientation of a beam, i.e. the phased array system is different from the mechanical scanning system with respect to implementation of the scanning, but they can be consolidated by certain process method.

Fig. 4.1 Field component in different pointing directions

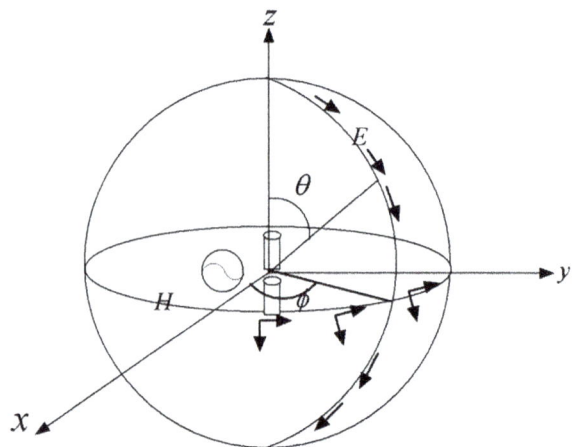

4.1.2   Description of Spatial Polarization Characteristics of Phased Array

According to definition of the radiation direction pattern: radiation field of the antenna is distribution (variation) trend of the radiated electric field at spherical surface of the antenna in view of the spatial angle (θ, φ). Hence, an electric field radiated by the antenna is usually defined by the spherical coordinate $\left(\hat{r}, \hat{\theta}, \hat{\varphi}\right)$. At far field area of the antenna, the radiation field E has no radial component, so it can be expressed as follows:

$$E\left(\hat{r}, \hat{\theta}, \hat{\varphi}\right) = E_\theta \cdot \hat{u}_\theta + E_\varphi \cdot \hat{u}_\varphi \tag{4.1.1}$$

where, $\hat{u}_\theta$ and $\hat{u}_\varphi$ are unit vector in tilt and orientation directions in the spherical coordinate system, respectively; E_θ and E_φ are electric field component in $\hat{u}_\theta$ and $\hat{u}_\varphi$ direction.

If xy plane is parallel with the ground, E_φ is the horizontal component of the wave (it is always parallel with xy plane, so it is parallel with the ground surface), $-E_\theta$ is the vertical component, $(\hat{u}_\varphi, -\hat{u}_\theta, \hat{u}_r)$ forms the right-hand coordinate system, and $\hat{\varphi}$ and $-\hat{\theta}$ can be defined as horizontal polarization basis $\hat{h}$ and $\hat{v}$, respectively, i.e.:

$$\begin{cases} \hat{h} = \hat{\varphi} \\ \hat{v} = -\hat{\theta} \end{cases} \tag{4.1.2}$$

When position of the measuring point is confirmed in relation to antenna position, it is possible to use radiation field of the antenna where the point is situated to define its polarization state. Jones vector for normalization of the antenna is expressed as follows:

$$h = \begin{bmatrix} E_H \\ E_V \end{bmatrix} = \frac{1}{\|E_{\max}\|} \cdot \begin{bmatrix} E_\varphi \\ -E_\theta \end{bmatrix} \tag{4.1.3}$$

With respect to the phased array antenna, when non-ideal factors including cross coupling are ignored, angle between scanning direction of its main lobe and x axis and between the scanning direction and y axis is α_s, β_s, respectively, which are mainly dependent on array factor function expressed as follows [123]

$$\mathbf{AF} = F(\alpha, \beta) = \sum_{m=0}^{M} \sum_{n=0}^{N} e^{jmk_0 dx(\cos \alpha - \cos \alpha_s) + jmk_0 dy(\cos \beta - \cos \beta_s)} \tag{4.1.4}$$

where M, N represents number of elements of planar matrix. Spacing between elements is d_x and d_y, respectively, and phase shift of element excitation current along the x and y axis is $\psi_x = k_0 d_x \cos \alpha_s$ and $\psi_y = k_0 d_y \cos \beta_s$, respectively.

The above-mentioned equation can be further written as follows:

$$\mathbf{AF} = F(\alpha, \beta) = \left| \frac{\sin\left[\frac{1}{2} M k_0 d_x \tau_x\right]}{\sin\left[\frac{1}{2} k_0 d_x \tau_x\right]} \right| \times \left| \frac{\sin\left[\frac{1}{2} M k_0 d_y \tau_y\right]}{\sin\left[\frac{1}{2} k_0 d_y \tau_y\right]} \right| \tag{4.1.5}$$

where $\tau_x = \cos \alpha - \cos \alpha_s$, $\tau_y = \cos \beta - \cos \beta_s$; at this time, scanning direction (α_s, β_s) of the beam depends on phase displacement ψ_x, ψ_y between neighboring elements, i.e.:

$$\cos \alpha_s = \frac{\psi_x}{k_0 d_x}, \cos \beta_s = \frac{\psi_y}{k_0 d_y} \tag{4.1.6}$$

Because the coordinate (α, β) has following relationship with the angular coordinate (θ, φ) under the spherical coordinate system of the normal line in the direction of $(\theta = 0)$ along the z axis,

$$\cos \alpha = \sin \theta \cos \varphi$$
$$\cos \beta = \sin \theta \sin \varphi$$
$$\sin^2 \theta = \cos^2 \alpha + \cos^2 \beta \tag{4.1.7}$$
$$\tan \varphi = \cos \beta / \cos \alpha$$

A resolution function for scalar quantity of the array factor can be obtained as follows:

$$\mathbf{AF}(\theta, \varphi) = F(\alpha, \beta)$$
$$= \left| \frac{\sin\left[\frac{1}{2} M (k_0 d_x \sin \theta \cos \varphi - \psi_x)\right]}{\sin\left[\frac{1}{2} (k_0 d_x \sin \theta \cos \varphi - \psi_x)\right]} \right|$$
$$\times \left| \frac{\sin\left[\frac{1}{2} M (k_0 d_y \sin \theta \sin \varphi - \psi_y)\right]}{\sin\left[\frac{1}{2} (k_0 d_y \sin \theta \sin \varphi - \psi_y)\right]} \right| \tag{4.1.8}$$

Hence, relationship between phase scanning and beam spatial direction under the spherical coordinate system can be obtained:

$$\varphi = \arctan\left(\frac{\psi_y}{\psi_x} \cdot \frac{d_x}{d_y}\right), \theta = \arcsin\sqrt{\frac{\psi_x^2}{k_0^2 d_x^2} + \frac{\psi_y^2}{k_0^2 d_y^2}} \tag{4.1.9}$$

It can be seen from the Eqs. (4.1.7)–(4.1.9) that any beam direction of the phased array is realized by varying the feed phase, and scanning direction (α_s, β_s) of the beam corresponds to the space angle coordinate. According to multiplying principle in the antenna array theory, overall radiation field of the phased array antenna is a product obtained by multiplying a polarization vector element direction pattern $f(\theta, \varphi)$ by a scaled array factor function $\mathbf{AF}(\theta, \varphi)$. Hence, it can be known from the above-mentioned expression that the array factor decides direction angle (α, β) of electric scanning of the beam, while scanning polarization characteristic of the phased array antenna can be decided and illustrated jointly by the polarization characteristic of the center radiation unit $f(\theta, \varphi)$ and observation direction (θ, φ).

In other words, the polarization characteristic can be jointly decided by the vector polarization direction pattern of element in array in observation direction (θ, φ) and the beam direction determined by the beam scanning; therefore, spatial polarization characterization used to describe the mechanical mechanism-scanning system antenna is still applicable for phased array system.

It is noted that, during scanning of the phased array antenna, structure of the polarization direction pattern will vary, and antenna radiation field at different beam scanning position will be different. Hence, polarization characteristic cannot be approximately described by variation law of the radiation field at several observation angle, but it is required to first consider distribution condition of the antenna radiation field at fixed scanning angle, then, spatial polarization characteristic of the phased array antenna can be described accurately only by summing up or fitting polarization characteristic of the main lobe of the radiation field at several scanning angles.

Supposing that $P(\theta, \varphi)$ represents space angle coordinate at measuring point, which is a two-dimensional vector of the orientation direction φ and elevation direction θ, abbreviated to P, and $P \in \Omega$, where Ω is spatial range scanned by the antenna. It can be known from the Eq. (4.1.3) that spatial polarization ratio of the antenna radiation field is as follows:

$$\rho(P) = \frac{E_V}{E_H} = -\frac{E_\theta}{E_\varphi} \tag{4.1.10}$$

Following relationship between the antenna spatial Jones vector and spatial polarization ratio exists:

$$h(P) = \frac{1}{\sqrt{1 + |\rho(P)|^2}} \begin{bmatrix} 1 \\ \rho(P) \end{bmatrix} \tag{4.1.11}$$

Spatial polarization ratio is the most illustrative characterization quantity of the antenna spatial polarization characteristic, meanwhile it is also possible to define spatial polarization phase descriptor $(\gamma(P), \phi(P))$, spatial polarization elliptic descriptor $(\varepsilon(P), \tau(P))$, spatial Stokes vector and other description methods.

At horizontal and vertical polarization basis $(\hat{h}, \hat{v})$, normalized Jones vector of the antenna can be represented as follows:

$$h = \begin{bmatrix} E_H \\ E_V \end{bmatrix} = \begin{bmatrix} a_H e^{j\varphi_H} \\ a_V e^{j\varphi_V} \end{bmatrix} \tag{4.1.12}$$

Spatial polarization ratio of the antenna can be represented as follows:

$$\rho(P) = \frac{E_V}{E_H} = tg\gamma(P)e^{j\varphi(P)}, (\gamma, \phi) \in [0, \pi] \times [0, 2\pi] \tag{4.1.13}$$

where, $\gamma(P) = \tan^{-1}\frac{a_V}{a_H}$ 和 $\phi(P) = \varphi_V - \varphi_H$ is spatial polarization phase descriptor of the antenna.

Spatial polarization ratio of the antenna and spatial polarization phase descriptor are the most common characterization quantity used to describe spatial polarization characteristic of the antenna. Additionally, it is possible to use "Antenna cross polarization discrimination XPD(P)" to characterize variation in displacement degree of real polarization of the antenna in relation to expected polarization in the space angle position [3]. Spatial cross polarization discrimination of the antenna is defined by following equation:

$$XPD(P) = 20\lg(|\rho(P)|) \text{ dB} \tag{4.1.14}$$

As discussed in Ref. [40], electric field can be resolved at any pair of orthogonal polarization basis, so a far field radiated by the antenna can also be resolved as follows:

$$E_{far} = E_1\hat{e}_1 + E_2\hat{e}_2 \tag{4.1.15}$$

Theoretically, the orthogonal polarization basis $(\hat{e}_1, \hat{e}_2)$ is selected at random. For example, polarization ratio in Eq. (4.1.10) is defined under polarization basis $(\hat{\phi}, -\hat{\theta})$, and in the spherical coordinate system, this is a most natural and most widely-used resolution method.

However, cross polarization discrimination XPD obtained based on the polarization ratio ρ defined in Eq. (2.2.10) cannot illustratively characterize degree of deviation of the antenna polarization from expected polarization at different space angle position. Therefore, it is possible to select different polarization basis in specific conditions, and in the polarization basis $(\hat{e}_{co}, \hat{e}_{cross})$, the far field E_{far} can be represented as follows:

$$E_{far} = E_{co}\hat{e}_{co} + E_{cross}\hat{e}_{cross} \tag{4.1.16}$$

where $\hat{e}_{co}$ is expected polarization direction of the antenna, and $\hat{e}_{cross}$ is its orthogonal polarization (or parasitical polarization), Further, it is inferred that spatial cross polarization discrimination XPD(P) of the antenna can be represented by following equation:

$$XPD(P) = 10 \lg \left(\frac{P_{cross}}{P_{co}} \right) = 20 \lg \left(\left| \frac{E_{cross}}{E_{co}} \right| \right) dB \qquad (4.1.17)$$

where P_{cross} and P_{co} represent parasitical polarization power and expected polarization power, respectively, and E_{cross} and E_{co} represent voltage amplitude of parasitical polarization and expected polarization.

It can be seen from the above-mentioned equation that the smaller XPD value, the less will be cross polarization component of the antenna and the higher will be the "Polarization purity". In this sense, the above-mentioned equation can also be used to define "Spatial polarization *Purity* (*P*)" of the antenna. Now, spatial polarization *Purity* (*P*) is consistent with the spatial polarization discrimination XPD(P), which is able to characterize cross polarization characteristic of the antenna in a very illustrative manner.

When expected polarization (co-polarization) of the antenna is a horizontal polarization $E_H = E_\varphi$, and parasitical polarization is vertical polarization $E_V = -E_\theta$, cross polarization discrimination defined in Eq. (4.1.17) is different from Eq. (4.1.14). When expected polarization of the antenna is in different polarization state, it is possible to select different polarization basis to define spatial cross polarization discrimination XPD(P) of the antenna.

For example, when expected polarization of the antenna is horizontal/vertical linear polarization, it is possible to select polarization basis $(\hat{h}, \hat{v})$; when expected polarization of the antenna is 45°/135° linear polarization, it is possible to select polarization basis $(\hat{e}_{45}, \hat{e}_{135})$; when expected polarization of the antenna is left-hand/right-hand circular polarization, it is possible to select polarization basis $(\hat{l}, \hat{r})$.

To better describe distribution condition of change in antenna polarization state during spatial scanning of the phased array antenna, including overall distribution trend of individual polarization states, center position of polarization distribution, scattering degree of space distribution of individual points and variation speed of individual space polarization states, it is possible to make a reference to characterization method of antenna spatial instantaneous polarization characteristics and definition of spatial instantaneous polarization of non-time harmonic electromagnetic wave in Ref. [3], and at this moment, different space point can be converted into angular function of the main lobe in relation to the array normal line during scanning of phased array. In practice, they are described by the same method, and can be characterized by use of such parameters as spatial instantaneous polarization projection set, spatial polarization cluster center and spatial polarization divergence.

4.2 Modelling and Simulation of Phased Array Spatial Polarization Characteristics

4.2.1 Antenna Element Polarization Characteristics

4.2.1.1 Polarization Characteristic of Dipole Element

Radiation field of the dipole can be obtained by the principle of superposition [139]. When observation point P is far away from the dipole, it is possible to take current on the dipole as line current on the axis, and due to distance between two internal end points $d \ll l$, it can be thought that the line current is continuously distributed from $z = -l$ to $z = l$.

The dipole is divided into numerous differential section with length of dz, each section equivalent to a current element. The whole dipole can be seen as combination of numerous current elements which are connected together. Based on the principle of superposition, the field where the dipole is situated at P point is superposition of the field where these current elements are at this point.

When $r \to \infty$:

$$R = |r - z| \approx r - z \cos \theta \tag{4.2.1}$$

Radiation field of the current source at z is represented as follows:

$$dE_0 = j \frac{W_0 I(z) dz}{2 \lambda r} \sin \theta e^{-jkr} e^{jkz \cos \theta} \tag{4.2.2}$$

Hence, it is inferred that radiation field of the dipole is represented as follows [43]:

$$
\begin{cases}
E_\theta = \displaystyle\int dE_0 = j \frac{W_0 I_m}{2 \lambda r} \sin \theta e^{-jkr} \int_{-l}^{l} \sin k(l - |z|) e^{jkz \cos \theta} dz \\[2mm]
\quad = j \frac{60 I_m}{r} \sin \theta e^{-jkr} \left[\dfrac{\cos(kl \cos \theta) - \cos kl}{\sin^2 \theta} \right] \\[4mm]
H_\phi = \dfrac{E_\theta}{W_0}
\end{cases}
\tag{4.2.3}
$$

When axis of the dipole is placed along the x axis, the radiation field has two components, i.e. E_θ and E_ϕ. During calculation, $\sin \theta$ in the above-mentioned equation is replaced by a directivity function $(-\hat{\theta} \cos \theta \cos \varphi + \hat{\varphi} \sin \varphi)$ of the current element on the x axis, which is obtained by replacing the $z \cdot \hat{r} = z \cos \theta$ by $x \cdot \hat{r} = x \sin \theta \cos \varphi$, i.e.:

$$E = j\frac{60\pi I_m}{\lambda\pi}\mathrm{e}^{-jkr}(-\hat{\theta}\cos\theta\cos\varphi + \hat{\varphi}\sin\varphi)\int_{-l}^{l}\sin k(l - |x|)\mathrm{e}^{jkx\sin\theta\cos\varphi}dx$$

$$= j\frac{60\pi I_m}{\lambda\pi}\mathrm{e}^{-jkr}(-\hat{\theta}\cos\theta\cos\varphi + \hat{\varphi}\sin\varphi)\times\left[\frac{\cos(kl\sin\theta\cos\varphi) - \cos kl}{\sqrt{1 - \sin^2\theta\cos^2\varphi}}\right]$$

$$(4.2.4)$$

Then

$$|E| = \sqrt{|E_\theta|^2 + |E_\varphi|^2}$$

$$= \frac{60I_m}{r}\left[\frac{\cos(kl\sin\theta\cos\varphi) - \cos kl}{\sqrt{1 - \sin^2\theta\cos^2\varphi}}\right]$$

$$(4.2.5)$$

where, $E_\theta = j\frac{60\pi I_m}{\lambda\pi}\mathrm{e}^{-jkr}\cos\theta\cos\varphi\left[\frac{\cos(kl\sin\theta\cos\varphi) - \cos kl}{\sqrt{1 - \sin^2\theta\cos^2\varphi}}\right]$

$$E_\varphi = j\frac{60\pi I_m\sin\varphi}{\lambda\pi}\mathrm{e}^{-jkr}\left[\frac{\cos(kl\sin\theta\cos\varphi) - \cos kl}{\sqrt{1 - \sin^2\theta\cos^2\varphi}}\right]$$

4.2.1.2 Polarization Characteristic of Wave-Guided Slot Elements

Real slot antenna is at a limited conductive plane. Figure 4.2 shows narrow and long slot on a rectangular conductive panel. Distance between the slot center line and both edges of the rectangular panel is d_1 and d_2, and there is an electric field which is polarized in x direction in the narrow slot, and distributed evenly along x direction and in a cosine manner along y direction, as well as the slot length is usually half-wave length.

Fig. 4.2 Slot on a rectangular conductive panel

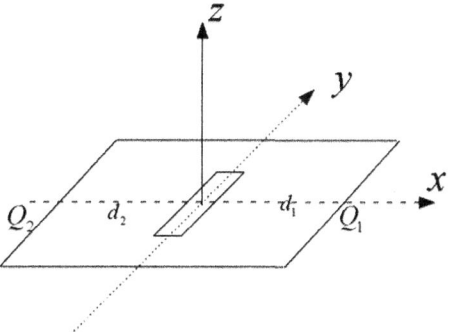

Maximum radiation direction of the slot is at xz plane, and axial radiation is zero. Resolution of the slot radiation field can be approximately simplified to obtain 2D radiation of the slot at a zone with infinite length, shown on Fig. 4.3. According to geometrical diffraction theory and in addition to direct radiation of the slot, there is diffraction of two edges parallel to the slot [123]. Therefore, with respect to far field point in the upper space, total field is the sum of direct ray emitted by the slot and diffraction ray emitted by both edges, ignoring secondary diffraction field passing through the one edge and then second edge and diffraction field at higher order.

Resolution can be made through Fourier transformation method. Fourier transformation of tangential component at electric field of the radiation field for any slot is as follows:

$$\tilde{E}_z(n, w) = \frac{1}{2\pi} \int_0^{2\pi} d\varphi \int_{-\infty}^{\infty} E_z(a, \varphi, z) e^{-jn\varphi} e^{-jwz} dz \qquad (4.2.6)$$

$$\tilde{E}_\varphi(n, w) = \frac{1}{2\pi} \int_0^{2\pi} d\varphi \int_{-\infty}^{\infty} E_\varphi(a, \varphi, z) e^{-jn\varphi} e^{-jwz} dz \qquad (4.2.7)$$

Its inverse transformation is as follows

$$E_z(a, \varphi, z) = \frac{1}{2\pi} \sum_{n=-\infty}^{\infty} e^{jn\varphi} \int_{-\infty}^{\infty} \tilde{E}_z(n, w) e^{jwz} dw \qquad (4.2.8)$$

$$E_\varphi(a, \varphi, z) = \frac{1}{2\pi} \sum_{n=-\infty}^{\infty} e^{jn\varphi} \int_{-\infty}^{\infty} \tilde{E}_\varphi(n, w) e^{jwz} dw \qquad (4.2.9)$$

Field at any observation point P can be represented as the sum of TE wave and TM wave. TE wave and TM wave can be calculated based on z component A_z, A_{mz} of magnetic vector and electric vector. In the spherical coordinate system, A_z, A_{mz} can be represented by superposition form of following wave function:

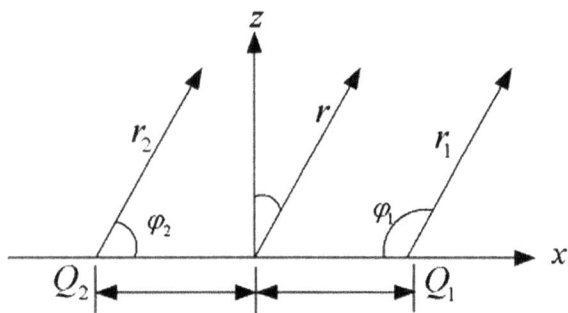

Fig. 4.3 Slot on rectangular conductive panel

$$A_z(\rho, \varphi, z) = \frac{1}{2\pi} \sum_{n=-\infty}^{\infty} e^{jn\varphi} \int_{-\infty}^{\infty} f_n(w) H_n^{(2)}\left(\rho\sqrt{k_0^2 - w^2}\right) e^{jwz} dw \qquad (4.2.10)$$

$$A_{mz}(\rho, \varphi, z) = \frac{1}{2\pi} \sum_{n=-\infty}^{\infty} e^{jn\varphi} \int_{-\infty}^{\infty} g_n(w) H_n^{(2)}\left(\rho\sqrt{k_0^2 - w^2}\right) e^{jwz} dw \qquad (4.2.11)$$

The cylindrical function is selected as second category of Hankel function $H_n^{(2)}$ to represent outward traveling wave. $f_n(w)$ and $g_n(w)$ are undetermined coefficients.

$$f_n(\omega) = \frac{j\omega\varepsilon_0 \hat{E}_z(n, \omega)}{(k_0^2 - \omega^2) H_n^{(2)}\left(a\sqrt{k_0^2 - \omega^2}\right)} \qquad (4.2.12)$$

$$g_n(\omega) = \frac{\varepsilon_0 \left[\hat{E}_\varphi(n, \omega) + \frac{n\omega}{a(k_0^2 - \omega^2)} \hat{E}_z(n, \omega)\right]}{\sqrt{k_0^2 - \omega^2} H_n^{(2)'}\left(a\sqrt{k_0^2 - \omega^2}\right)} \qquad (4.2.13)$$

They can be determined by following method: Based on in-homogeneous Helmholtz equation, field component at any point of the electromagnetic field is solved by use of A_z and A_{mz} to calculate an infinite integral, so as to obtain an approximate solution.

$$E_\theta = jw\mu_0 \frac{e^{-jk_0 r}}{\pi r} \sin\theta \sum_{n=-\infty}^{\infty} e^{jn\varphi} j^{n+1} f_n(-k_0 \cos\theta) \qquad (4.2.14)$$

$$E_\varphi = -jk_0 \frac{e^{-jk_0 r}}{\pi r} \sin\theta \sum_{n=-\infty}^{\infty} e^{jn\varphi} j^{n+1} g_n(-k_0 \cos\theta) \qquad (4.2.15)$$

4.2.2   Phased Array Antenna Spatial Polarization Characteristics Modelling

Considering a 2D $M \times N$ planar array [140], each array element has same directional pattern, and θ_i', φ_i' are elevation angle and direction angle in local coordinate system of the array element. To investigate cross polarization characteristics of the array, it is required to calculate polarization component under given polarization basis in local coordinate system of the array element, and then through coordinate transformation, transform the polarization component in local coordinate system of the array element to the total coordinate system for superposition. When building the array, the coordinate system where the array element is situated can be assigned with different position under the spherical coordinate system, and 3D coordinate transformation can be achieved through Euler matrix, hence the key to model is rotation transformation of polarization component of individual elements by following process:

(1) Build coordinate of the array element vector in the global coordinate system;
(2) According to specific geometrical structure of the planar array antenna and position relation of individual elements, build coordinate of local coordinate system for individual array elements in the global coordinate system;
(3) Build coordinate of directional unit vector for array elements in the local rectangular coordinate system of individual array element;
(4) Calculate projection of unit vector of the array element on its local coordinate axis in the global coordinate system;
(5) By use of coordinate of directional unit vector of the array elements in the local rectangular coordinate system of individual array elements, and relationship with projection of unit vector of the array element on local coordinate axis of individual array element, solve transformation relation between elevation angle and directional angle in the global coordinate system and local coordinate system of individual array elements, and complete transformation of polarization component of array elements in the global coordinate system.

As shown in Fig. 4.4, ε_X, ε_Y, ε_Z represents three rotating angles for transformation of 3D rectangular coordinate transformation, also called Euler angle. When rotating around x axis at angle of ε_X, the rotation matrix is defined as $[R_1]$; when rotating around y axis at angle of ε_Y, the rotation matrix is defined as $[R_2]$; when rotating around z axis at angle of ε_Z, the rotation matrix is defined as $[R_3]$

Therefore, the rotation matrix can be represented as $R_0 = [R_3][R_2][R_1]$, which is expanded as follows:

Fig. 4.4 Rotation of coordinate system and definition of rotation matrix

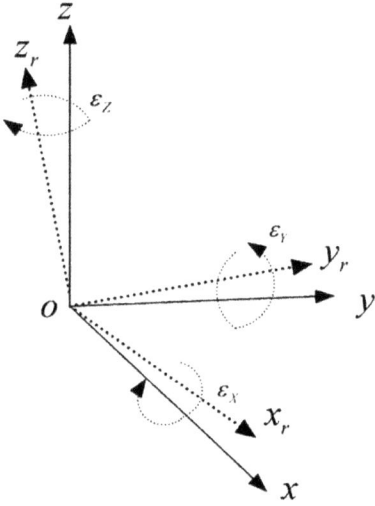

$$R_0 = \begin{bmatrix} \cos \varepsilon_Z & \sin \varepsilon_Z & 0 \\ -\sin \varepsilon_Z & \cos \varepsilon_Z & 0 \\ 0 & 0 & 1 \end{bmatrix} \begin{bmatrix} \cos \varepsilon_Y & 0 & -\sin \varepsilon_Y \\ 0 & 1 & 0 \\ \sin \varepsilon_Y & 0 & \cos \varepsilon_Y \end{bmatrix} \begin{bmatrix} 1 & 0 & 0 \\ 0 & \cos \varepsilon_X & \sin \varepsilon_X \\ 0 & -\sin \varepsilon_X & \cos \varepsilon_X \end{bmatrix}$$

$$= \begin{bmatrix} \cos \varepsilon_Y \cos \varepsilon_Z & \cos \varepsilon_Y \sin \varepsilon_Z & -\sin \varepsilon_Y \\ -\cos \varepsilon_X \sin \varepsilon_Z + \sin \varepsilon_X \sin \varepsilon_Y \cos \varepsilon_Z & \cos \varepsilon_X \cos \varepsilon_Z + \sin \varepsilon_X \sin \varepsilon_Y \sin \varepsilon_Z & \sin \varepsilon_X \cos \varepsilon_Y \\ \sin \varepsilon_X \sin \varepsilon_Z + \cos \varepsilon_X \sin \varepsilon_Y \cos \varepsilon_Z & -\sin \varepsilon_X \cos \varepsilon_Z + \cos \varepsilon_X \sin \varepsilon_Y \sin \varepsilon_Z & \cos \varepsilon_X \cos \varepsilon_Y \end{bmatrix}.$$

$$(4.2.16)$$

Unit vector of $[\theta, \varphi]$ in the spherical coordinate system is $\hat{n} = (\sin \theta \cos \varphi, \sin \theta \sin \varphi, \cos \theta)$, so the Cartesian coordinate that is transformed from global coordinate system to local coordinate system of the array element is represented as follows:

$$[\tilde{x}, \tilde{y}, \tilde{z}]^T = R_0 (\sin \theta \cos \varphi, \sin \theta \sin \varphi, \cos \theta)^T \quad (4.2.17)$$

$$\tilde{\theta} = \arccos(\tilde{z}) \quad (4.2.18)$$

$$\tilde{\varphi} = \arctan\left(\frac{\tilde{y}}{\tilde{x}}\right) \quad (4.2.19)$$

So, when electric field of the array element is represented as $\bar{E}_i(\theta, \varphi)$ in spherical coordinate system, and two orthogonal polarization components are $\bar{E}_{i\theta}$, $\bar{E}_{i\varphi}$, it can be represented in the rectangular coordinate system $(\tilde{x}, \tilde{y}, \tilde{z})$ of the array element:

$$\begin{cases} \bar{E}_{i\tilde{x}} = \bar{E}_{i\theta} \cos \tilde{\theta} \cos \tilde{\varphi} - \bar{E}_{i\varphi} \sin \tilde{\varphi} \\ \bar{E}_{i\tilde{y}} = \bar{E}_{i\theta} \cos \tilde{\theta} \sin \tilde{\varphi} - \bar{E}_{i\varphi} \cos \tilde{\varphi} \\ \bar{E}_{i\tilde{z}} = -\bar{E}_{i\theta} \sin \tilde{\theta} \end{cases} \quad (4.2.20)$$

Representation of individual components in the local rectangular coordinate system $(\tilde{x}, \tilde{y}, \tilde{z})$ is transformed to representation in the global rectangular coordinate system [141].

$$[E_{ix} \quad E_{iy} \quad E_{iz}]^T = R_0^{-1}(\varepsilon_X, \varepsilon_Y, \varepsilon_Z) \cdot [\bar{E}_{i\tilde{x}} \quad \bar{E}_{i\tilde{y}} \quad \bar{E}_{i\tilde{z}}] \quad (4.2.21)$$

Two polarization components of individual array elements can be determined under polarization basis of the global coordinate system (θ, φ):

$$E_{i\theta}(\theta, \varphi) = \frac{E_{ix} \cos \varphi + E_{iy} \sin \varphi}{\cos \theta}, \quad E_{i\varphi}(\theta, \varphi) = -E_{ix} \sin \varphi + E_{iy} \cos \varphi \quad (4.2.22)$$

Hence, the far-zone radiation field of the planar array antenna can be represented as follows:

$$E_{Array}(\theta, \varphi) = \sum_i^N G_i(\theta, \varphi)E_i(\theta, \varphi) = \sum A_i e^{jP_i(\theta,\varphi)} \cdot E_i(\theta, \varphi)$$
$$= \sum A_i e^{jk_0 \bar{r}_i \cdot \hat{n}} \cdot E_i(\theta, \varphi) \tag{4.2.23}$$

where $\bar{r}_i = (x_i, y_i, z_i)$ represents position of the ith array element in the global coordinate system; A_i represents excitation current of the ith array element; and $\bar{E}_i(\theta, \varphi)$ represents directional pattern of the ith array element.

To simplify analysis, it is supposed that when the planar arrays are arranged in xoy plane at equal distance of $M \times N$, distance between elements is d_x, d_y, respectively, and electric field at the antenna aperture is distributed evenly, array factor of horizontal and longitudinal electric field and field under polarization basis $[\theta, \varphi]$ are represented as follows:

$$
\begin{aligned}
F_\theta(\theta, \varphi) &= \frac{1}{\sqrt{M}} \sum_{i=0}^{M-1} E_{i\theta}(\theta, \varphi) e^{jk_0(i-\frac{M-1}{2})d_x \cos\varphi \sin\theta} \frac{1}{\sqrt{N}} \sum_{i=0}^{N-1} E_{i\theta}(\theta, \varphi) e^{jk_0(i-\frac{N-1}{2})d_y \cos\varphi \sin\theta} \\
&= \frac{E_\theta(\theta, \varphi)}{\sqrt{MN}} \frac{\sin\frac{M}{2}(k_0 d_x \cos\varphi \sin\theta - \frac{2\pi}{\lambda}d_x \sin\theta_B)}{\sin\frac{1}{2}(k_0 d_x \cos\varphi \sin\theta - \frac{2\pi}{\lambda}d_x \sin\theta_B)} \\
&\quad \times \frac{\sin\frac{N}{2}(k_0 d_y \sin\varphi \sin\theta - \frac{2\pi}{\lambda}d_y \cos\varphi_B \sin\theta_B)}{\sin\frac{1}{2}(k_0 d_y \sin\varphi \sin\theta - \frac{2\pi}{\lambda}d_y \cos\varphi_B \sin\theta_B)}
\end{aligned} \tag{4.2.24}
$$

$$
\begin{aligned}
F_\varphi(\theta, \varphi) &= \frac{1}{\sqrt{M}} \sum_{i=0}^{M-1} E_{i\varphi}(\theta, \varphi) e^{jk_0(i-\frac{M-1}{2})d_x \cos\varphi \sin\theta} \frac{1}{\sqrt{N}} \sum_{i=0}^{N-1} E_{i\varphi}(\theta, \varphi) e^{jk_0(i-\frac{N-1}{2})d_y \cos\varphi \sin\theta} \\
&= \frac{E_\varphi(\theta, \varphi)}{\sqrt{MN}} \frac{\sin\frac{M}{2}(k_0 d_x \cos\varphi \sin\theta - \frac{2\pi}{\lambda}d_x \sin\theta_B)}{\sin\frac{1}{2}(k_0 d_x \cos\varphi \sin\theta - \frac{2\pi}{\lambda}d_x \sin\theta_B)} \\
&\quad \times \frac{\sin\frac{N}{2}(k_0 d_y \sin\varphi \sin\theta - \frac{2\pi}{\lambda}d_y \cos\varphi_B \sin\theta_B)}{\sin\frac{1}{2}(k_0 d_y \sin\varphi \sin\theta - \frac{2\pi}{\lambda}d_y \cos\varphi_B \sin\theta_B)}
\end{aligned} \tag{4.2.25}
$$

where, $\beta = \frac{2\pi}{\lambda}d_x \sin\theta_B$ is horizontal phase displacement when spatial direction of the beam is θ_B, and $\alpha = \frac{2\pi}{\lambda}d_y \cos\varphi_B \sin\theta_B$ is longitudinal phase displacement when spatial direction of the beam is φ_B.

Hence, $T(\theta) = \begin{bmatrix} F_\theta(\theta, \varphi) \\ F_\varphi(\theta, \varphi) \end{bmatrix}$ is directional pattern vector produced by current distribution (amplitude and phase) on the array element at far zone, which is able to reflect polarization characteristics of the radiation field. When M, N are big enough, main lobe width, minor lobe level and other radiation characteristics are mainly depends on the array factor.

It can be seen from spatial polarization ratio of this kind of planar array antenna obtained by the above-mentioned equation that, if polarization ratio or cross polarization discrimination is used to characterize a spatial polarization character-istics of a constant-amplitude phased array antenna, its expression is a function about array element field, and has little relationship with number of array elements, arrangement method, and array element distance. These factors (mainly array fac-tor) will influence beam width, shape of antenna directional pattern, but will not influence relative variation rate of cross polarization component in different scan-ning spatial with respect to co-polarization component. Polarization characteristics of the array mainly depends on polarization characteristics of the array element and polarization state of scanning. For any array antenna, radiation field at any scanning angle is required to multiply array factor for co-polarization and cross polarization, so polarization characteristics of the whole array can be described by polarization characteristics of a single array element in an equivalent manner. For density weighing, amplitude weighing and conformal array antenna, theoretical analysis will be more complicated when considering cross coupling between several array elements, whose in-depth calculation and analysis will be carried out in Sect. 2.4.

4.2.3 Phased Array Antenna Spatial Polarization Characteristics Simulation Analysis

1. Evenly-distributed rectangular dipole array

Simulation calculation is carried out by taking planar dipole array antenna as an example [141], in which the array is arranged in an 5×5 equivalent-distance manner, the element direction is parallel to x axis; as shown in Fig. 4.5, distance in xy direction is 0.7λ, resonant frequency is 10 GHz, dipole length is 1.5 cm, and height above the ground is 0.75 cm, considering following conditions:

- Analysis of polarization characteristic at different observation angle when main beam of the phased array is in direction of the normal line;
- Analysis of polarization characteristic at different elevation observation angle when main beam of the phased array is in direction of the normal line;
- Analysis of polarization characteristic nearby the main lobe when main beam of the phased array conducts scanning at directional plane;
- Analysis of polarization characteristic nearby the main lobe when main beam of the phased array conducts scanning at elevation plane;

Fig. 4.5 Geometrical layout
of evenly-distributed
rectangular dipole array

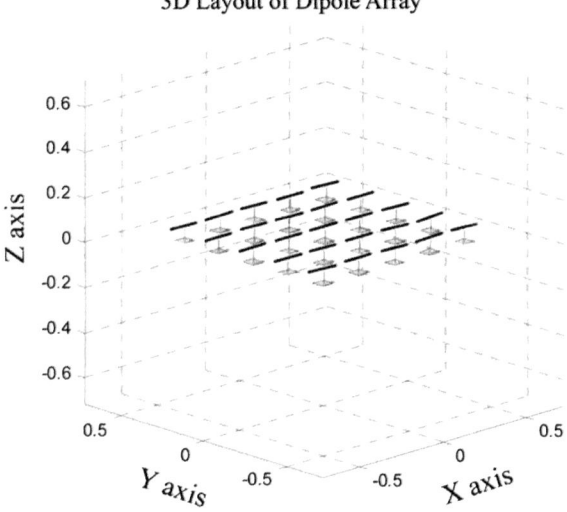

Through calculation and simulation analysis, some meaningful conclusions are obtained, and polarization characteristics of the phased array antenna are closely related to three quantities.

First, maximum radiation direction determined by mechanical structure of the antenna, i.e. direction of normal line;

Second, maximum radiation beam direction during spatial scanning which is usually considered center direction of the main beam, also called electric axis direction;

Third, difference between both above-mentioned values, which is defined here ass "Beam steering angle" (Defined as beam steering by IEEE), hereafter abbreviated as "Beam steering angle", means angle of the beam deviating from the normal line during spatial scanning.

Table 2.1 lists variation law of polarization characteristics during scanning of observation point in the azimuth and elevation direction when the antenna beam is in normal line direction of the array plane. The following gives specific analysis results.

(1) Polarization characteristics of azimuth and elevation direction

When the antenna beam is in normal line direction of the array plane, Fig. 4.6 gives azimuth direction polarization pattern of the antenna at several elevation tangential planes. It can be seen that cross polarization radiation is smallest and co-polarization radiation is biggest in azimuth center direction of the beam. When observation direction deviates from the azimuth center, cross polarization component that is obviously increased in the man lobe area of the antenna at this moment is subject to decrease in polarization purity, indicating that electric wave polarization state received will vary in the deviation direction and antenna azimuth

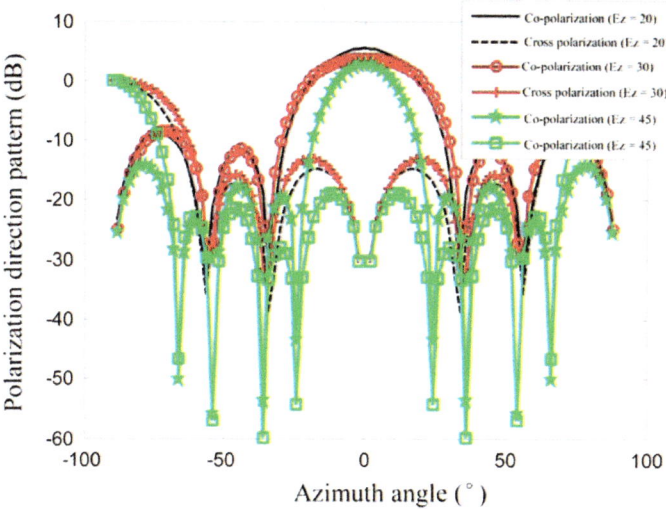

Fig. 4.6 Azimuth direction pattern in normal line direction

polarization characteristics at tangential plane of several elevation angles have this feature. As elevation angle (Ez) of the observation point increases, the narrower the beam width, the lower the antenna gain is; Fig. 4.7 gives antenna elevation polarization pattern at tangential plane of several elevation angles. It can be seen that co-polarization direction pattern of the elevation direction is similar to cross polarization direction pattern, and in elevation center direction, the co-polarization

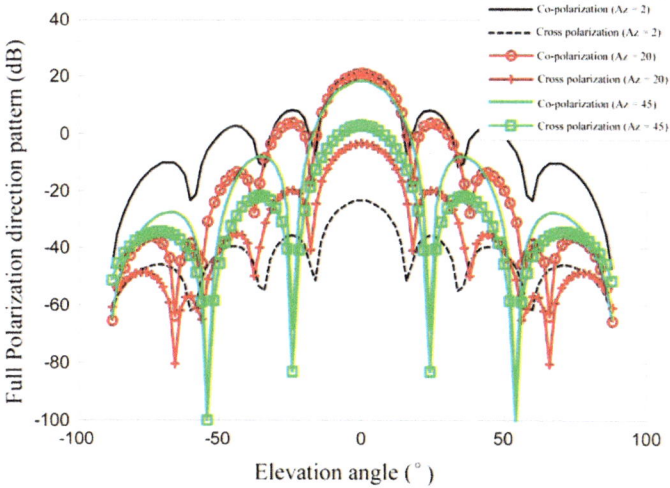

Fig. 4.7 Elevation direction pattern in normal line direction

Table 4.1 Quantitative description of polarization characteristics during scanning of observation point in normal line direction

Angle of beam deviating from the elevation angle at observation point in normal line direction	Polarization purity in azimuth direction	Remarks
Deviation of observation point from elevation angle of the electric axis by 10°	−18 dB	Copol = 15 dB, Cxpol = −3 dB
Deviation of observation point from elevation angle of the electric axis by 20°	−18 dB	Copol = 5.5 dB, Cxpol = −14 dB
Deviation of observation point from elevation angle of the electric axis by 30°	−16 dB	Copol = 3.7 dB, Cxpol = −13 dB
Deviation of observation point deviating from elevation angle of the electric axis by 45°	−20 dB	Copol = 2.6 dB, Cxpol = −19 dB
Deviation of observation point from elevation angle of the electric axis by 0°	−∞	V polarization of co-polarization, H polarization of cross polarization
Deviation of observation point from elevation angle of the electric axis by 2°	−44 dB	Co-pol = 21 dB, Cx-pol = −23 dB
Deviation of observation point from elevation angle of the electric axis by 20°	−24 dB	Co-pol = 21 dB, Cx-pol = −3 dB
Deviation of observation point from elevation angle of the electric axis by 45°	−16 dB	Co-pol = 19 dB, Cx-pol = 3 dB
Deviation of observation point from elevation angle of the electric axis by 90°	Cross polarization only	

radiation is biggest, and cross polarization is stronger, however, when deviation of observation direction from the center angle increases, beam width of the direction pattern, and radiation gain will not be subject to significant change, and at this moment cross polarization component increases significantly, and the polarization purity decreases (Table 4.1).

(2) When main beam scans at spatial azimuth plane and elevation plane, polarization characteristics of the azimuth direction and polarization characteristics of elevation direction

When antenna beam is scanning in the spatial, the direction of the electric axis will deviate normal line direction of the array plane, so that a beam steering angle exists between the electric axis direction and normal line direction. At this moment, structure of the phased array antenna radiation direction pattern will vary. At different azimuth different beam steering angle and elevation beam steering angle, polarization purity of electromagnetic wave radiated or received in main lobe direction of the antenna will decrease, and at this moment, co-polarization component does not only exist in main lobe area of the antenna, but also higher cross polarization component exist. It can be seen from Fig. 4.8 that when elevation angle and observation angle remain fixed and beam steering angle of the azimuth

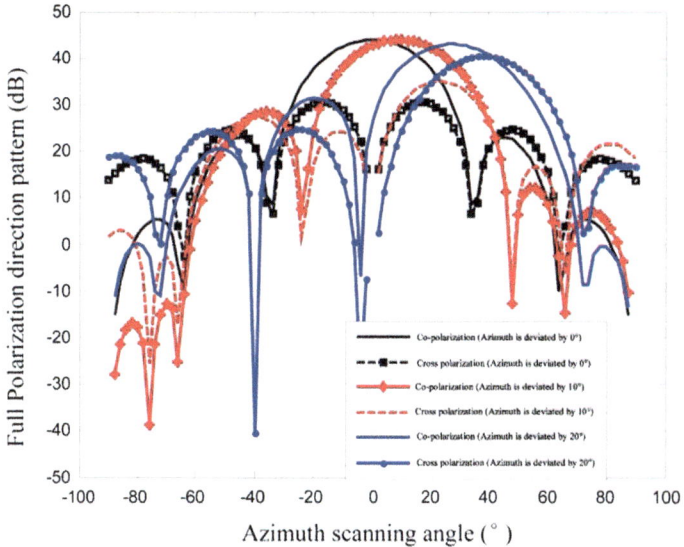

Fig. 4.8 Azimuth scanning direction pattern at several azimuth off-axis angles

scanning increases, cross polarization level increases significantly, and polarization purity and azimuth scanning decreases. Particularly, when the elevation beam steering angle is 30° and azimuth beam steering angle is 45°, orthogonal polarization components of the antenna are very nearly equal.

Figure 4.9 gives polarization characteristics of the beam during scanning in elevation direction when the azimuth beam steering angle and observation angle remain fixed. It can be seen that change in polarization characteristics in elevation direction is not obvious, structure of the antenna direction pattern varies significantly, electric level of the minor lobe remains high and is unsymmetrical to the main lobe, as well as main lobe gain is declined slightly. Table 4.2 gives quantitative description of polarization characteristics during spatial Beam Steering.

2. Wave-guided slot phased array

The following takes a slot array having 25 elements as an example to analyze spatial polarization characteristics of the antenna, where working frequency of the antenna is 1 GHz, and slot length and width is 0.1λ and 0.5λ respectively, and slot gap on the rectangular panel is 0.25λ and 0.8λ respectively. Figure 4.10 gives a 3D geometrical structure diagram for simulated slot array. Value simulation is carried out based on following three conditions to give polarization direction pattern at different observation direction and elevation direction, and polarization direction pattern of beam scanning within small scope of beam scanning and wide-angle azimuth during equivalent-amplitude and equidirectional feed and when scanning angle is 0°. The above calculation doesn't take consider of array element coupling

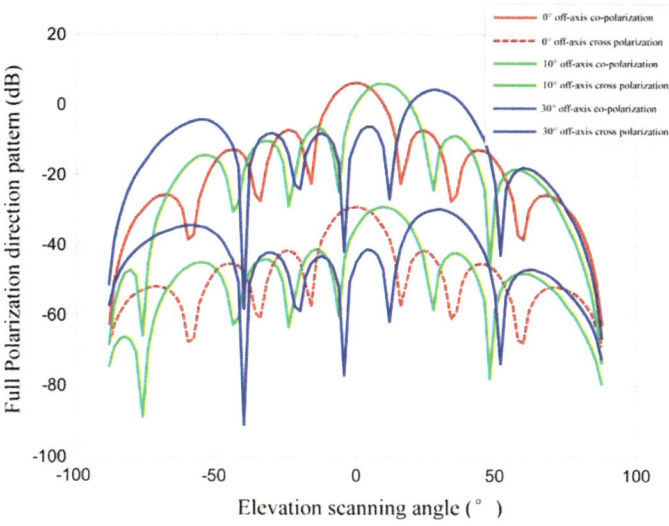

Fig. 4.9 Elevation scanning direction pattern at several elevation off-axis angle

Table 4.2 Quantitative description of polarization characteristics during spatial beam steering

When beam scanning in spatial, it deviates from normal line direction The azimuth beam steering angle remains fixed, and elevation beam steering angle increases	Polarization purity in elevation direction	Remarks
Azimuth angle 1°, elevation angle 0°	−35 dB	Structure of cross polarization direction pattern is similar to that of co-polarization direction pattern
Azimuth angle 1°, elevation angle 10°	−35 dB	
Azimuth angle 1°, elevation angle 20°	−35 dB	Polarization purity of elevation scanning is not varied
Azimuth angle 1°, elevation angle 30°	−35 dB	
Azimuth angle 0°, elevation angle 30°	−15 dB	As beam azimuth scanning angle increases, cross polarization component increases and azimuth scanning polarization purity decreases
Azimuth angle 10°, elevation angle 30°	−10 dB	
Azimuth angle 20°, elevation angle 30°	−7 dB	
Azimuth angle 30°, elevation angle 30°	−5 dB	

and mismatch effect, whose purpose is to determine the influence of beam scanning on the antenna polarization characteristics of slot array.

It can be seen from 3D co-polarization and cross polarization direction pattern shown in Fig. 4.11 that E-side direction pattern for co-polarization is wide, covering a large azimuth angle, and minor lobe of the azimuth direction pattern is below −20 dB, so that a funnel-shaped direction pattern structure is formed. Figure 4.12 gives elevation direction pattern at different observation angle when scanning angle is 0°; and when observation angle is 0°, only co-polarization

Geometrical layout of the array

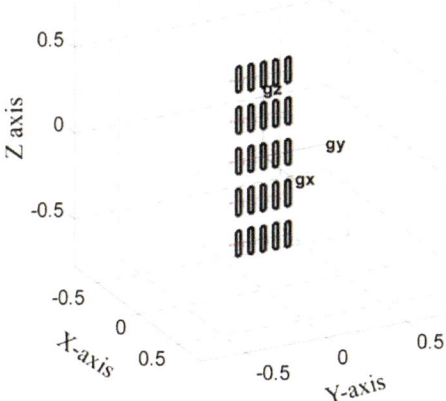

Fig. 4.10 3D geometrical structure diagram of slot array

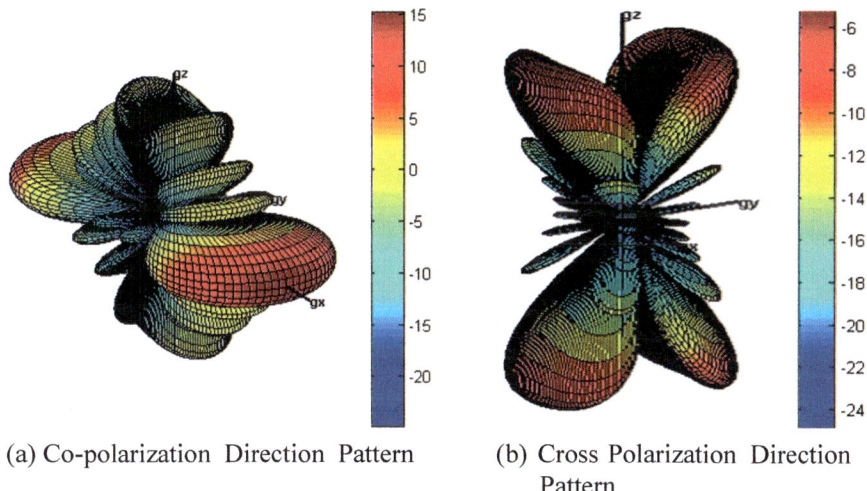

(a) Co-polarization Direction Pattern

(b) Cross Polarization Direction Pattern

Fig. 4.11 Polarization direction pattern of slot array

component exists, without cross polarization component; when observation angle is deviated by 30° and 60°, co-polarization component gain declines, while cross polarization component increases. Similarly, as shown in Fig. 4.13, when elevation observation angle is 0°, only co-polarization component exists; and as elevation observation angle increases, cross polarization component increases and beam width gets narrow. This indicates that when no beam scanning is done, the antenna radiation efficiency is biggest; and when measuring target direction is deviated from

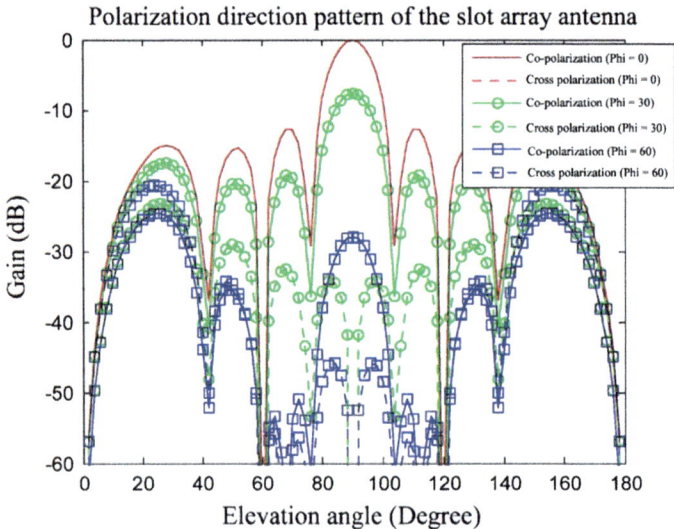

Fig. 4.12 Polarization direction pattern at different azimuth tangent planes

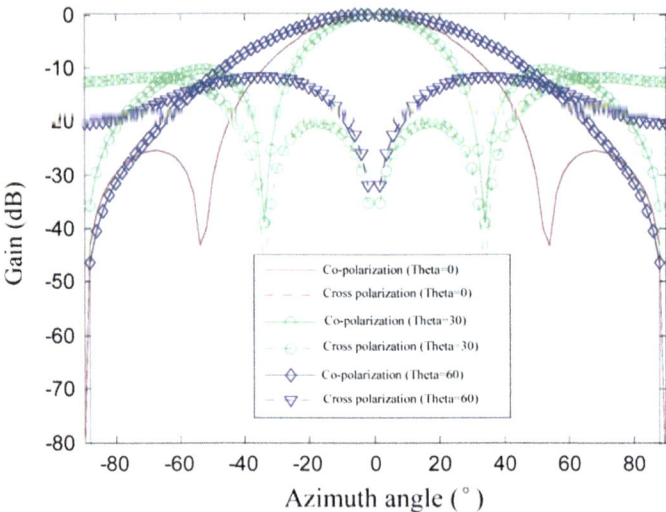

Fig. 4.13 Polarization direction pattern at different elevation tangent planes

electric axis direction of the antenna, electric wave polarization state received will vary with deviation from the electric axis direction and elevation angle. Figure 4.14a gives elevation direction pattern at 1° azimuth scanning angle and 0°, 10° and 25° elevation scanning angles. It can be seen that when no elevation scanning is not done, cross polarization is below −50 dB; however, when elevation

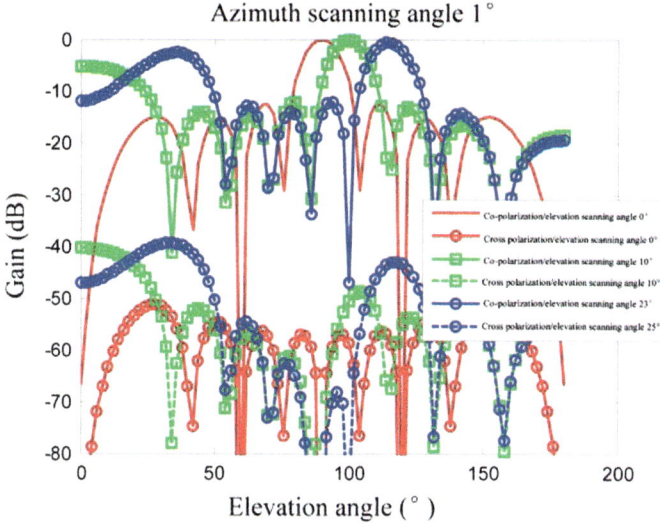

(a) Elevation Polarization Direction Pattern during Beam Steering

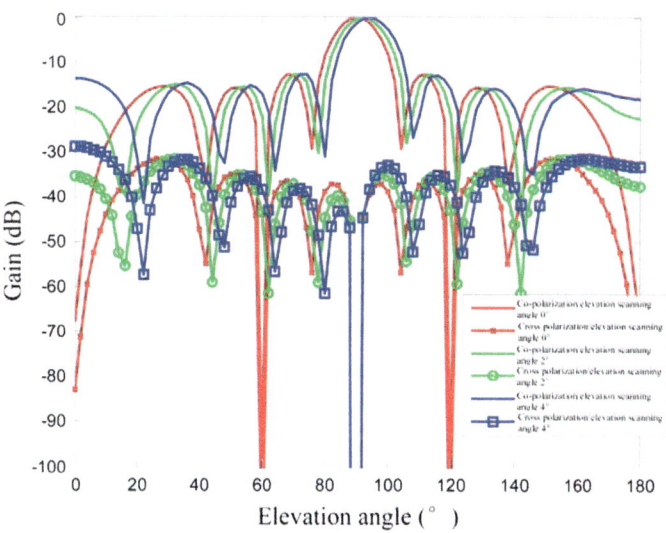

(b) Elevation Polarization Direction Pattern during Beam Steering

Fig. 4.14 Azimuth and elevation polarization direction pattern during beam steering

scanning angle is 10°, structure of cross polarization direction pattern is varied, similar to co-polarization direction pattern, and is increased by more than 10 dB. And, as elevation scanning angle increases, electric level of the minor lobe increases, and the cross polarization gain also gets bigger. When azimuth scanning angle is increased to 10° (as shown in Fig. 4.14b) and elevation scanning angle is

small (0°, 2°, 4°), cross polarization level increases significantly, and cross polarization discrimination (XPD) is increased to about −30 dB. When azimuth scanning
angle is increased to 30° (as shown in Fig. 4.15a), XPD is increased to about
−20 dB. To investigate scanning polarization characteristics at azimuth plane,
simulation analysis is carried out, either. Figure 4.16b gives azimuth direction
pattern at 2° elevation scanning angle and 20° azimuth scanning angle. When
electric level of the minor lobe is increased from −25 to −15 dB, shape of the
direction pattern is also distorted; and when electric level of the minor lobe is
unsymmetrical, electric level of cross polarization is increased from −42 to −35 dB.
It can be seen from Fig. 4.16 that when the beam scans the area with 30° elevation
angle and 20° azimuth angle, electric level of cross polarization in the main lobe is
increased to −12 dB. To sum up, when co-polarization of the slot array is horizontal
polarization, wide-angle scanning in azimuth and elevation direction significantly
changes polarization characteristics of the beam. In addition to influence on antenna
radiation characteristics (beam broaden, gain decrease), polarization characteristics
in the main lobe also changes significantly. Increase in electric level of cross
polarization results in increase in ratio between two orthogonal polarization components and decrease in polarization purity. Through statistics of several groups of
simulation result, Table 4.3 gives variation law of polarization characteristics
during spatial scanning of the beam by the antenna.

This Section analyzes spatial polarization characteristics of evenly-distributed
dipole array and wave-guided slot array at different angles, and gives corresponding
analysis results. However, there are many types of array elements in actual application, and other factors including amplitude of aperture electric field and phase
weighing, feed method, and individual elements of conformal array have different
steering direction, so it is inevitable to bring new problems for and influence on
analyzing polarization characteristics, and it is required to carry out further research.
However, as a preliminary investigation of spatial polarization characteristics of the
phased array antenna, it has certain theoretical and practical meaning, and can be
applied to reduce influence of antenna polarization characteristics during evaluation
of radar power and accuracy in out-field test, correct simulation model of phased
array radar antenna and enrich in-field simulation theory. More accurate modeling
and calculation carried out by making use of advanced finite difference time-domain
electromagnetic calculation method (XFDTD) will be the content for further study.

4.3 Phased Array Antenna Design and Spatial Polarization Characteristics Analysis Based on XFDTD

Based on simplified theoretical model, Sect. 4.3 analyzes and researches spatial
polarization characteristics of two kinds of phased array antenna in a detailed
manner by taking spatial direction of the antenna as variation factor [142], so as to

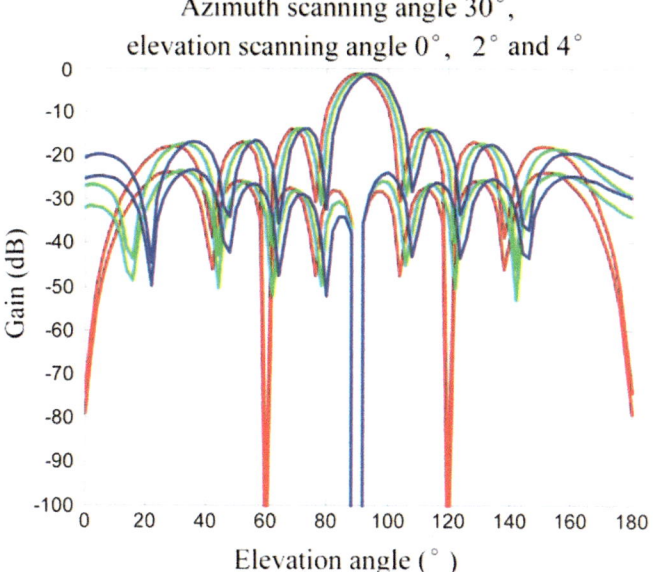

(a) Elevation Direction Pattern during Beam Steering

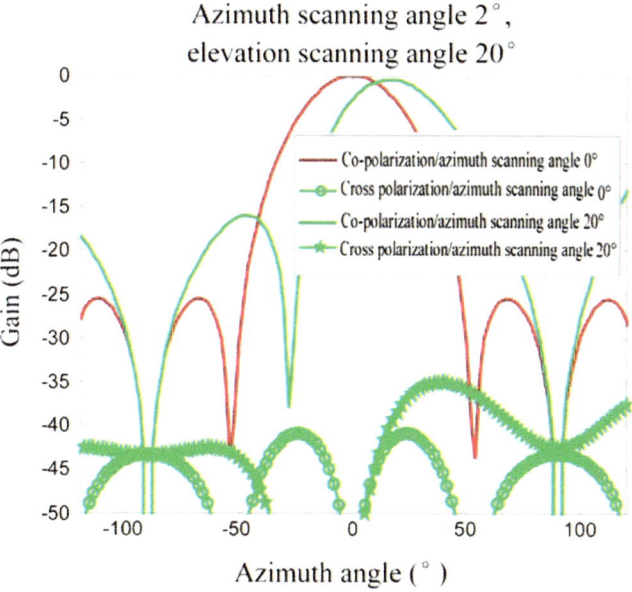

(b) Azimuth Direction Pattern during Beam Steering

Fig. 4.15 Elevation polarization direction pattern and azimuth polarization direction pattern during beam steering

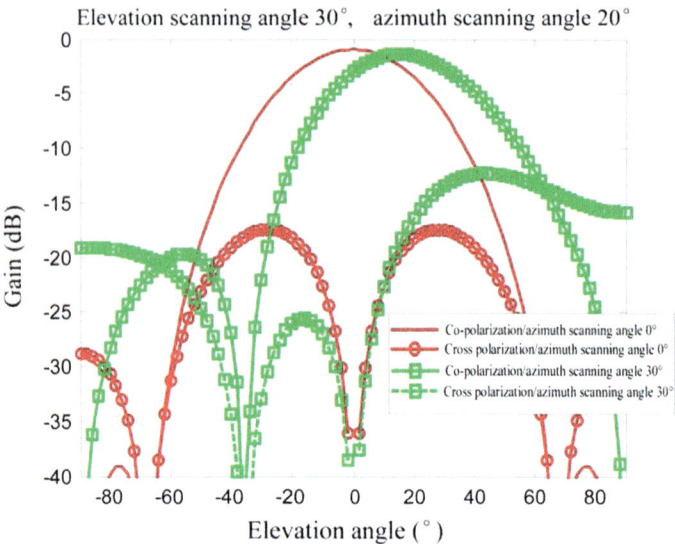

Fig. 4.16 Azimuth polarization direction pattern during beam steering

Table 4.3 Quantitative analysis of polarization characteristics during spatial scanning of the beam

Azimuth scanning and elevation scanning of beam in spatial	Polarization purity (dB)
Azimuth scanning angle 1°, elevation scanning angle 0°	−58
Azimuth scanning angle 1°, elevation scanning angle 10°	−48
Azimuth scanning angle 1°, elevation scanning angle 25°	−40
Azimuth scanning angle 10°, elevation scanning angle 0°	−37
Azimuth scanning angle 10°, elevation scanning angle 2°	−35
Azimuth scanning angle 10°, elevation scanning angle 4°	−33
Azimuth scanning angle 30°, elevation scanning angle 0°	−28
Azimuth scanning angle 30°, elevation scanning angle 2°	−25.8
Azimuth scanning angle 30°, elevation scanning angle 4°	−23.9
Azimuth scanning angle 20°, elevation scanning angle 0°	−41
Azimuth scanning angle 20°, elevation scanning angle 2°	−35
Azimuth scanning angle 0°, elevation scanning angle 30°	−18
Azimuth scanning angle 20°, elevation scanning angle 30°	−12
Azimuth scanning angle 30°, elevation scanning angle 30°	−8

obtain a preliminarily-qualitative conclusion on change in polarization characteristics occurred during beam steering in spatial. To further verify this conclusion, this Section gives more accurate analysis, and adopts electromagnetic field simulation and analysis software based on full-wave 3D finite difference time-domain method-XFDTD to design a planar rectangular phased array antenna having 625 elements, and through electromagnetic calculation, co-polarization and cross polarization component data are obtain for far field of the antenna, so as to utilize spatial polarization descriptor to obtain variation law of polarization characteristics of typical phased array antenna at different beam steering angle. The results show that analysis on polarization characteristics of the phased array antenna is more complicated than mechanical scanning system, and point out difference in spatial polarization characteristics between phased array antenna and mechanical scanning antenna. Change its polarization state depends on non-ideal factors including scanning angle, array element characteristics and array element coupling, and changes approximately in a linear manner with scanning angle. This conclusion coincides with theoretical analysis results, providing theoretical basis and practical basis for application of spatial polarization characteristics of the phased array antenna.

4.3.1 Uniformly Distributed Phased Array Antenna Design and Simulation

4.3.1.1 Geometrical Modeling

XFDTD is a full-wave 3D high-frequency electromagnetic simulation analysis software based on finite difference time domain method (FDTD) developed by REMCOM company, which is mainly applied to antenna analysis and design, microwave circuit design, bioelectromagnetics, electromagnetic compatibility analysis, electromagnetic scatter calculation, photonics research and other fields [174–178]. XFDTD is taken as a kind of high-end antenna simulation and calculation tool which has been widely recognized by many internationally-distinguished antenna research and development centers and scholars, and whose calculation results have approximately reached practical measuring results with respect to their accuracy. This Section will, with the help of this tool, designs a kind of typical phased array antenna, and carries out analysis on polarization characteristics in several spatial scanning directions.

As shown in Fig. 4.17a, the antenna element patch is square, with edge length of 64 mm, distance of 12 mm between feed point and center position, and thickness of 3 mm, whose substrate material is polytetrafluoroethylene with dielectric constant of 2.6, and floor slab size is 90 × 90 mm. Taking the antenna as an array element, Fig. 4.17b gives a 25 × 25 square array designed, in which array elements are spaced at 112 mm, and distributed evenly, as well as individual elements are symmetrical to axis of the center element. The floor slab is square, with edge of 2832 × 2832 mm, and element edge is at distance of 40 mm from the edge of the floor slab. To obtain characteristics of co-polarization and cross polarization direction patterns of the array whose beam is at normal direction, individual elements of the array are excited at the same amplitude and in the same phase,

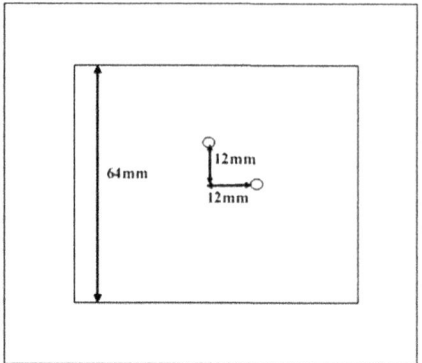 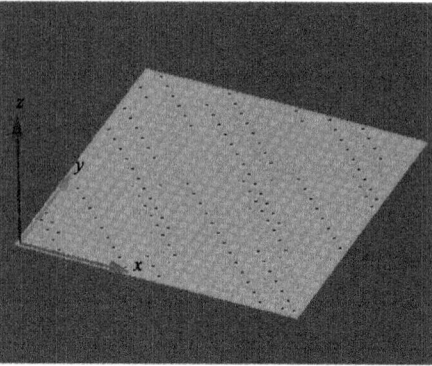

(a) Array Element Model Designed by   (b) Array Model Designed by Use
     Use of XFDTD   of XFDTD

Fig. 4.17 Array element model and array model designed by use of XFDTD

respectively. To enable the beam to be deviated from normal direction, it is required to scan polarization characteristics at different spatial directions along the azimuth axis, and amplitude of excitation signal for individual array elements remains unchanged, but phase displacement of the 25×25 square array is modified to $90°$, $127.3°$, $155.9°$ and $178.9°$, respectively, so as to compare polarization character-istics of the main lobe at different space position displacement.

Simulation excitation source use a Gaussian pulse with pulse width of 32 (basic time step) [174], using Liao boundary condition. Simulation waveform uses 1362 MHz sine wave.

Then, above-mentioned antenna elements are used to form 25×25 square array, with element spacing of 112 mm, and floor slab is a square with edge of 2832 mm. Direction pattern and axial ratio characteristics of the array center ele-ment are simulated, and cross coupling curve between elements is given (frequency sweep can be made between 1300–1400 MHz).

Characteristics of 25×25 square array is calculated (direction pattern). Phase difference of each row of array elements is modified to enable space scanning direction of the main lobe is at $0°$, $5°$, $10°$, $15°$, $30°$, $45°$, $60°$ and $75°$.

Computer used for simulation is configured as follows:

CPU: Intel(R) Pentium(R) 4, 2.40 GHz
Memory: 2048 MB (DDR SDRAM)

4.3.1.2 Mesh Subdivision

As shown in Fig. 4.18, unit mesh in X, Y and Z direction is 1 mm, and 29.80 MB memory space is required [174].

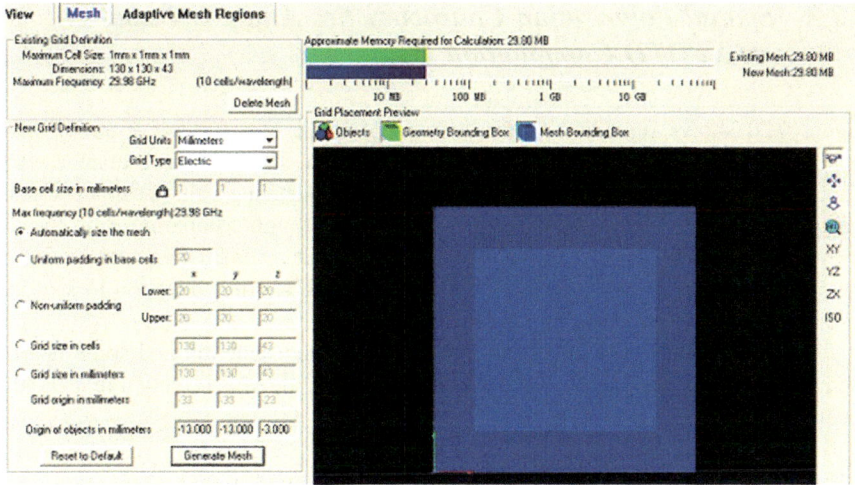

Fig. 4.18 Mesh subdivision of unit antenna

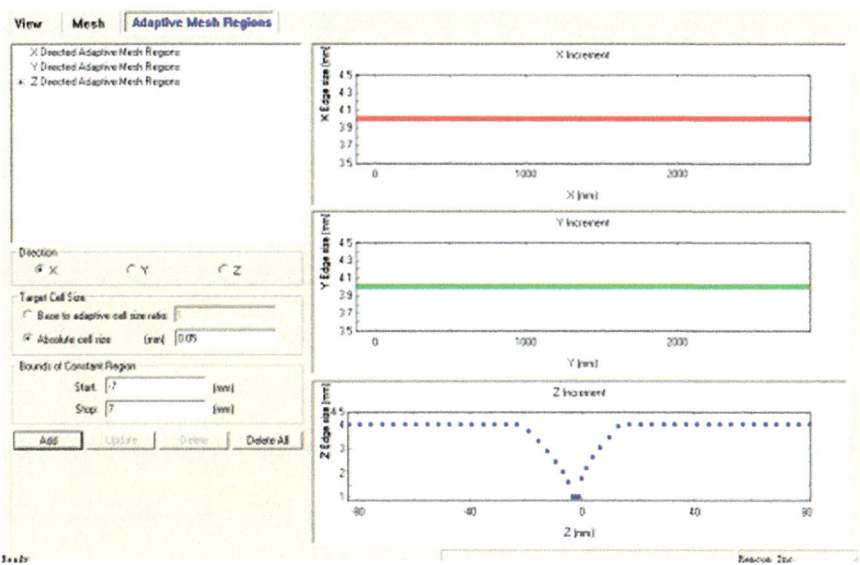

Fig. 4.19 Use of Adaptive mesh in Z Direction

Figure 4.19 shows mesh subdivision of array antenna, with basic unit mesh being (4,4,4) (expressed in mm), and an adaptive mesh is used at point (−3,0) in Z direction. Simulation requires 716.22 MB memory space in total.

4.3.2 Spatial Polarization Characteristics Analysis Based on XFDTD Computation Data

Figure 4.20 shows impedance characteristic of input port for linear polarization center element. When the reactance is 0, the antenna is subject to resonance, with resonant frequency of 1332 MHz. The maximum resistance value at this resonant frequency point is 100 Ω. As shown in Fig. 4.21, through comparing echo loss of array center element and unit antenna, it is found that resonant frequency point of the array center element antenna is shifted by 30 MHz. And, insertion loss of the resonance point is increased by about 6 dB.

Figure 4.22 gives coupling curve for center element antenna and some nearby element antennas, and it can be seen that coupling degree is very big near the resonance point (-14 dB at most).

Fig. 4.20 Impedance characteristics of input port for center element

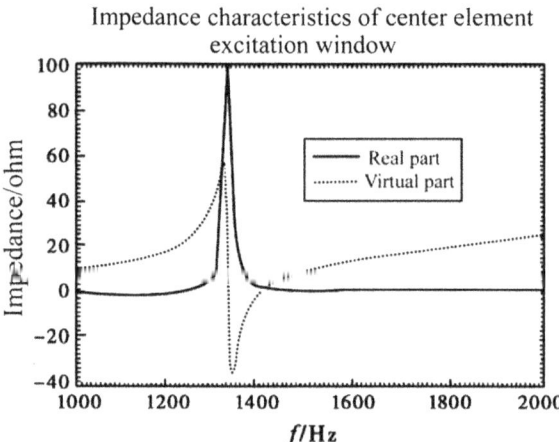

Fig. 4.21 Echo loss of input port

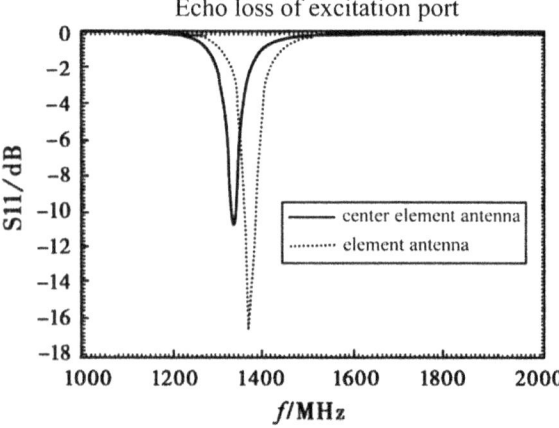

Fig. 4.22 Coupling curve of antennas for linear polarization center element and other elements

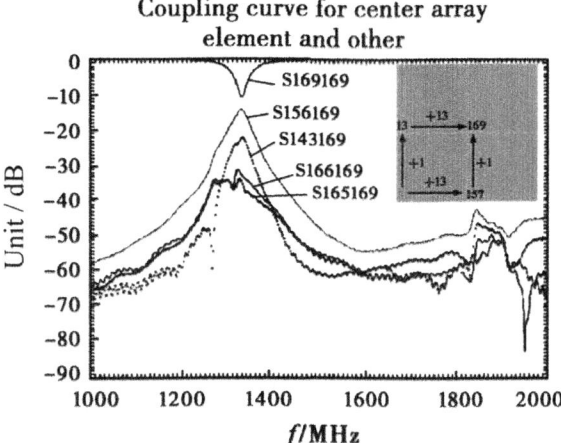

Through modeling and electromagnetic calculation of the above-mentioned phased array antenna, a great deal of radiation field data is obtained. Description method for spatial polarization characteristics of the antenna mentioned in Sect. 4.2 is used to analyze and process calculation results in a detailed manner to obtain polarization direction pattern of the phased array antenna in different main beam scanning directions, phase direction pattern, distribution of complex polarization ratio, variation curve of cross polarization discrimination, distribution condition of polarization descriptor, distribution results of instantaneous polarization projection vector (IPPV). The following gives some typical processing results.

Figures 4.23, 4.24, 4.25, 4.26, 4.27, 4.28 gives 3D polarization direction pattern of the phased array antenna at 0° and 60° spatial scanning angle, and Fig. 4.29 illustratively gives polarization direction pattern of the phased array antenna at several scanning angles, and it can be seen obviously that co-polarization, i.e. vertical polarization direction pattern, is broaden gradually with increase in scanning angle, the antenna gain is declined slightly, and the minor lobe shows unsymmetrical structure; meanwhile, cross polarization electric level is increased gradually from $-\infty$ to about 30 dB in electric axis direction.

For the reason that scanning characteristics of the phased array antenna will vary, single-direction polarization pattern is difficult to describe complete spatial polarization characteristics. So, through calculation of several waves and then through insertion fitting, it is possible to obtain polarization characteristics restructuring results at 0–60 spatial scanning angles, as shown in Fig. 4.30. It can be seen that polarization characteristic of several waves is varied in an approximately linear manner during phased array antenna beam scanning. Figure 4.31 gives phase direction pattern of the beam in normal line direction of the array, which is obviously different from characteristics of conventional parabolic reflector antenna. Figure 4.32 gives distribution of spatial Strokes vector during antenna scanning. The above-mentioned analysis results show that polarization

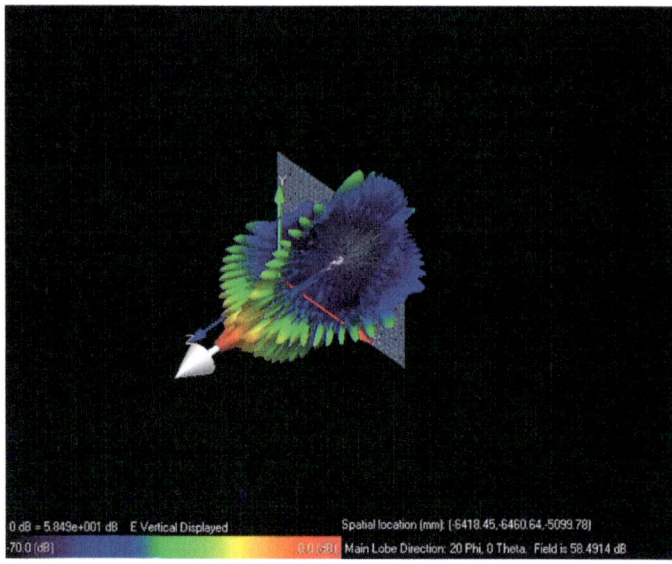

Fig. 4.23 Co-polarization direction pattern in normal line direction

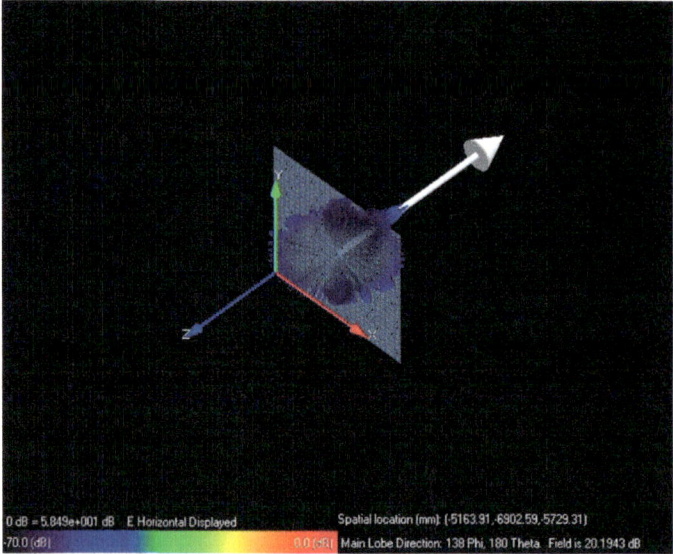

Fig. 4.24 Cross polarization direction pattern in normal line direction obtained through electromagnetic calculation

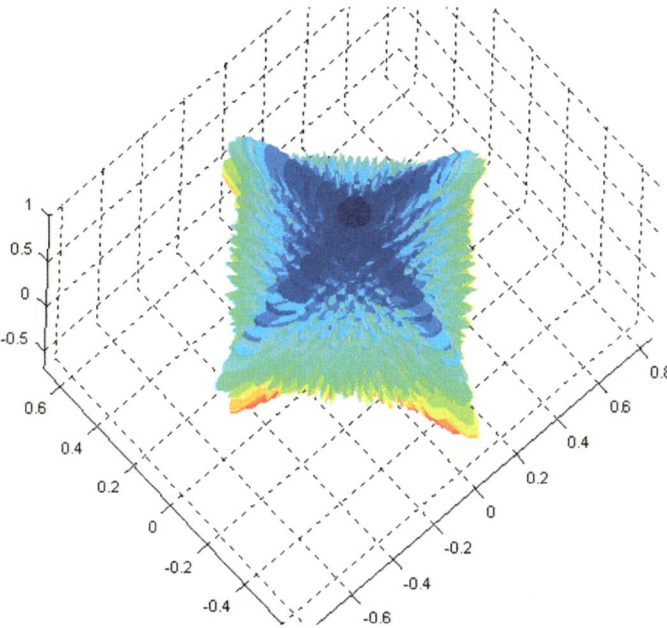

Fig. 4.25   Co-polarization direction pattern of main beam in normal line direction

Fig. 4.26   Cross polarization
direction pattern of main
beam in normal line direction

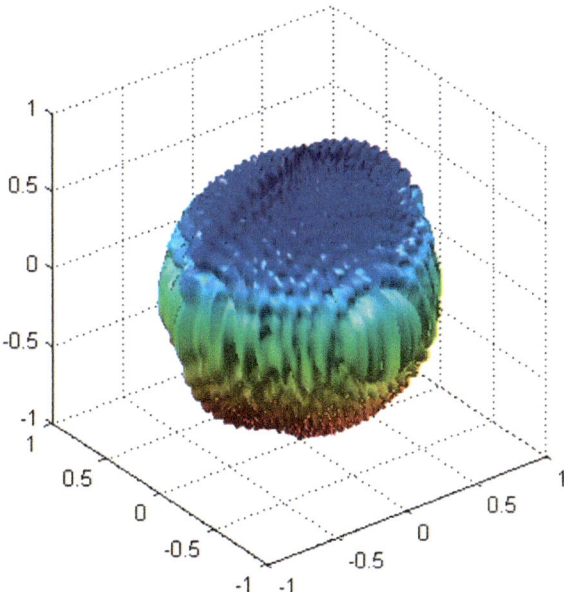

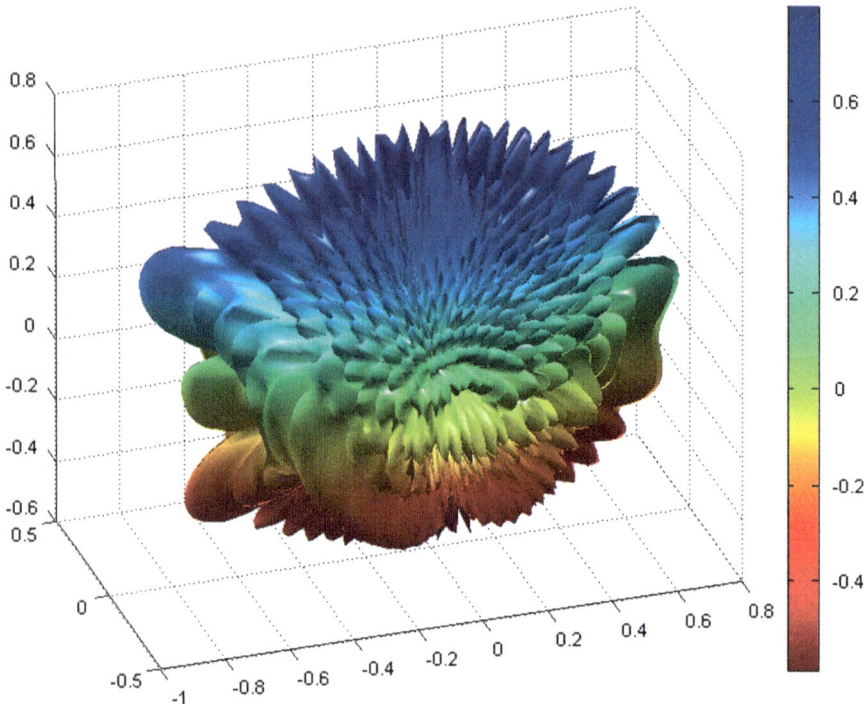

Fig. 4.27 Main polarization direction pattern of main beam at 60° scanning angle

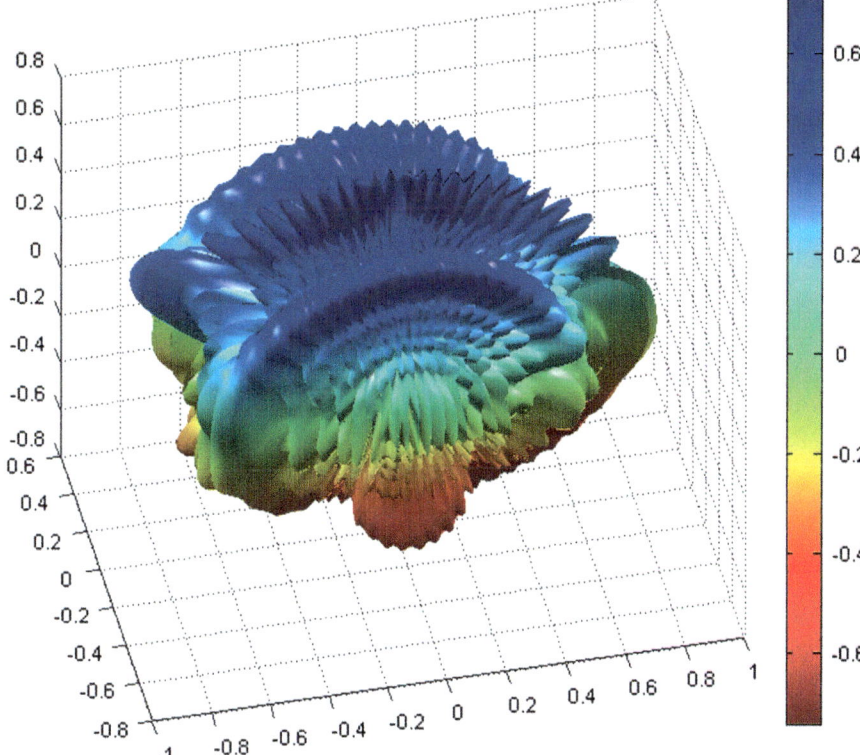

Fig. 4.28 Cross polarization direction pattern of main beam at 60° scanning angle

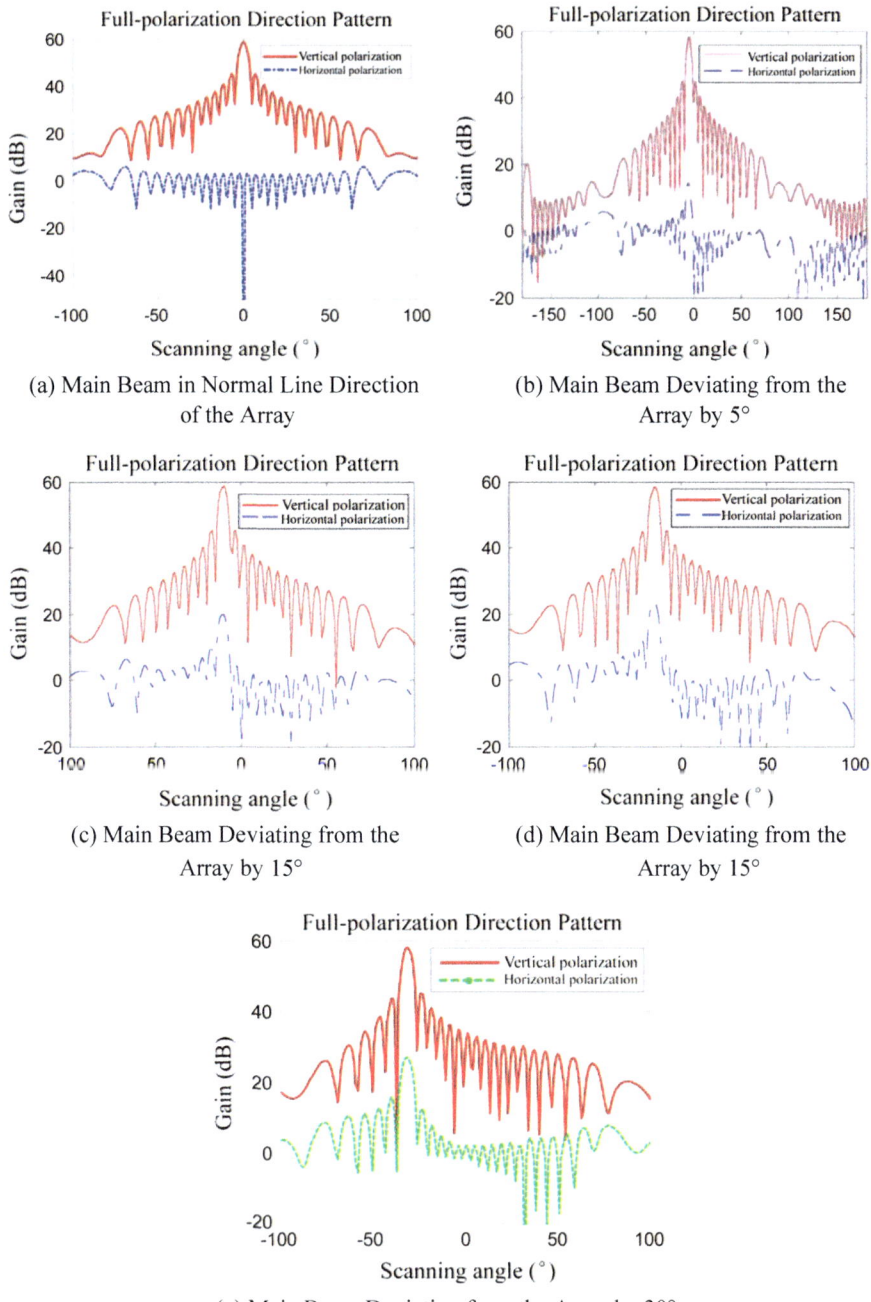

Fig. 4.29 Polarization direction pattern of phased array antenna beam at several scanning angles

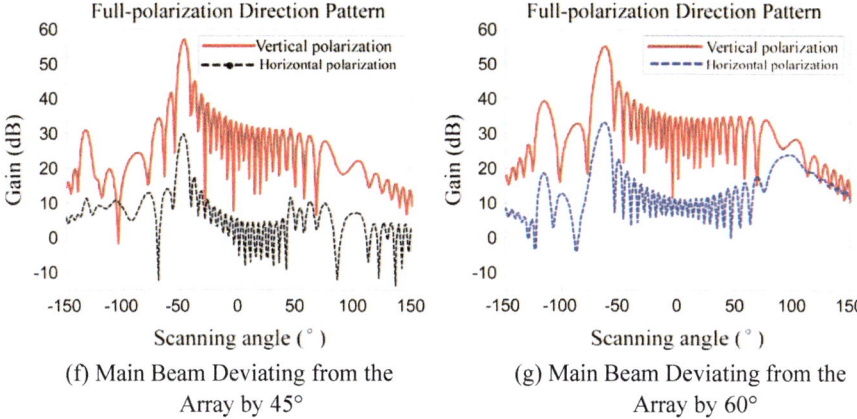

(f) Main Beam Deviating from the
Array by 45°

(g) Main Beam Deviating from the
Array by 60°

Fig. 4.29 (continued)

Fig. 4.30 Variation curve of
spatial polarization purity at
different scanning angles

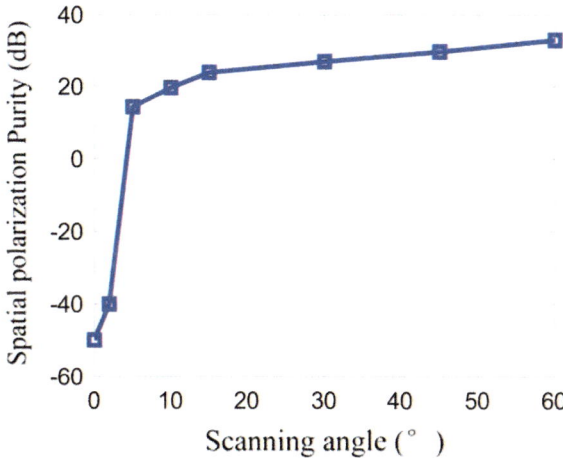

Fig. 4.31 Phase direction
pattern of the antenna

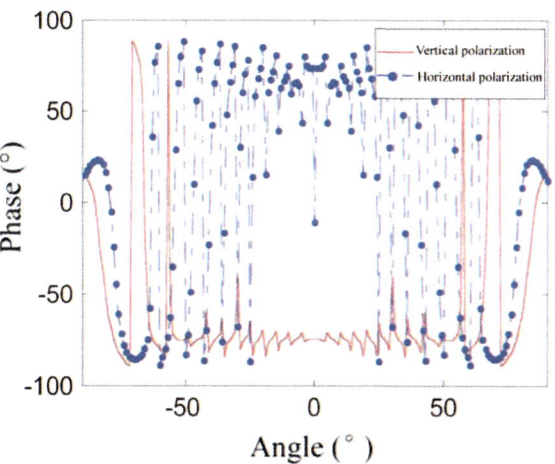

Fig. 4.32 IPPV pattern for
phased array beam scanning

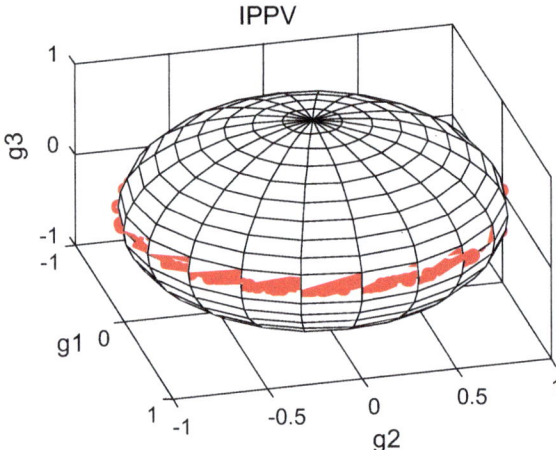

characteristics of the phased array antenna is subject to certain regular variation during continuous spatial scanning of the beam, the antenna polarization state is deviated from expected state, change in polarization state depends on such non-ideal factors as scanning angle, scanning axis, array element characteristics and array element coupling. This conclusion coincides with results of theoretical analysis.

Chapter 5
Antenna Spatial Polarization Characteristics Measurement and Calibration

Measurement of target polarization characteristics is an important basic issue for recognition of radar target characteristics and target identification, and is widely concerned for a long time. Radar target can be considered a polarization changer, which is usually represented by a second-order complex matrix, and is solved through "receiving and measuring two target echo signal under orthogonal polarization electromagnetic wave excitation on two orthogonal polarization basis in a simultaneous or time-sharing manner". However, from the prospect of engineering implementation, there are many technical difficulties to develop special target polarization characteristics measurement radar, such as improving orthogonal polarization channel separation degree, maintaining consistency in amplitude and phase of several polarization channels, and solving calibration of polarization measurement system. Reference [7] proposes a new method for measuring target polarization scattering matrix by use of spatial polarization characteristics of the single polarization antenna, which has certain application prospect. But, it brings new problems-problem in outfield measurement of actual spatial polarization characteristics of the radar antenna. Reference [143–149] propose some solutions for acquiring polarization characteristics of the antenna, but complicated electromagnetic calculation and compact field measurement and other methods are difficult to simulate actual utilization environment in a real manner, reflect influence of non-ideal factors in cascading radio frequency head and reception channel on polarization characteristics of the antenna, and meet requirements for practical application. Outfield measurement is closest to actual working condition of the radar antenna, and more illustratively reflects polarization characteristics of receive and transmit beam of the antenna, as well as the most efficient means and approach to adjust and balance antenna characteristics and describe spatial polarization characteristics of the antenna. Additionally, influence of such factors as non-ideal conditions and manual operation in outfield measurement, causes polarization basis

© National Defense Industry Press, Beijing and Springer Nature Singapore Pte Ltd. 2019   173
H. Dai et al., *Spatial Polarization Characteristics of Radar Antenna*,
https://doi.org/10.1007/978-981-10-8794-3_5

of "inspecting the antenna" not to completely align with two groups of orthogonal polarization basis where "antenna to be tested" is situated, i.e. certain deviation exists, so that certain measurement error is resulted from energy projection leakage of co-polarization component to cross polarization component at deviation angle; meanwhile, influence by acquisition frequency of the antenna measurement data, signal source frequency shift, and phase center shift will also result in measurement error. Hence, it is required to carry out subsequent correction of measurement data to obtain accurate description of the antenna polarization characteristics.

This Chapter mainly studies measurement method of the antenna spatial polarization characteristics and measurement error correction technology [150−152]. Section 5.1 builds model by converting the polarization modulation characteristics of the radar incident wave to time variable and space variable characteristics of echo signal of the radar receiving target, take metallic ball and dihedral angle as radar target for observation, respectively, derives a measurement equation for antenna spatial polarization characteristics, and through sampling and solution of the echo voltage, it is possible to obtain a polarization direction pattern for circular polarization antenna and linear polarization antenna. Measurement equation of antenna spatial polarization characteristics, and through sample processing and calculation of the echo voltage, it is possible to obtain polarization direction pattern for circular antenna and linear polarization antenna. This method wisely makes use of spatial polarization time-variable characteristics during antenna scanning, and no orthogonal polarization antenna and two-way receiver and data acquisition system are required, so that it is possible to reduce quantity of equipment and operation complexity, avoid polarization error of the antenna and amplitude and phase measurement error resulted from inconsistency in channels. So, it has certain inspirational value and application prospect. Section 5.2 studies the model for measurement of antenna polarization characteristics, derives a correction restriction model for measurement of the cross polarization direction pattern, gives optical receive and transmit polarization basis, and carries out outfield measurement experiment for actual radar antenna. A restriction model is used to effectively correct polarization direction pattern measured. Section 5.3 takes measured data in the radar antenna polarization direction pattern as a basis to analyze main difference of measurement results and ideal antenna radiation field, discuss causes of measurement error in antenna phase direction pattern during measurement, and points out that data sampling frequency, signal source frequency shift, and phase center shift are main factors restricting measurement accuracy. By combining measured data, the above-mentioned measurement error is corrected, and the correction results prove correctness of theoretical analysis and effectiveness of correction method, and have certain reference value.

5.1 Antenna Spatial Polarization Characteristics Measurement Method Based on Target Character of Calibrator

5.1.1 Base Transformation Computation of Polarization Scattering Matrix

Radar scattering cross-section is a quantity used to describe target electromagnetic scattering efficiency, which only characterizes radar target scattering amplitude characteristics, lacking of characterization of target characteristics such as polarization and phase characteristics. Polarization scattering matrix (PSM) characterizes all information of the radar target electromagnetic scattering. Usually, scattering matrix is in the form of complex number, and varies with working frequency and target altitude [2]. At any altitude of the target, scattering of different polarization waves differs, enabling polarization of the radiation field to be different from polarization of the incident field. Hence, this Section attempts to use a standard object whose polarization scattering characteristics have been known, and through measuring echo wave reflected from the standard target which is radiated by the radar, obtain polarization characteristics of the radar antenna by inverse method. The calculation is derived as follows:

It has been known that scattering matrix of the standard metallic ball at $(\hat{h}, \hat{v})$ polarization basis is $S_{HV} = \begin{bmatrix} 1 & 0 \\ 0 & 1 \end{bmatrix}$, which is often used for calibration of co-polarization component in engineering [192, 193]. Polarization scattering matrix of 0° dihedral angle is $S_2 = \begin{bmatrix} -1 & 0 \\ 0 & 1 \end{bmatrix}$, after being rotated by 45°, its polarization scattering matrix is $S_3 = \begin{bmatrix} 0 & 1 \\ 1 & 0 \end{bmatrix}$, which can be used for calibration of scattering matrix cross polarization component.

The dihedral angle is formed by folding two metallic apertures, whose reflection component includes primary and secondary refection components of the flat plate. Geometrical relationship between two kinds of calibration objects is shown in following Fig. 5.1.

According to conversion relationship between horizontal and vertical polarization basis $(\hat{h}, \hat{v})$ and left-hand and right-hand circular polarization basis $(\hat{l}, \hat{r})$, it can be known:

$$(\hat{l}, \hat{r}) = (\hat{h}, \hat{v}) \cdot U \tag{5.1.1}$$

where, $U = \frac{1}{\sqrt{2}} \begin{bmatrix} 1 & 1 \\ j & -j \end{bmatrix}$.

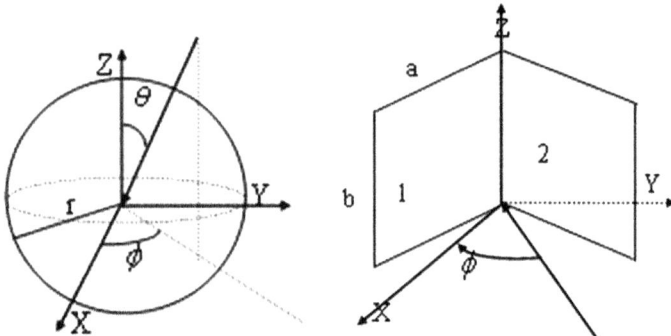

Fig. 5.1 Geometrical structure of metallic ball and rectangular metallic dihedral angle

Voltage equation for polarization basis conversion of the target is as follows:

$$S_V(\hat{l}, \hat{r}) = U^T \cdot S_{HV} \cdot U \tag{5.1.2}$$

The scattering wave from the target is transmitted in left-hand system. To be converted back to right-hand system and ensure the matrix calculation is correct, an additional matrix is introduced as follows:

$$S_V(\hat{l}, \hat{r}) = U^T \cdot \begin{bmatrix} 1 & 0 \\ 0 & -1 \end{bmatrix} \cdot S_{HV} \cdot U \tag{5.1.3}$$

Then, scattering matrix of the metallic ball at left- and right-hand circular polarization basis $(\hat{l}, \hat{r})$ can be written as follows:

$$\begin{aligned}
S_{V1}(\hat{l}, \hat{r}) &= \frac{1}{2} \begin{bmatrix} S_{HH} + jS_{HV} + jS_{VH} - S_{VV} & S_{HH} - jS_{HV} + jS_{VH} + S_{VV} \\ S_{HH} + jS_{HV} - jS_{VH} + S_{VV} & S_{HH} - jS_{HV} - jS_{VH} - S_{VV} \end{bmatrix} \\
&= \begin{bmatrix} 0 & 1 \\ 1 & 0 \end{bmatrix}
\end{aligned} \tag{5.1.4}$$

It has been known that theoretical value of the polarization scattering matrix at $45°$ dihedral angle is $S_2 = \beta \begin{bmatrix} 0 & 1 \\ 1 & 0 \end{bmatrix}$; similarly, after the dihedral angle of the metallic ball is turned $45°$, scattering matrix at $(\hat{l}, \hat{r})$ polarization basis is calculated as follows:

$$\begin{aligned}
S_{V2}(\hat{l}, \hat{r}) &= \frac{1}{2} \begin{bmatrix} S_{HH} + jS_{HV} + jS_{VH} - S_{VV} & S_{HH} - jS_{HV} + jS_{VH} + S_{VV} \\ S_{HH} + jS_{HV} - jS_{VH} + S_{VV} & S_{HH} - jS_{HV} - jS_{VH} - S_{VV} \end{bmatrix} \\
&= \begin{bmatrix} j & 0 \\ 0 & -j \end{bmatrix}
\end{aligned} \tag{5.1.5}$$

As discussed above, theoretical value of the metallic ball's polarization scattering matrix is $S_{V1} = \alpha \begin{bmatrix} 0 & 1 \\ 1 & 0 \end{bmatrix}$, and theoretical value of polarization scattering matrix at 45° dihedral angle is $S_{V2} = \beta \begin{bmatrix} j & 0 \\ 0 & -j \end{bmatrix}$, where α, β is square root value of backward scattering cross-section (RCS) of two calibration objects, respectively.

5.1.2 Measurement Algorithm of Circular Polarized Antenna

Supposing that spatial polarization characteristic is $\boldsymbol{h}(\varphi_i, \theta_k) = \begin{bmatrix} h_R(\varphi_i, \theta_k) \\ h_L(\varphi_i, \theta_k) \end{bmatrix}$ at elevation angle θ_i of circular polarization antenna and azimuth angle φ_i, it can be known from the section above that polarization scattering matrix of standard target at $(\hat{l}, \hat{r})$ polarization basis is $S = \begin{bmatrix} s_{RR} & s_{RL} \\ s_{LR} & s_{LL} \end{bmatrix}$, and where reciprocal conditions are met, $s_{RL} = s_{LR}$, and the target distance is r_0.

When a metallic ball target is placed on a high tower, a right-hand circular polarization continuous signal is used to excite the same target, and the antenna scans at azimuth and receives target echo, so when radar antenna is at a different azimuth angle φ_k, $k = -N, \ldots, 0, \ldots, N$ echo complex voltage sampling value can be represented as follows:

$$
\begin{aligned}
V_1(\theta_i, \varphi_k) &= \boldsymbol{h}^T(\theta_i, \varphi_k) \cdot S_{V1} \cdot \boldsymbol{h}(\theta_i, \varphi_k) = A \cdot \frac{\alpha \cdot e^{-j2kr_0}}{4\pi r_0^2} \cdot \begin{bmatrix} h_R & h_L \end{bmatrix} \begin{bmatrix} 0 & 1 \\ 1 & 0 \end{bmatrix} \begin{bmatrix} h_R \\ h_L \end{bmatrix} \\
&= A \cdot \frac{\alpha \cdot e^{-j2kr_0}}{4\pi r_0^2} \cdot 2h_R(\theta_i, \varphi_k) h_L(\theta_i, \varphi_k)
\end{aligned}
$$

$$(5.1.6)$$

where, A is a coefficient decided by parameters of individual elements in radar equation (including: maximum gain of the transmit antenna, range, transmitting power, transmitting loss, transmission loss).

Theoretically, if inner production $\langle h_R, h_L \rangle = 0$ of orthogonal polarization component of the antenna, then $V_1(\theta_i, \varphi_k) = 0$. This shows circular polarization signal will produce circular polarization echo wave after being scattered by the metallic ball, so that antenna's receive voltage is 0. This corresponds to principle of removing rain clutter by use of circular polarization signal in meteorological radar. However, no absolute ideal circular polarization exists virtually, and as observation azimuth (i.e. scanning azimuth angle) increases, antenna's orthogonal polarization component increases and depolarization effect also increases significantly, as well as electromagnetic will deviate circular polarization, showing certain elliptical polarization, which is discussed in Refs. [58–60], either. Hence, echo wave signal can be collected.

Similarly, when 45° dihedral angle is place on the high tower, the same dihedral angle is excited by use of left-hand circular polarization continuous wave signal, and the antenna scans at azimuth to receive target echo wave; so, $k = -N, \ldots 0, \ldots, N$ echo voltage sampling value can be represented as follows at different azimuth angle φ_k of the radar antenna:

$$
\begin{aligned}
V_2(\theta_i, \varphi_k) &= \boldsymbol{h}^T(\theta_i, \varphi_k) \cdot \boldsymbol{S}_{V2} \cdot \boldsymbol{h}(\theta_i, \varphi_k) = A \cdot \frac{\beta \cdot e^{-j2kr_0}}{4\pi r_0^2} \cdot [h_R \quad h_L] \begin{bmatrix} j & 0 \\ 0 & -j \end{bmatrix} \begin{bmatrix} h_R \\ h_L \end{bmatrix} \\
&= A \cdot \frac{\beta \cdot e^{-j2kr_0}}{4\pi r_0^2} \cdot \left[j \cdot h_R^2(\theta_i, \varphi_k) - j \cdot h_L^2(\theta_i, \varphi_k) \right]
\end{aligned}
$$

$$(5.1.7)$$

Because cross polarization component of the antenna at major axis direction of the direction pattern is very low ($\leq -20\,\text{dB}$) and can be ignored, voltage sampling value of the standard target is as follows where $\varphi_0 = 0°$:

$$
V_{2,\max}(\theta_i, \varphi_0) = A \cdot \frac{\beta \cdot e^{-j2kr_0}}{4\pi r_0^2} \cdot j \cdot h_R^2(\theta_i, \varphi_k) \tag{5.1.8}
$$

Through normalization of complex voltage sampling sequence (5.1.6) and (5.1.7) sampled by the radar receiver for two times by Eq. (5.1.8), it is possible to obtain:

$$
V_{2,norm}(\theta_i, \varphi_k) = j \cdot h_{R,norm}^2(\theta_i, \varphi_k) - j \cdot h_{L,norm}^2(\theta_i, \varphi_k) \tag{5.1.9}
$$

$$
V_{1,norm}(\theta_i, \varphi_k) = \frac{\alpha}{\beta} 2h_R(\theta_i, \varphi_k)h_L(\theta_i, \varphi_k) \tag{5.1.10}
$$

To simplify calculation and operate easily, it is possible to suppose that scattering cross section of the metallic ball equals to metallic dihedral angle, then

$$
V_{1,norm}(\theta_i, \varphi_k) = 2h_R(\theta_i, \varphi_k)h_L(\theta_i, \varphi_k) \tag{5.1.11}
$$

By combining Eqs. (5.1.9) and (5.1.10) and after fuzzy processing, it is possible to solve normalized spatial polarization characteristics of the circular polarization antenna:

$$
h_{R,norm}^2(\theta_i, \varphi_k) = \frac{-j \cdot V_2(\theta_i, \varphi_k) - \sqrt{V_2^2(\theta_i, \varphi_k) - V_1^2(\theta_i, \varphi_k)}}{2j} \tag{5.1.12}
$$

$$
h_{L,norm}^2(\theta_i, \varphi_k) = \frac{-V_2(\theta_i, \varphi_k) + \sqrt{V_2^2(\theta_i, \varphi_k) - V_1^2(\theta_i, \varphi_k)}}{2j} \tag{5.1.13}
$$

The circular polarization antenna is a special example for elliptical polarization antenna. Through derivation, it is found that the above-mentioned processing method is also applicable to characteristics measurement of the elliptical polarization antenna, and is not described any more.

5.1.3 Measurement Algorithm of Linear Polarization Antenna

5.1.3.1 Measurement Algorithm Combining Metallic Ball and 0° Dihedral Angle

Supposing that spatial polarization characteristics of the radar antenna at elevation angle φ_i and azimuth angle θ_i is $\boldsymbol{h}(\theta_i, \varphi_k) = \begin{bmatrix} h_H(\theta_i, \varphi_k) \\ h_V(\theta_i, \varphi_k) \end{bmatrix}$, polarization scattering matrix of standard target at $(\hat{h}, \hat{v})$ polarization basis is $\boldsymbol{S} = \begin{bmatrix} s_{HH} & s_{HV} \\ s_{VH} & s_{VV} \end{bmatrix}$. Where reciprocal conditions are met [10], $s_{HV} = s_{VH}$, target distance is r_o, and complex voltage measurement of receive antenna induction current is as follows:

$$v(\theta_i, \varphi_k) = A \cdot \frac{e^{-j2kr_0}}{4\pi r_0^2} \cdot \boldsymbol{h}^T(\theta_i, \varphi_k) \cdot \boldsymbol{S} \cdot \boldsymbol{h}(\theta_i, \varphi_k) \qquad (5.1.14)$$

By substituting parameters, a specific expression is obtained:

$$v(\theta_i, \varphi_k) = A \cdot \frac{e^{-j2kr_0}}{4\pi r_0^2} \cdot \left[s_{HH} h_H^2(\theta_i, \varphi_k) + 2s_{HV} h_H(\theta_i, \varphi_k) h_V(\theta_i, \varphi_k) \right.$$
$$\left. + s_{VV} h_V^2(\theta_i, \varphi_k) \right] \qquad (5.1.15)$$

where, A is gain coefficient determined by parameters of individual elements in radar equation (including: maximum gain of the transmit antenna, range, transmitting power, transmitting loss, transmission loss).

Supposing that polarization scattering matrix of the metallic ball is $S_1 = \alpha \begin{bmatrix} 1 & 0 \\ 0 & 1 \end{bmatrix}$, polarization scattering matrix of 0° dihedral angle $S_2 = \beta \begin{bmatrix} -1 & 0 \\ 0 & 1 \end{bmatrix}$, and measuring range is r_0.

Voltage sequence sampling value of the target echo is represented as follows at different elevation angle θ_i and azimuth angle $\varphi_k(k = -N, \ldots, 0, \ldots, N)$ of the antenna to be measured:

$$v_1(\theta_i, \varphi_k) = \boldsymbol{h}^T(\theta_i, \varphi_k) \cdot \boldsymbol{S}_1 \cdot \boldsymbol{h}(\theta_i, \varphi_k) = A \cdot \frac{\alpha \cdot e^{-j2kr_0}}{4\pi r_0^2} \cdot \left[h_H^2(\theta_i, \varphi_k) + h_V^2(\theta_i, \varphi_k)\right]$$

(5.1.16)

Particularly, when $\theta_0 = 0°$ in electric axis direction, co-polarization component is much larger than cross polarization (more than 20 dB, it is here supposed that co-polarization of the antenna is horizontal polarization), so influence of cross polarization component can be ignored in this direction. The measurement result obtained through the equation above can be written as follows:

$$v_1(\theta_i, \varphi_0) \approx A \cdot \frac{\alpha \cdot e^{-j2kr_0}}{4\pi r_0^2} \cdot h_H^2(\theta_i, \varphi_0)$$

(5.1.17)

Measurement result of Eq. (5.1.16) obtained from Eq. (5.1.17) is normalized, so as to eliminate influence of unknown parameters.

$$\begin{aligned} v_{1,norm}(\theta_i, \varphi_k) &= \frac{v_1(\theta_i, \varphi_k)}{v_1(\theta_i, \varphi_0)} = \left(\frac{h_H(\theta_i, \varphi_k)}{h_H(\theta_i, \varphi_0)}\right)^2 + \left(\frac{h_V(\theta_i, \varphi_k)}{h_H(\theta_i, \varphi_0)}\right)^2 \\ &= h_{H,\mathrm{norm}}^2(\theta_i, \varphi_k) + h_{V,\mathrm{norm}}^2(\theta_i, \varphi_k) \end{aligned}$$

(5.1.18)

where, $h_{H,\mathrm{norm}}(\theta_i, \varphi_k) = \frac{h_H(\theta_i, \varphi_k)}{h_H(\theta_i, \varphi_0)}$, $h_{V,\mathrm{norm}}(\theta_i, \varphi_k) = \frac{h_V(\theta_i, \varphi_k)}{h_H(\theta_i, \varphi_0)}$.

In next step, the radar radiation target is replaced by $0°$ dihedral angle. By repeating the above-mentioned process, the receive antenna measures voltage sampling value at different elevation angle θ_i and azimuth scanning angle φ_k $(k = -N, \ldots, 0, \ldots, N)$, which is represented as follows:

$$v_2(\theta_i, \varphi_k) = A \cdot \frac{\beta \cdot e^{-j2kr_0}}{4\pi r_0^2} \cdot \left[-h_H^2(\theta_i, \varphi_k) + h_V^2(\theta_i, \varphi_k)\right]$$

(5.1.19)

Similarly, cross polarization component is ignored in direction of $\varphi_k = 0$, and measuring result is simplified as follows:

$$v_2(\theta_i, \varphi_0) = -A \cdot \frac{\beta \cdot e^{-j2kr_0}}{4\pi r_0^2} \cdot h_H^2(\theta_i, \varphi_0)$$

(5.1.20)

Through normalizing measurement value of Eq. (5.1.19) by use of Eq. (5.1.20), obtain following equation:

$$v_{2,\mathrm{norm}}(\theta_i, \varphi_k) = h_{H,\mathrm{norm}}^2(\theta_i, \varphi_k) - h_{V,\mathrm{norm}}^2(\theta_i, \varphi_k)$$

(5.1.21)

By combining Eqs. (5.1.18) and (5.1.21), it can be derived that normalized spatial polarization characteristic of the radar antenna is as follows at angle of (θ_i, φ_k):

$$h^2_{H,\text{norm}}(\theta_i, \varphi_k) = \frac{v_{1,\text{norm}}(\theta_i, \varphi_k) + v_{2,\text{norm}}(\theta_i, \varphi_k)}{2} \tag{5.1.22}$$

$$h^2_{V,\text{norm}}(\theta_i, \varphi_k) = \frac{v_{1,\text{norm}}(\theta_i, \varphi_k) - v_{2,\text{norm}}(\theta_i, \varphi_k)}{2} \tag{5.1.23}$$

Because theoretical value of the metallic ball RCS, and standard target range r_0 are known quantities, it is possible to solve co-polarization gain of the antenna in electric axis direction, i.e. at angle (θ_i, φ_0):

$$A \cdot h^2_H(\theta_i, \varphi_0) = B \cdot v_1(\theta_i, \varphi_0) \tag{5.1.24}$$

where, B is a gain compensation factor.

Similarly, polarization characteristic is measured in azimuth direction by changing elevation angle θ_i and repeating the above-mentioned step, so spatial polarization characteristics of the radar antenna are obtained at different elevation angle $\theta_{\min} \leq \theta_i \leq \theta_{\max}$.

Block diagram for calculation steps is shown in Fig. 5.2.

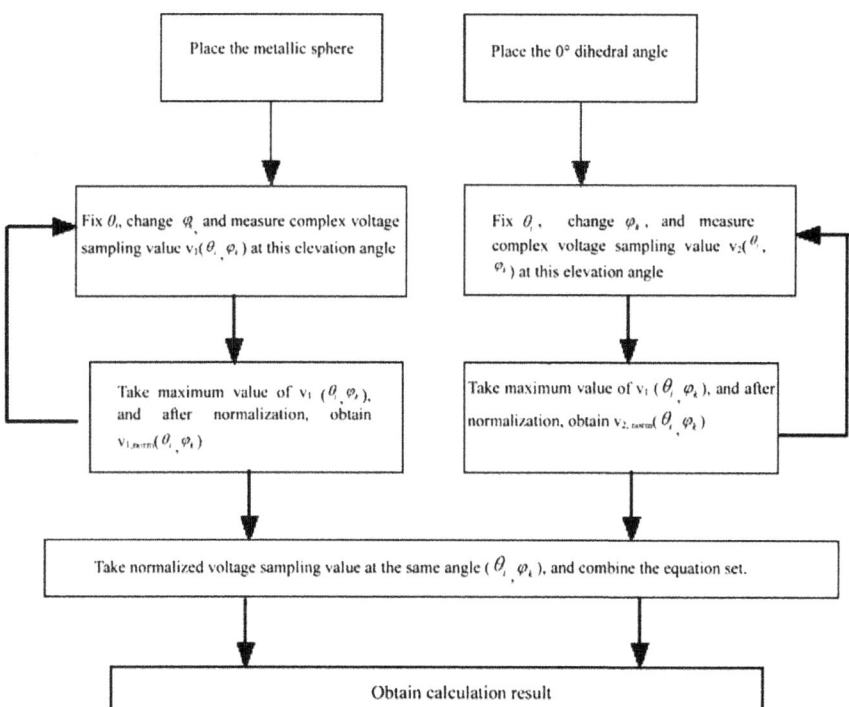

Fig. 5.2 Block diagram for passive measurement steps of antenna spatial polarization characteristics

5.1.3.2 Measurement Algorithm Combining 0°/45° Dihedral Angles

It has been known that theoretical value of polarization scattering matrix of 0° dihedral angle is $S_1 = \beta \begin{bmatrix} -1 & 0 \\ 0 & 1 \end{bmatrix}$, and theoretical value of polarization scattering matrix of 45° dihedral angle is $S_2 = \beta \begin{bmatrix} 0 & 1 \\ 1 & 0 \end{bmatrix}$, visible range from antenna to be measured to standard object is r_0.

Step 1: Place the 0° dihedral angle, fix the elevation angle θ_i, and measure echo voltage sampling value at different azimuth angle φ_k, with $k = -N, \ldots 0, \ldots, N$.

$$
\begin{aligned}
v_1(\theta_i, \varphi_k) &= \boldsymbol{h}^T(\theta_i, \varphi_k) \cdot \boldsymbol{S}_1 \cdot \boldsymbol{h}(\theta_i, \varphi_k) \\
&= A \cdot \frac{\beta \cdot e^{-j2kr_0}}{4\pi r_0^2} \cdot \left[-h_H^2(\theta_i, \varphi_k) + h_V^2(\theta_i, \varphi_k) \right]
\end{aligned}
\tag{5.1.25}
$$

Similarly, supposing that cross polarization component of the antenna in major axis direction of the direction pattern is very low and can be ignored, then voltage sampling value of the standard target in direction of $\varphi_0 = 0°$ is as follows:

$$
v_1(\theta_i, \varphi_0) = -A \cdot \frac{\beta \cdot e^{-j2kr_0}}{4\pi r_0^2} h_H^2(\theta_i, \varphi_0)
\tag{5.1.26}
$$

Complex voltage sampling sequence is normalized by this quantity to obtain:

$$
v_{1,\text{norm}}(\theta_i, \varphi_k) = h_{H,\text{norm}}^2(\theta_i, \varphi_k) - h_{V,\text{norm}}^2(\theta_i, \varphi_k)
\tag{5.1.27}
$$

Step 2: Place the 45° dihedral angle, fix the elevation angle θ_i, and measure echo voltage sampling value at different azimuth angle φ_k, with $k = -N, \ldots 0, \ldots, N$.

$$
v_2(\theta_i, \varphi_k) = \boldsymbol{h}^T(\theta_i, \varphi_k) \cdot \boldsymbol{S}_2 \cdot \boldsymbol{h}(\theta_i, \varphi_k) = A \cdot \frac{\beta \cdot e^{-j2kr_0}}{4\pi r_0^2} \cdot \left[2h_H(\theta_i, \varphi_k)h_V(\theta_i, \varphi_k) \right]
\tag{5.1.28}
$$

The Eq. (3.2.28) is normalized by Eq. (3.2.26) to obtain:

$$
v_{2,\text{norm}}(\theta_i, \varphi_k) = -2h_{H,\text{norm}}(\theta_i, \varphi_k) \cdot h_{V,\text{norm}}(\theta_i, \varphi_k)
\tag{5.1.29}
$$

Step 3: By combining Eqs. (5.1.26) and (5.1.29) and after fuzzy processing, normalized spatial polarization characteristic of the antenna is obtained:

$$
h_{H,\text{norm}}^2(\theta_i, \varphi_k) = \frac{v_1(\theta_i, \varphi_k) + \sqrt{v_1^2(\theta_i, \varphi_k) + v_2^2(\theta_i, \varphi_k)}}{2}
\tag{5.1.30}
$$

$$h_{V,\text{norm}}^2(\theta_i, \varphi_k) = \frac{-v_1(\theta_i, \varphi_k) + \sqrt{v_1^2(\theta_i, \varphi_k) + v_2^2(\theta_i, \varphi_k)}}{2} \qquad (5.1.31)$$

Step 4: If RCS and distance of the dihedral angle have been known, gain of the co-polarization component in electric axis direction (i.e. (θ_i, φ_0) angle) can be obtained by Eq. 5.1.26, i.e.:

$$A \cdot h_H^2(\theta_i, \varphi_0) = B \cdot v_1(\theta_i, \varphi_0) \qquad (5.1.32)$$

Step 5: Within wide elevation angle $\theta_{\min} \leq \theta_i \leq \theta_{\max}$, total measurement result for spatial polarization characteristics of the radar antenna is obtained by changing the antenna's elevation angle and repeating step 1–4 (Fig. 5.3).

5.1.4 Outfield Experiment Design

(1) Outfield experiment layout

The experiment layout is mainly composed of flat test field and calibration test tower where the radar to be measured is arranged at an open position, measuring

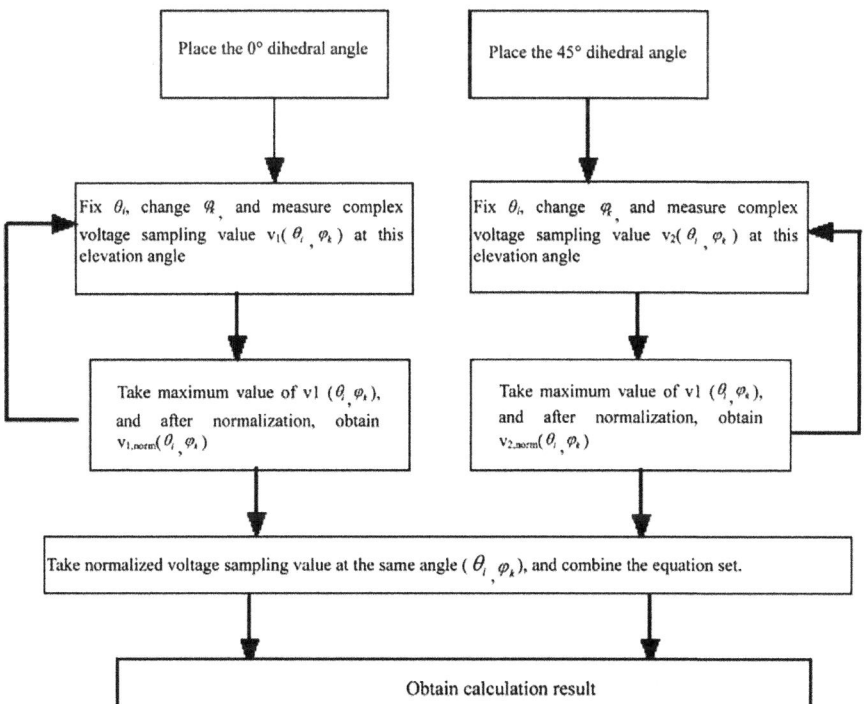

Fig. 5.3 Block diagram for passive measurement of antenna spatial polarization characteristics

distance l from the calibration test tower meets far field equation ($l \geq \frac{2D^2}{\lambda}$, D is longitudinal distance of the antenna aperture, and λ is length of working wave), calibration object is placed at top of the test tower, and altitude angle of the calibration object can be controlled accurately. The experiment layout is shown in Fig. 5.4. The radar antenna scans and radiates the target in a circumferential manner in azimuth direction, and high-speed data acquisition of the echo signal is carried out by use of data acquisition system.

Another kind of experiment layout can be considered, either, as shown in Fig. 5.5. Supposing that height (equivalent to the radar vehicle height plus the antenna serve seat) of the radar antenna (which is usually placed on the radar vehicle) is h_1, and signal source is connected to standard transmit antenna and is beside the radar antenna, with erection height being basically identical to the radar antenna.

In the experiment, the standard signal source transmits continuous wave signal, with frequency being consistent with working frequency of the radar. The transmit antenna is aimed at the target for continuous radiation. Meanwhile, the radar transmitter is stopped, and works in receiving condition only. The antenna conducts circumferential scanning in azimuth direction. Then, by stopping automatic gain control of the radar, I, Q signal of the intermediate frequency output of the radar are connected with the data acquisition system to acquire intermediate frequency after down conversion and I, Q processing. In the experiment, cascading relationship of individual systems is shown in Fig. 5.6.

(2) Calculation and analysis of outfield experiment conditions

According to outfield experiment layout shown in Sect. 5.1.4, it can be supposed that height of the tower is h_2. The standard target is placed at top of the tower; at this moment, elevation angle of the radar antenna is θ, distance from the radar to the tower is d, main lobe width of the antenna at zero point is B_w, maximum size of the antenna aperture is D_1 and D_2, respectively, and working wave length of the radar is λ, as well as k is a constant chosen randomly.

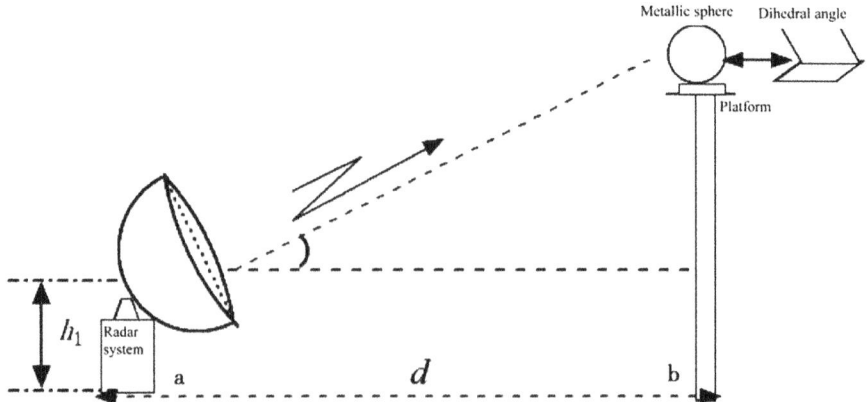

Fig. 5.4 Experiment layout

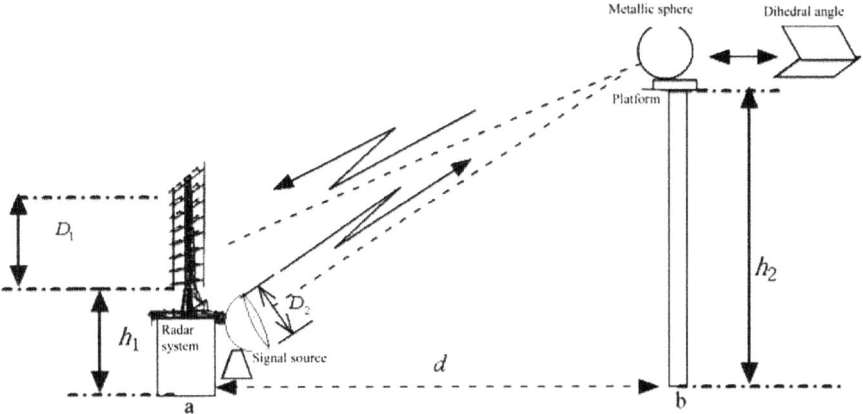

Fig. 5.5 Experiment layout

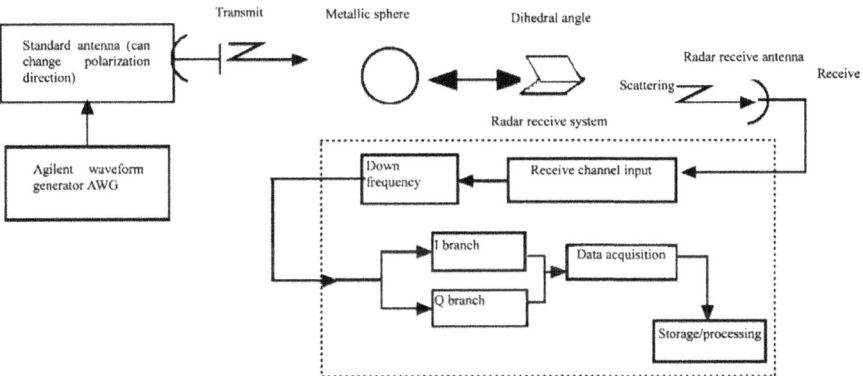

Fig. 5.6 Diagram for connection of individual systems in the experiment

- To meet far field conditions in experiment, distance between the high tower and antenna to be measured should meet:

$$d \geq k \frac{(D_1 + D_2)^2}{\lambda} \qquad (5.1.33)$$

Usually, if it is allowed that phase shift at aperture of the antenna is $\pi/8$, k is 2, and relative error in field strength tested is 2.5%; the bigger k value, the higher the phase measurement accuracy is; if this condition is met, influence of induction field can be ignored.

- To avoid multi-path effect, elevation angle θ of the radar antenna must meet:

$$\theta \geq B_w/2 + \arctan(h_1/d) \tag{5.1.34}$$

- When direction of the electric axis is aligned with phase center where the calibration body is located, elevation angle of the antenna should meet following conditions:

$$\theta_1 = \arctan\left(\frac{h_2 - h_1}{d}\right) \quad \text{is} \quad h_2 = d \cdot \tan \theta_1 + h_1 \tag{5.1.35}$$

- Considering spatial polarization characteristics of azimuth scanning and elevation scanning within main lobe area of the antenna measured, elevation angle of the antenna should be restricted to following measuring zone:

$$\arctan\left(\frac{h_2 - h_1}{d}\right) \leq \theta < \frac{B_w}{2} + \arctan\left(\frac{h_2 - h_1}{d}\right) \tag{5.1.36}$$

As discussed above, it is possible to derive value scope of the distance between the antenna and tower to enable the antenna to meet far field conditions, and meet measurement scope in elevation direction of the main lobe and phase measurement accuracy of electric field, as well as ensure that main lobe radiation of the antenna does not cause multi-path effect.

$$k\frac{(D_1 + D_2)'}{\lambda} \leq d < \frac{h_2 - h_1}{\tan \theta} \tag{5.1.37}$$

At this moment, θ value should be $\theta = B_w/2 + \arctan(h_1/d)$.

The antenna rack is 3 m high, the tower is 120 m high, the antenna aperture size is 4 m, the signal source aperture size is 0.5 m, and the antenna beam width is $4° \times 4°$. According to above-mentioned measurement conditions, it is possible to derive variation curve for scope of distance from the antenna to the tower with respect to radar frequency, as shown in Figs. 5.7 and 5.8. It can be seen that the maximum distance from the antenna to the tower is not more than 3150 m. Meanwhile, it is possible to derive elevation variation scope of the antenna when aiming at the target in this beam width condition, and variation scope of the maximum elevation angle when measuring spatial polarization characteristics within the main lobe, and the minimum elevation angle of the antenna for the purpose of avoiding multi-path effect.

Azimuth sampling interval is one of factors influencing measurement performance of the antenna. As shown in Fig. 5.9, scanning period of the radar antenna can be interfered manually, i.e. conducted at slowest speed; meanwhile, pulse repetition frequency of the capture card is increased as far as possible, so as to ensure that azimuth sampling interval should be as small as possible.

Fig. 5.7 Curve of distance equivalence derivation for outfield measurement layout

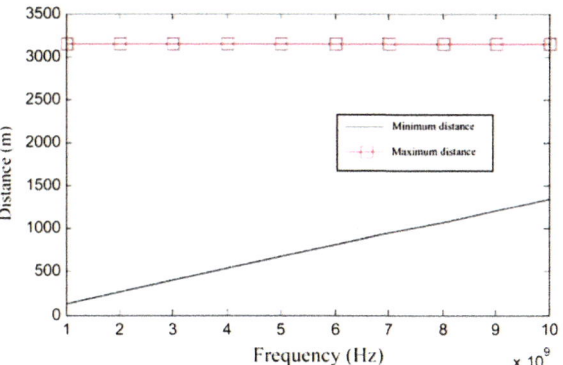

Fig. 5.8 Value range of antenna elevation angle for outfield measurement layout

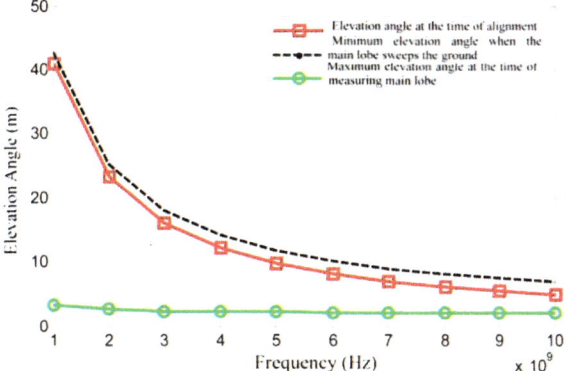

5.1.5 Antenna Property Measurement Experiment and Analysis

(1) Measurement experiment based on radar antenna data

To verify effectiveness of the algorithm, this Section simulates target echo by use of signal source radiation signal received during mechanical scanning in azimuth direction for the array antenna of a radar system, so as to obtain normalized co-polarization and cross polarization direction pattern of the antenna by use of the algorithm proposed when the antenna works at center frequency.

Figure 5.10 gives distribution of echo signal complex voltage when the radar antenna sweeps the metallic sphere and dihedral target respectively. It can be seen that when main lobe of the antenna aims at the calibration body and receives target echo signal, the signal is strong, and there is obvious echo signal envelope. Because minor lobe of the antenna is very low, the echo wave is very weak at other scanning angles.

Figure 5.11 shows a normalized direction pattern of the antenna, and Fig. 5.12 shows antenna polarization direction pattern obtained by proposed method

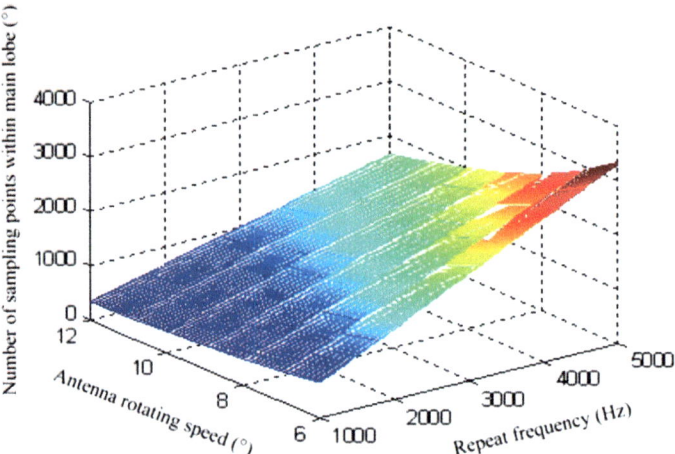

Fig. 5.9 Relationship between sampling frequency (PRF), antenna rotating speed and azimuth sampling interval

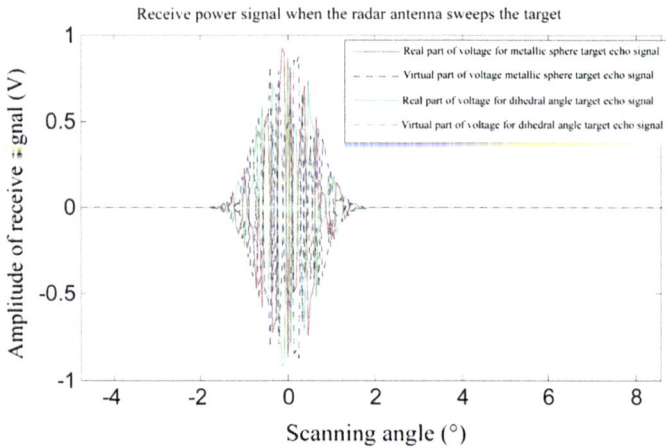

Fig. 5.10 Voltage signal received when the radar antenna sweeps calibration body

discussed in this section. In the figure, the horizontal ordinate is azimuth angle, and the vertical coordinate represents antenna gain in dB. It can be seen from the figure that the measured direction pattern well coincides with estimated direction pattern, being able to accurately estimate real spatial polarization characteristics of the antenna, stating that estimate performance of the proposed method is excellent, and inversion results proves that the antenna is a narrow beam antenna. In beam center direction, cross polarization component of the antenna remains high, and cross polarization discrimination reaches about −30 dB.

Fig. 5.11 Actual antenna spatial polarization characteristics

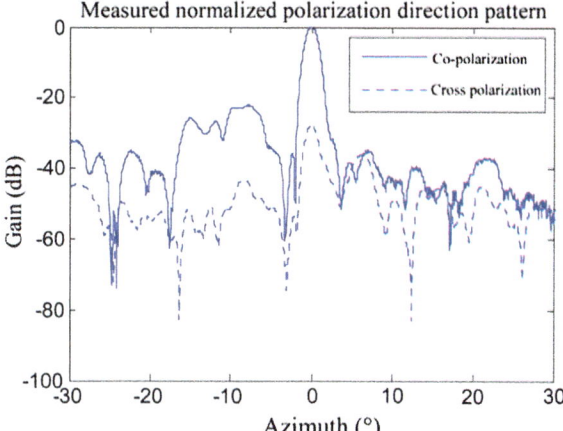

Fig. 5.12 Estimate results of antenna spatial polarization characteristics

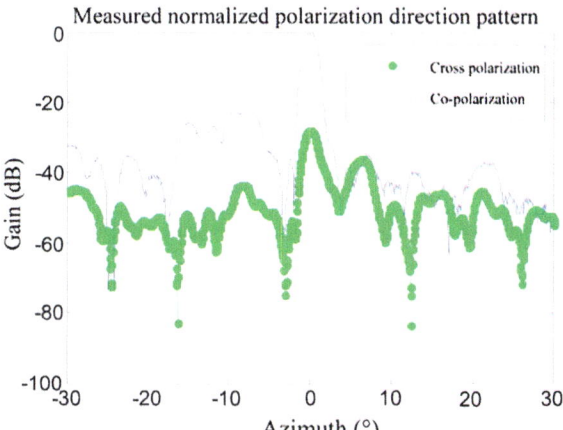

(2) Measurement experiment based on microwave dark room measurement data

In the light of dark room measurement data of a real C jammer antenna in C wave band, a target echo is simulated. The dark room measurement mode is as follows: ① Azimuth direction scanning range: −60° to +60°, scanning interval of 0.5°; ② Elevation direction scanning range: −45°to +45°, scanning interval of 5°; ③ Working frequency range: 3.9–6.2 GHz.

Figure 5.13 gives distribution of echo wave signal complex voltage when the radar antenna sweeps the metallic sphere and dihedral angle target, respectively. Figure 5.14 gives one-dimension azimuth scanning polarization direction pattern of the antenna estimated by proposed algorithm. At this moment, the antenna is working at center frequency of $f = 5.05$ GHz $\pm$ 12.2 MHz, and there is certain gap

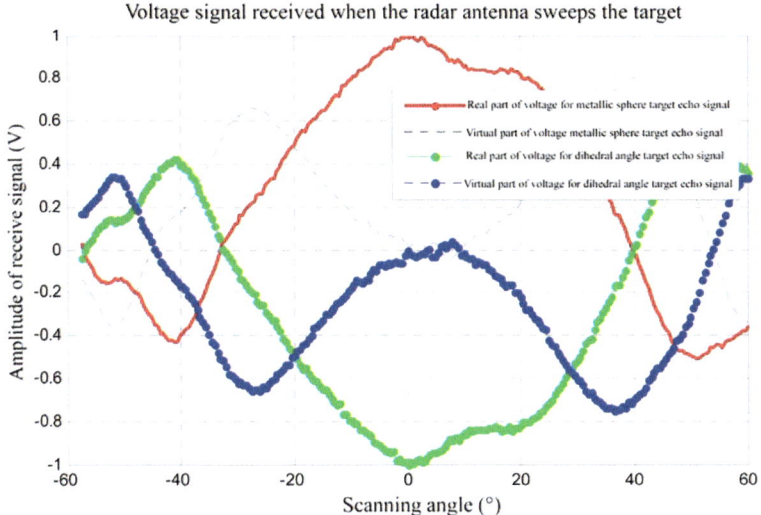

Fig. 5.13 Voltage signal received when radar antenna sweep the target

between co-polarization and cross polarization direction pattern of the antenna estimated before gain compensation and real polarization characteristics. Figure 5.15 gives antenna polarization direction pattern estimated after gain compensation, the horizontal coordinate is azimuth angle, the vertical coordinate represents voltage amplitude of the antenna, and two curves are coincided with each

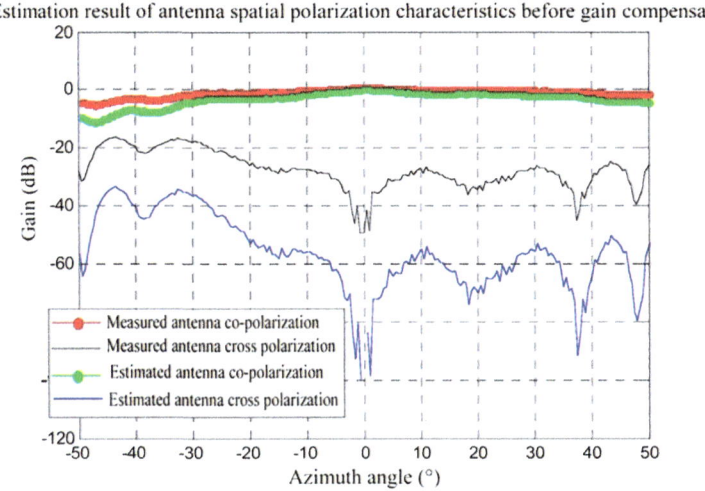

Fig. 5.14 Estimation results for antenna spatial polarization characteristics before gain compensation

other well, being able to accurately estimate real spatial polarization characteristics of the antenna. Meanwhile, inversion results prove that the jammer antenna is a wide beam antenna. When the antenna sweeps in azimuth and elevation direction, the polarization characteristics vary according to certain law. At zero-point position of the direction patter, as the antenna polarization purity increases, the cross polarization discrimination is lowered to −50 dB. It can be seen from Fig. 5.16 that as direction of the antenna beam is deviated from the center position, its cross polarization component increases gradually, and the polarization purity decreases. When the antenna is placed horizontally, and elevation angle is 5°, cross polarization discrimination variation curve in relation to azimuth differs slightly, stating smaller elevation measurement error does not influence measurement of the cross polarization component, so alignment error in vertical (V) polarization basis of two antennas does not bring significant influence on measurement data of the azimuth scanning polarization characteristics. It is noted that, due to different antenna polarization direction and antenna type, different law in degree of sensitivity of the polarization purity with respect to azimuth and elevation is different. This will be considered during actual measurement.

In the light of reliability and credibility of measurement data, the method proposed in this section requires that polarization scattering matrix of the target is consistent with theoretical value at observation altitude of the radar; however, change in actual PSM in relation to its altitude is sensitive very much. Use of algorithm mentioned in Sect. 5.1.3.1 can reduce polarization characteristic measurement error resulted from altitude control error. But, the calibration body is placed a relatively large 3D spatial area directed by the beam, so it is difficult to align the antenna center with geometrical center of the standard target, which will result in certain measurement error. Therefore, the next study is to reduce the error

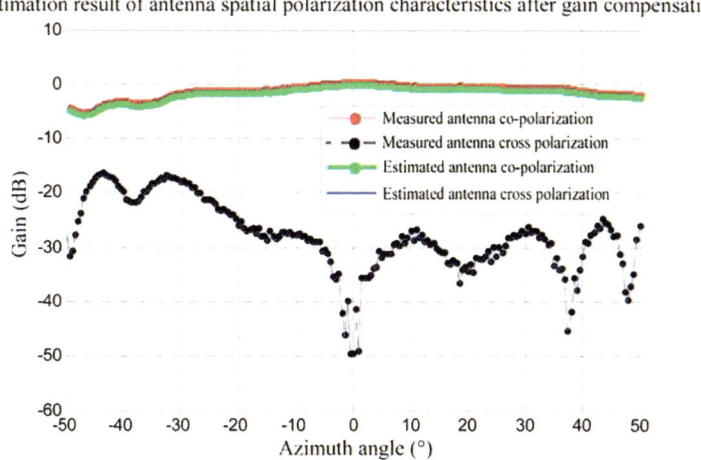

Fig. 5.15 Estimation result of antenna spatial polarization characteristics after gain compensation

Fig. 5.16 Spatial distribution chart for real measured jammer antenna polarization purity

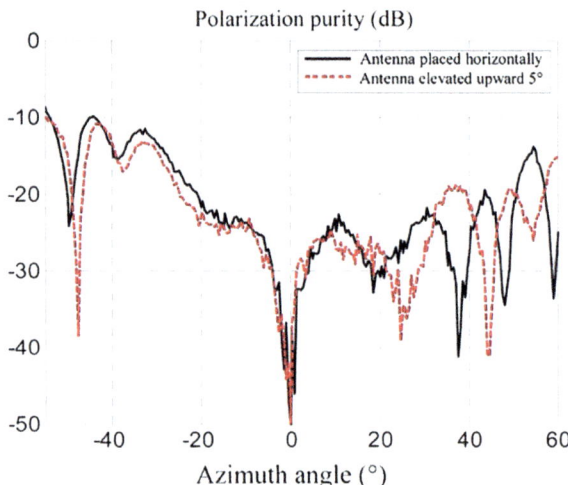

at most and propose an improvement method. In spite of this, this method is close to actual working condition of the radar antenna, and more illustrative to reflect direction characteristics of transmitting/receiving beam by the antenna, which has certain practical significance.

During measurement of antenna polarization characteristics, analysis on possible errors is more complicated than the polarization problem, including: (1) polarization error of the reference antenna; (2) amplitude measurement error; (3) phase measurement error; (4) measurement error resulted from jamming signal; (5) measurement error in polarization characteristics due to failure to align the antenna phase center, and align the polarization basis; (6) Measurement error resulted from multi-path produced by minor radiation. Sections 5.2 and 5.3 will focus on antenna polarization characteristic measurement data and correction technologies, and give practical handling results.

5.2 Polarization Base Mismatch Effect and Calibration in Antenna Property Measurement

Polarization characteristics of the antenna are important parameters which are used to describe the antenna performance, and are basic problem in radar target characteristics measurement field. With respect to mechanical scanning antenna and active phased array antenna, outfield measurement of its polarization characteristics is a complicated and tremendous job. With respect to radar, the antenna can be considered as space polarization filter, which can be described and defined completely in five definition fields, such as space, frequency, amplitude, phase and polarization [34, 35].

Usually, the antenna can be represented by a second-order complex vector, whose polarization characteristics usually can be used to solve amplitude and phase response at its azimuth and frequency through "receiving signal based on two orthogonal polarization basis". "Nominal" polarization of the antenna, i.e. interesting and useful polarization, is often called "co-polarization", and polarization component which is not required during orthogonal intersection with co-polarization is called "cross polarization". If co-polarization and cross polarization direction pattern is measured within designated interested spatial, it is practically effective polarization vector of the antenna in individual directions, so it is possible to propose several rules to evaluate cross polarization characteristics of the "antenna to be measured", and determine polarization basis, correspondingly. In practical application, outfield measurement can be closest to actual working condition of the "antenna to measured", and more illustrative to reflect polarization characteristics of the antenna under direction of the fixed beam, so it is the most effective means and approach to balance the antenna characteristics and describe antenna spatial polarization characteristics. But, certain deviation angle may exist because polarization basis of the "inspection antenna" is not completely aligned with two groups of orthogonal polarization basis where the "antenna to be measured" is situated, enabling energy projection of the co-polarization component at the deviation angle to be leaked to the cross polarization component, resulting in certain measurement error.

This section first builds an antenna polarization characteristics measurement model [153], derives two restriction conditions for cross polarization measurement correction; first: enable "co-polarization" voltage (or power) at measured antenna end to reach maximum optimum polarization basis; second, enable cross polarization receiving power of the measured antenna at place of the maximum co-polarization to reach maximum optimum polarization basis with minimum value or partial small value. Meanwhile, cross polarization characteristics should also be symmetrical for the antenna that is symmetrical in zero azimuth direction. This section focuses on outfield measurement experiment for actual radar antenna, verifies the above-mentioned conclusion by use of measurement handling results, correct polarization direction pattern by use of restriction model, and obtains polarization direction pattern of the radar antenna at optimum receive/transmit polarization basis.

5.2.1 Radar Antenna Spatial Polarization Characteristics Measurement Model

Supposing that the radar antenna scans in azimuth direction at rotating speed of ω_s, sampled pulse repeat frequency of the continuous wave signal is f_r, and radar azimuth sampling interval of the target is $\Delta\varphi_s$ (where $\Delta\varphi_s = \omega_s/f_r$). After "inspection antenna" is connected to the signal source and transmit signal is sampled, receiving voltage sequence $v_{\varphi_i}(t)$ is obtained.

Supposing that "antenna to be measured" peak gain is G_r, co-polarization is horizontal polarization (H), and cross polarization is vertical polarization (V), there are $2M + 1$ azimuth sampling points in total within azimuth scanning spatial $[-\varphi_{\max}, \varphi_{\max}]$, which are $\varphi_{-M}, \ldots, \varphi_0, \ldots, \varphi_M$, respectively. Supposing that co-polarization normalization direction pattern sample of the antenna is $g_H(\varphi_i)$ and cross polarization normalization direction pattern sample of the antenna is $g_V(\varphi_i)$, the antenna spatial polarization vector is written as follows:

$$\boldsymbol{g}(\varphi_i) = G_r \cdot \begin{bmatrix} g_H(\varphi_i) \\ g_V(\varphi_i) \end{bmatrix}, \quad i = -M, \ldots, 0, \ldots, M \tag{5.2.1}$$

For the reason that "inspection antenna" polarization state keeps unchanged during scanning period of the radar antenna, $\boldsymbol{S} = \begin{bmatrix} s_H & s_V \end{bmatrix}^T$, and $\|\boldsymbol{S}\|^2 = 1$; supposing that the inspection antenna gain is G_t, transmit signal power is P_s and signal model is $s(t)$, signal received at each spatial scanning angle φ_i is down-converted to obtain following intermediate frequency signal:

$$\begin{aligned} v_{\varphi_i}(t) &= m_A \big[G_t \cdot P_S \cdot G_r \cdot \boldsymbol{g}^T(\varphi_i) \cdot \boldsymbol{S} \big] \cdot s(t) + n_{\varphi_i}(t) \\ &= m_A \big[G_t \cdot P_S \cdot G_r \cdot (g_H(\varphi_i) \cdot s_H + g_V(\varphi_i) \cdot s_V) \big] \exp(j2\pi f_0 t + j2\pi f_0/f_r + \phi_0) \\ &\quad + n_{\varphi_i}(t) \end{aligned}$$

$$\tag{5.2.2}$$

Where f_0 is intermediate frequency, f_r is repeat frequency sampled by sampling equipment for intermediate frequency signal, ϕ_0 is initial phase of the transmit signal, $n_{\varphi_i}(t)$ is receiver's channel noise, the variance is δ_n^2, and peak signal/noise ratio is defined as $\mathrm{SNR}_{\max} = \frac{G_r \cdot P_S}{\delta_n^2}$, $m_A = \frac{\lambda^2}{(4\pi R)^2 \cdot 10^L}$ is a loss coefficient that takes account of such factors as electromagnetic wave space propagation and loss of measurement system antenna feed line and devices [195]. When receiving horizontal polarization and vertical polarization signal radiated by the signal source, respectively, this parameter is kept the same, and can be ignored in following analysis.

During sampling of the receive signal, supposing that number of sampling points is N, voltage sampling sequence obtained is $v_{\varphi_i}(k)$ (where $k = 1, \ldots, N$), and k mainly depends on sampling depth of the sampling card. The above-mentioned equation is written in the form of linear equation set as follows:

$$\begin{cases} v_{\varphi_{-M}}(k) = [G_t \cdot P_S \cdot G_r \cdot (g_H(\varphi_{-M}) \cdot s_H + g_V(\varphi_{-M}) \cdot s_V)]s(t) + n_{\varphi_{-M}}(k) \\ \qquad\qquad\qquad\qquad \cdots \\ v_{\varphi_0}(k) = [G_t \cdot P_S \cdot G_r \cdot (g_H(\varphi_0) \cdot s_H + g_V(\varphi_0) \cdot s_V)]s(t) + n_{\varphi_0}(k) \\ \qquad\qquad\qquad\qquad \cdots \\ v_{\varphi_M}(k) = [G_t \cdot P_S \cdot G_r \cdot (g_H(\varphi_M) \cdot s_H + g_V(\varphi_M) \cdot s_V)]s(t) + n_{\varphi_M}(k) \end{cases}$$
$$(5.2.3)$$

Equation (5.2.3) is written in the form of matrix operation as follows:

$$G_t \cdot P_S \cdot G(\varphi) \cdot S + n_\varphi(k) = V_\varphi(k) \qquad (5.2.4)$$

Where, $G(\varphi) = G_r \begin{bmatrix} g_H(\varphi_{-M}) & g_V(\varphi_{-M}) \\ & \cdots \\ g_H(\varphi_0) & g_V(\varphi_0) \\ & \cdots \\ g_H(\varphi_M) & g_V(\varphi_M) \end{bmatrix}$, $V_\varphi(k) = \begin{bmatrix} v_{\varphi_{-M}}(k) & \cdots \end{bmatrix}$

$v_{\varphi_0}(k) \ldots v_{\varphi_M}(k)]^{\mathrm{T}}$

$$n_\varphi(k) = \begin{bmatrix} n_{\varphi_{-M}}(k) & \cdots & n_{\varphi_0}(k) & \cdots & v_{\varphi_M}(k) \end{bmatrix}^{\mathrm{T}}$$

Therefore, for the single-frequency continuous wave signal transmitted by the "inspection antenna" and signal source, the radar "antenna to be measured" and receiver process signal received at each azimuth scanning angle, intermediate frequency of the receiver is sampled by the sampling system at certain sampling frequency, sampling repeat frequency and sampling depth, and the sampled data, i.e. sampled voltage signal, is subject to spectrum analysis, i.e. fast time sampling sequence $v_{\varphi_i}(k_0)$ at each azimuth sampling interval is subject to Fourier transformation to obtain spatial spectrum as follows:

$$V_{\varphi_i}(f) = FFT[v_{\varphi_i}(k)] \qquad (5.2.5)$$

To extract amplitude and phase at each azimuth frequency point can obtain polarization direction pattern within scanning period of the antenna.

5.2.2 Antenna Polarization Measurement Optimizing and Calibration

In ideal conditions, Eq. (5.2.2) is rewritten as follows when the radar antenna is receiving a horizontal polarization signal with high polarization purity and being completely matched:

$$v_{\varphi_i}(t) = m_A \left[G_t \cdot P_S \cdot G_r \cdot \boldsymbol{g}^T(\varphi_i) \cdot \boldsymbol{S} \right] \cdot s(t) + n_{\varphi_i}(t)$$
$$= m_A [G_t \cdot P_S \cdot G_r \cdot g_H(\varphi_i)] \exp(j2\pi f_0 t + j2\pi f_0/f_r + \phi_0) \qquad (5.2.6)$$
$$+ n_{\varphi_i}(t)$$

At this moment, according to Eqs. (5.2.2)–(5.2.5), it is possible to obtain horizontal polarization direction pattern of the antenna.

Accordingly, when the radar antenna is receiving a vertical polarization signal with high polarization purity and orthogonal polarization, the Eq. (3.3.2) is rewritten as follows:

$$v_{\varphi_i}(t) = m_A \left[G_t \cdot P_S \cdot G_r \cdot \boldsymbol{g}^T(\varphi_i) \cdot \boldsymbol{S} \right] \cdot s(t) + n_{\varphi_i}(t)$$
$$= m_A [G_t \cdot P_S \cdot G_r \cdot g_V(\varphi_i)] \exp(j2\pi f_0 t + j2\pi f_0/f_r + \phi_0) \qquad (5.2.7)$$
$$+ n_{\varphi_i}(t)$$

By repeating the process of Eqs. (5.2.2)–(5.2.5), it is possible to measure and obtain the antenna's vertical polarization direction pattern. Through normalization, it is possible to calculate the antenna spatial polarization vector of the Eq. (5.2.1).

But, the tower is far away from the radar in practical operation, an big apparent error exists, and deviation of the radiation direction and even polarization direction of the "radar antenna to be measured" and "inspection antenna" exists, enabling their polarization basis not be aligned with each other, i.e. there is certain displacement error τ, as shown in Fig. 5.17. Supposing that co-polarization of the antenna to be measured is horizontal (H) polarization, H polarization component projection of the polarization basis where the inspection antenna is located in H polarization direction of the polarization basis of the antenna to be measured is H polarization direction pattern that we measured actually by us, under influence of the error, while projection of the polarization basis where the inspection antenna is located combines with the V polarization component of the antenna to be measured to form V polarization direction pattern measured actually, i.e. because polarization basis alignment error results in co-polarization component leakage to cross polarization component, producing certain measurement error.

To correct the measurement error, the optimum polarization basis is defined. According to the radar polarization optimum theory [8], target echo power or signal power density at the receive antenna end reaches maximum or minimum value. This antenna transmits or receive polarization is called "maximum polarization", while this matches the optimum polarization, and enable the echo to match the radar antenna to be measured. The process of solving optimum polarization can be simplified to a process of confirming peak value point of the co-polarization and cross polarization. When polarization of the "transmit antenna" is cross polarization zero point or co-polarization zero point, cross polarization receive power of the "antenna to be measured" is lowest, and the receive power reaches maximum value or local extreme.

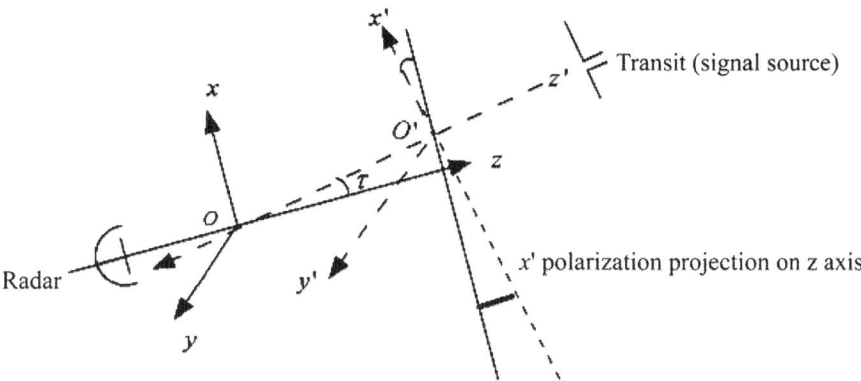

Fig. 5.17 Diagram for antenna polarization measurement coordinate system

Therefore, the optimum polarization basis has two restriction conditions: first, enable "co-polarization" voltage (or power) of the antenna to be measured to reach maximum optimum polarization basis; second, enable cross polarization receive power to reach optimum polarization basis with the minimum value or local minimum when the antenna is at maximum of the co-polarization. That's to say, co-polarization h_m is biggest and cross polarization component h_c is smallest in beam center direction. When the above-mentioned restriction conditions are met, it is possible that the antenna obtains optimum receive polarization characteristics. The restriction is similar to optimum polarization of the target in the radar polarization theory [1].

Supposing that the polarization basis mismatch error angle is τ, transition matrix of the polarization basis can be represented as follows:

$$R(\tau) = \begin{bmatrix} \cos \tau & -\sin \tau \\ \sin \tau & \cos \tau \end{bmatrix} \tag{5.2.8}$$

Supposing that normalized polarization direction pattern measured at horizontal and vertical polarization basis of the antenna is written as $J_{HV}(\varphi_i) = G \cdot \begin{bmatrix} g_H(\varphi_i) \\ g_V(\varphi_i) \end{bmatrix}$, effective polarization vector of the antenna co-polarization is written as $h_m = \begin{bmatrix} \cos \gamma_m \\ \sin \gamma_m e^{j\phi_m} \end{bmatrix}$, and effective polarization vector of the cross polarization can be written as follows:

$$h_c = \begin{bmatrix} \cos \gamma_c \\ \sin \gamma_c e^{j\phi_c} \end{bmatrix} = \begin{bmatrix} \cos(\frac{\pi}{2} - \gamma_m) \\ \sin(\frac{\pi}{2} - \gamma_m)e^{j(\pi - \phi_m)} \end{bmatrix} = \begin{bmatrix} \sin \gamma_m \\ -\cos \gamma_m e^{-j\phi_m} \end{bmatrix} \tag{5.2.9}$$

In the above-mentioned restriction conditions, measurement value of the antenna polarization characteristics has following mathematical relationship with real value:

$$\begin{bmatrix} \boldsymbol{h}_m \\ \boldsymbol{h}_c \end{bmatrix}^T = \boldsymbol{R}(\tau) \cdot \boldsymbol{J}_{HV}(\varphi_i) \tag{5.2.9}$$

i.e.,

$$\begin{bmatrix} \boldsymbol{h}_m \\ \boldsymbol{h}_c \end{bmatrix}^T = \begin{bmatrix} \cos\tau & -\sin\tau \\ \sin\tau & \cos\tau \end{bmatrix} \cdot \boldsymbol{J}_{HV}(\varphi_i) \tag{5.2.10}$$

The above-mentioned equation can be detailed as follows:

$$\begin{cases} \cos\tau \cdot J_H - \sin\tau \cdot J_V = \boldsymbol{h}_m \\ \sin\tau \cdot J_H + \cos\tau \cdot J_V = \boldsymbol{h}_c \end{cases} \tag{5.2.11}$$

The following equation can be obtained when the optimum polarization basis has two restriction conditions:

$$\frac{J_V(0)}{J_H(0)} = -tg(\tau) \tag{5.2.12}$$

i.e., $\tau = \tan^{-1}\left(\frac{J_V(0)}{J_H(0)}\right)$, τ is obtained by the above-mentioned equation; then, calculated rotation matrix is used to correct direction pattern measured at each azimuth angle, i.e., it is possible to obtain antenna polarization characteristics defined under optimum receive polarization basis.

The process of measuring orthogonal polarization also can be described by another description method. It can be known from the above-mentioned analysis that orthogonal polarization description model of the antenna can be represented as follows:

$$\boldsymbol{H}_{MC}(\varphi) = \begin{bmatrix} G_M(\varphi) \\ G_C(\varphi)\exp(j\phi) \end{bmatrix} = \begin{bmatrix} \boldsymbol{h}_M^T \boldsymbol{J}_{HV}(\varphi) \\ \boldsymbol{h}_C^T \boldsymbol{J}_{HV}(\varphi) \end{bmatrix} \tag{5.2.13}$$

Where, ϕ is phase difference between co-polarization component and cross polarization component.

The above-mentioned equation can be detailed as follows:

$$\begin{cases} G_M(\varphi) = \cos\gamma_m\sqrt{G_H(\varphi)} + \sin\gamma_m\sqrt{G_V(\varphi)}\exp(j(\phi_m + \varphi)) \\ G_C(\varphi) = \sin\gamma_m\sqrt{G_H(\varphi)} + \cos\gamma_m\sqrt{G_V(\varphi)}\exp(j(\phi_m + \pi - \varphi)) \end{cases} \tag{5.2.14}$$

So, the process of measuring cross polarization characteristics is essentially to find optimum ϕ and φ_m, enabling $G_M(\varphi)$, $G_C(\varphi)$ and ϕ to meet optimum cross polarization characteristics of the antenna. Because optimum cross polarization characteristics of the antenna has following characteristics: cross polarization is smallest when the co-polarization is biggest; i.e., only co-polarization echo power exists in main lobe direction φ_0 of the beam. This restriction condition can be represented as follows:

$$
\begin{cases}
\max\limits_{\theta,\gamma_m,\phi_m} & G_M(\varphi_0) \\
s.t. & G_C(\varphi_0) = 0
\end{cases}
\tag{5.2.15}
$$

It is expanded to obtain:

$$
\begin{cases}
\max\limits_{\theta,\gamma_m,\phi_m} & \left| \cos\gamma_m \sqrt{G_H(\varphi)} + \sin\gamma_m \sqrt{G_V(\varphi)}\exp(j(\phi+\varphi_m)) \right| \\
s.t. & \begin{cases} \sin\gamma_m \sqrt{G_H(\varphi)} + \cos\gamma_m \sqrt{G_V(\varphi)}\cos(\pi+\phi-\varphi_m) = 0 \\ \cos\gamma_m \sqrt{G_V(\varphi)}\sin(\pi+\phi-\varphi_m) = 0 \end{cases}
\end{cases}
\tag{5.2.16}
$$

By solution, it is possible to obtain:

$$
\begin{cases}
\pi+\phi-\varphi_m = 0 \\
\gamma_m = -\tan^{-1}\left(\dfrac{\sqrt{G_V(\varphi_0)}}{\sqrt{G_H(\varphi_0)}} \right)
\end{cases}
\tag{5.2.17}
$$

$$
\text{Or}
\begin{cases}
\varphi_m = \phi \\
\gamma_m = \tan^{-1}\left(\dfrac{\sqrt{G_V(\varphi_0)}}{\sqrt{G_H(\varphi_0)}} \right)
\end{cases}
\tag{5.2.18}
$$

Where, results obtained by Eqs. (5.2.17) and (5.2.12) are kept consistent.

5.2.3 Antenna Measurement Experiment and Analysis

5.2.3.1 Antenna Characteristic Measurement Experiment Layout

Outfield test adopts inclined antenna test field, i.e. a kind of test field whose erection height of the "antenna to be measured" is different from that of "inspection antenna". We erect standard linear polarization antenna on a high non-metallic test tower, and this standard antenna altitude can be controlled accurately, and transmit polarization is changed by adjusting direction of the antenna. The test scene is shown in Fig. 5.18. The antenna is connected to a standard signal source to send spot frequency continuous wave signal. The radar to be measured is placed on the ground, and certain elevation angle exists between the antenna to be measured and the inspection antenna of the test tower, and the antenna to be measured rotates in azimuth direction at speed of 4 r/min. By properly adjusting azimuth and elevation of the inspection antenna, the maximum value of lobe at elevation plane of the

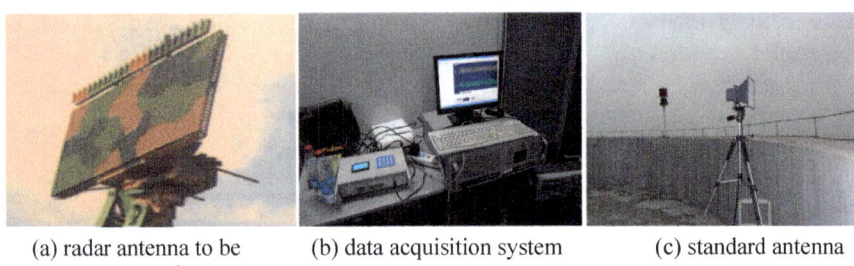

| (a) radar antenna to be | (b) data acquisition system | (c) standard antenna |
| measured | | |

Fig. 5.18 Layout for antenna polarization characteristics measurement test

antenna to be measured aims at phase center of the inspection antenna. Based on this, it works in receive condition only after stopping the radar transmitter. To accurately measure amplitude response of the signal at each azimuth angle, it is required to stop automatic gain control (AGC) of the radar receiver.

5.2.3.2 Steps for Antenna Characteristics Measurement Experiment

Step 1: As shown in Fig. 5.19, mechanical scanning of the radar antenna to be measured is carried out in azimuth direction, first by having the maximum gain direction of the inspection antenna on the tower directed at aperture direction of the radar antenna on the ground through a high-powered telescope. By connecting the signal source, transmit power of the signal source is targeted to working frequency point of the radar receiver, the signal power is adjusted to meet sampling card's dynamic range, and it is possible to measure signal response amplitude at individual azimuths, and the inspection antenna transmits signal that matches and intersects radar polarization.

Step 2: Antenna beam scanning is controlled by the antenna servo system to enable the antenna beam to scan at desired scanning speed. The independently-developed sampling card is connected to the radar receiver, so as to enable it to acquire intermediate frequency orthogonal signal of I and Q channels (I and Q signal has high channel gain and bigger signal/noise ratio). Because two kinds of polarization signal transmitted by the tower enter the same receive channel, amplitude and phase of the output signal has excellent consistency, which can ensure accuracy of the signal polarization state measurement.

Step 3: Fast time sampling at each azimuth scanning angle has 4000 points, and the azimuth sampling interval is less than 0.1°, as well as sampling frequency is 20 M. Sampling signal of each PRI should be subject to Fourier transformation, so as to obtain complex polarization direction pattern of the antenna. By modulo operation of the complex direction pattern, it is possible to obtain polarization amplitude direction pattern, and calculating anti-tangent of the virtual part to the real part of the complex direction pattern can obtain phase direction pattern.

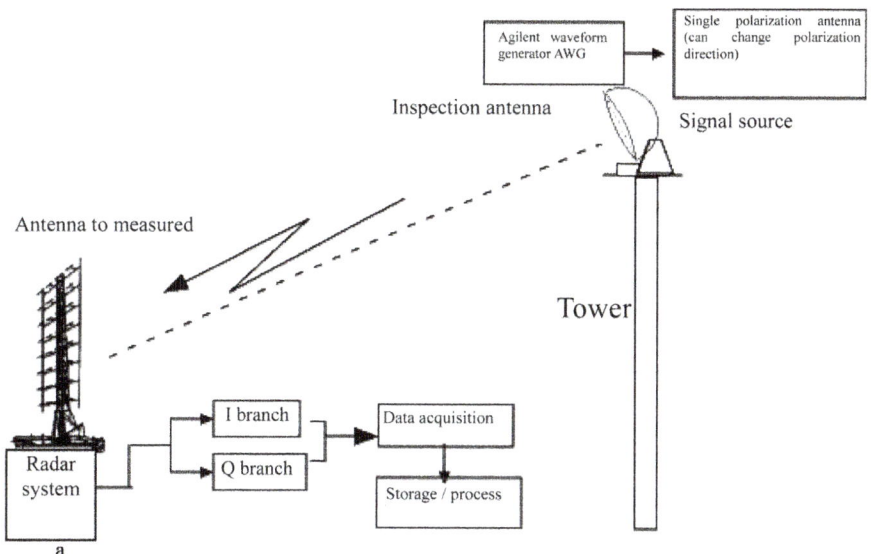

Fig. 5.19 Layout for experiment of antenna polarization characteristics measurement

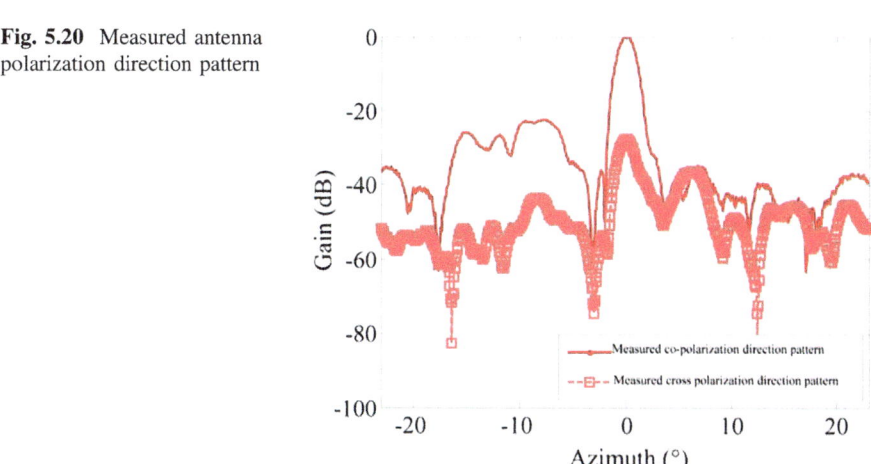

Fig. 5.20 Measured antenna polarization direction pattern

5.2.3.3   Calibration Results of Antenna Characteristics Measurement Data

When measuring co-polarization direction pattern of the antenna, the signal source is connected to the horizontal polarization antenna, with transmit power of −10 dBm. When measuring cross polarization component of the antenna, the signal source is connected to the vertical polarization antenna, with transmit power of 14 dBm. The signal source transmits spot frequency continuous wave signal, keeping consistency with working frequency of the radar. The sampling frequency of the sampling card is 20 M, and it is used to acquire intermediate frequency signal, whose dynamic azimuth is set to the maximum value. Number of sampling points for each pulse is 2000, storage depth is 4000, sampling card sampling frequency PRF is 280, and dynamic range of the sampling card is set to ±200 mV. The antenna rotates at speed of 3 r/min, so sampling interval is 0.0643°. By making use of processing method set forth in Sect. 3.3.1, polarization direction pattern of the antenna is obtained, as shown in Fig. 3.20. It can be seen from the figure that semi-power beam width of the antenna is about 2°, the first minor lobe is about −40 dB, and because non-ideal factors exist, the direction pattern shows an unsymmetrical structure; cross polarization discrimination of the antenna is about −30 dB, and cross polarization direction pattern is similar to that of the co-polarization with respect to their shape.

According to optimum theory of the radar polarization and by combining analysis in Sect. 5.2.2, it can be known that cross polarization receive power should theoretically reach minimum value or local minimum when the antenna to be measured receives maximum co-polarization echo power. This is different from actual measurement result. This phenomenon proves that cross polarization characteristics measurement of the antenna depends on selection of the polarization basis, and at different polarization basis, cross polarization performance obtained is different. In addition, polarization basis mismatch in non-ideal condition results in failure of actual measurement conditions to meet theoretical calculation conditions, which also leads to difference between actual engineering result and theoretical calculation result.

According to method set forth in Sect. 5.2.2, calculated polarization basis alignment error τ is approx. 2.3262°, and under restriction of the optimum receive polarization basis, it is possible to obtain corrected direction pattern, as shown in Fig. 5.21. It can be seen that measured co-polarization direction pattern is very close to corrected co-polarization direction pattern, and shape of cross polarization direction pattern shows significant difference before and after correction. After correction, cross polarization discrimination of the cross polarization direction pattern in the major axis direction is below −50 dB, and co-polarization power is biggest and cross polarization power is smallest at center position of the main lobe. Meanwhile, as observation azimuth increases, the cross polarization component increases, the polarization purity also decreases, and polarization characteristics of the antenna is subject to certain regular change. This coincides with conclusion set forth in Refs. [49, 155] (Fig. 5.20).

Optimum receive/transmit polarization basis is defined/derived through analyzing error in antenna polarization characteristics measurement, and in polarization measurement restriction conditions defined at the polarization basis, even if optimum polarization basis for which "co-polarization" gain of the antenna is biggest and "cross polarization" is lowest. Through actual outfield measurement of the radar antenna, the measurement processing result verifies correctness of the theoretical analysis. It proves antenna polarization characteristics evaluation result is different at different base. The polarization direction pattern is corrected by use of

Fig. 5.21 Corrected antenna polarization direction pattern

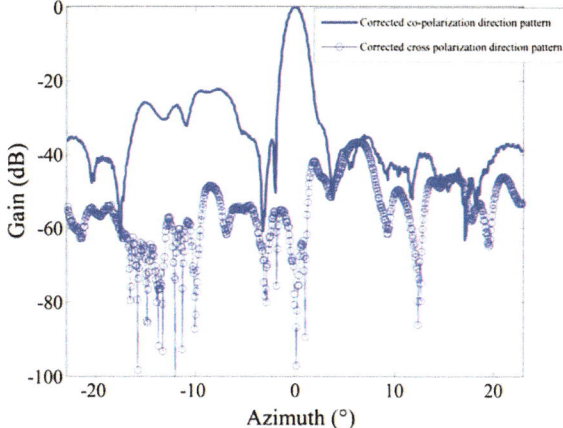

restriction model proposed, so that polarization direction pattern of the radar antenna is obtained at the optimum receive/transmit polarization basis. This has certain inspiring and application value for correcting error that may exist in antenna polarization characteristic measurement, and evaluation of polarization characteristics of the antenna to be measured.

5.3   Actual Radar Antenna Spatial Polarization Characteristics Measurement and Calibration

Section 5.2 builds an antenna polarization characteristics measurement model [154], providing a theoretical basis for subsequent outfield measurement data processing. This section first analyzes theoretical characteristics of the parabolic antenna co-polarization and cross polarization amplitude direction pattern and phase direction pattern; second, analyzes main difference between radiation characteristics in measurement data and ideal antenna radiation field, and discusses causes of producing antenna phase direction pattern measurement error, specifies data sampling frequency, signal source frequency shift, main factors of influencing phase direction patter measurement by phase center shift; lastly, proves correctness of error analysis and correction method by compensating the above-mentioned error to eliminate influence of non-ideal factors and due to the fact that measured antenna characteristics measurement result well coincides with theoretical value.

5.3.1 Prior Knowledge of Parabolic Antenna Spatial Polarization Characteristics

To obtain priori information on parabolic antenna spatial polarization characteristics, we adopt a special software, GRASP9, developed by TICAR Company in Denmark, which is used to analyze parabolic antenna. For a C wave band parabolic antenna with focus/aperture ratio F/D = 0.5, offset angle $\psi_0 = 44°$, we carry out calculation to obtain far zone radiation field data of co-polarization and cross-polarization. Through data analysis, we obtain amplitude characteristics and phase characteristics of 3D space radiation field of the antenna co-polarization and cross polarization.

In the calculation, coordinate system of the antenna is built as shown in Fig. 5.22. Origin of the coordinate system $O_r X_r Y_r Z_r$ is at phase center of the antenna, and the X axis is vertically upward, and XOZ plane is a vertical tangent plane of the antenna normal line. Supposing that angular coordinate of the electric wave propagation direction of the antenna is (θ, φ) in the coordinate system $O_r X_r Y_r Z_r$, where θ is angle between electric wave propagation direction and Z_r axis, φ is projection of the electric wave propagation direction in $X_r O_r Y_r$ plane, and polarization basis is $\hat{\theta}$ and $\hat{\varphi}$, respectively, as well as polarization of the electric wave can be represented at this polarization basis:

$$\boldsymbol{E} = E_1 \hat{\theta} + E_2 \hat{\varphi} \tag{5.3.1}$$

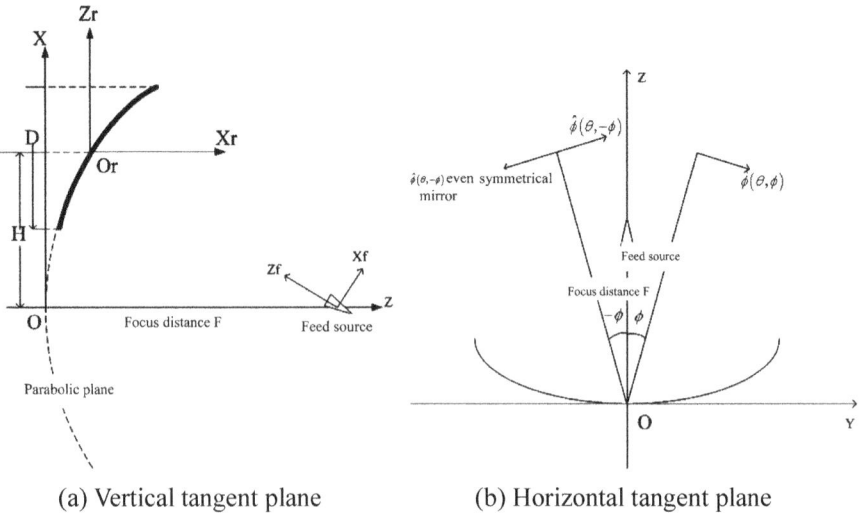

(a) Vertical tangent plane   (b) Horizontal tangent plane

Fig. 5.22 Offset feed type parabolic antenna diagram

Usually, the electric wave polarization is represented by following polarization ratio:

$$\rho = |\rho|e^{j\phi} = \tan \gamma e^{j\phi} \tag{5.3.2}$$

For majority of mechanical scanning radars, their antennas are always symmetrical to vertical tangent plane including antenna normal line, i.e. the antenna is symmetrical to the plane $\varphi = 0$ in a mirror manner, so their radiation field polarization is also mirror-symmetrical to the plane $\varphi = 0$ in a mirror manner. Because:

$$\hat{\phi}(\theta, -\varphi) = -\text{Mirror}\left[\hat{\phi}(\theta, \varphi)\right] \tag{5.3.3}$$

Where Mirror represents solution of the mirror. By use of symmetry, it can be known that the radiation field polarization essentially meets following equation as long as the antenna is symmetrical to the XOZ plane.

$$\begin{aligned} E_1(\theta, \varphi) &= E_1(\theta, -\varphi) \\ E_2(\theta, \varphi) &= -E_2(\theta, -\varphi) \end{aligned} \tag{5.3.4}$$

In electromagnetic calculation, the antenna polarization is $\hat{\theta}$ polarization, cross polarization is $\hat{\phi}$ polarization, E_1 and E_2 are function of the (θ, φ). Calculation results of the antenna polarization characteristics are shown in Fig. 5.23.

It can be seen from Fig. 5.23 co-polarization amplitude and phase are symmetrical to $\varphi = 0$ plane, i.e. $E_v(\varphi) = E_v(-\varphi)$. For the polarization ratio, its amplitude is symmetrical to $\varphi = 0$ plane, i.e. $\gamma(\varphi) = \gamma(-\varphi)$, whose phase meets: $\phi(-\varphi) = \phi(\varphi) + \pi$. In main lobe, following condition is met approximately:

$$\phi \approx \begin{cases} 90° & 0 < \varphi < \varphi_0 \\ -90° & -\varphi_0 < \varphi < 0 \end{cases}.$$

5.3.2 Actual Radar Antenna Spatial Polarization Characteristics Analysis and Error Calibration

5.3.2.1 Experiment Scene for Antenna Characteristic Measurement

Outfield experiment adopts inclined antenna test field, i.e. a kind of test field whose receive antenna is different from transmit antenna with respect to their erection height. The standard linear polarization antenna is erected at a higher non-metallic test tower, and altitude of this standard antenna can be controlled accurately, and transmit polarization can be varied by adjusting direction of the antenna. The antenna is connected to the standard signal source to send spot frequency

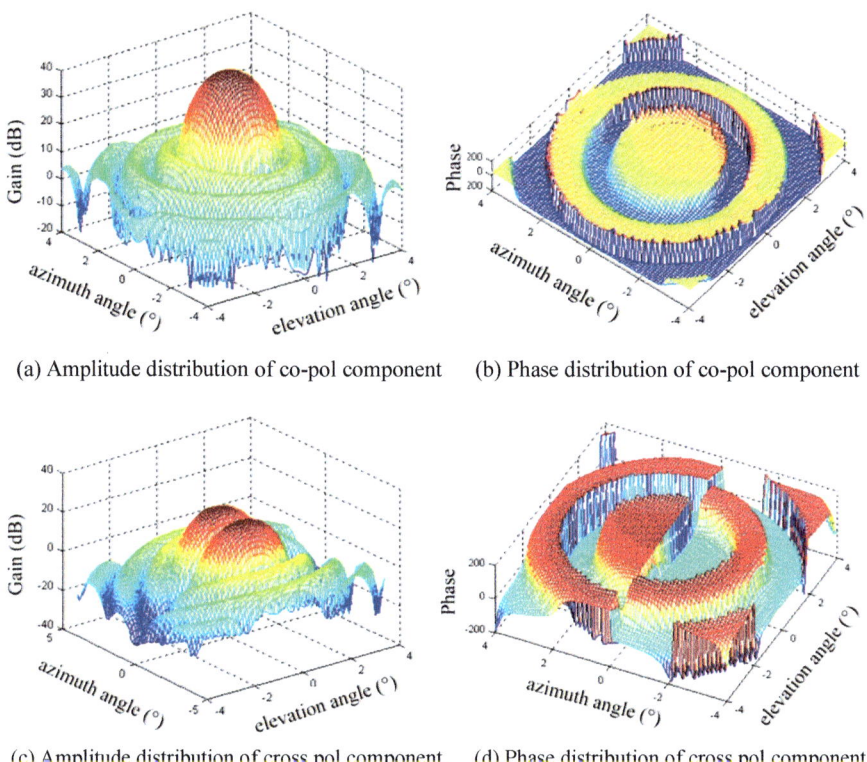

(a) Amplitude distribution of co-pol component   (b) Phase distribution of co-pol component

(c) Amplitude distribution of cross pol component   (d) Phase distribution of cross pol component

Fig. 5.23 Offset feed type parabolic antenna spatial polarization characteristics diagram

continuous wave signal. After measurement radar is placed on the ground, an elevation angle is formed between the antenna and the test tower, and the antenna is rotated upward by 360° at rate of 4 round/minute. Direction and elevation angle of the transmit antenna on the test tower are properly adjusted to enable maximum radiation direction of the free space direction pattern is aligned with the aperture center of the measurement radar antenna. By turning off the radar transmitter, the radar only works in receive condition. To measure amplitude response of signal at each azimuth angle, it is required to deactivate automatic gain control (AGC) of the radar receiver (Fig. 5.24).

5.3.2.2 Phase Measurement Error Model

It can be known from Sect. 5.2.1 that it can be represented when receiving horizontal polarization signal and vertical polarization signal:

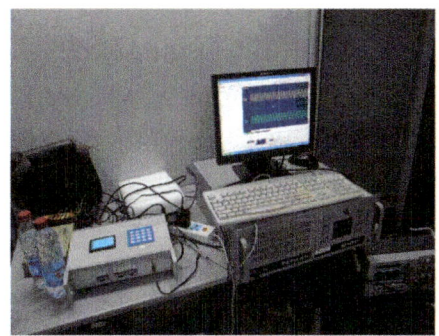 

(a) Independently-developed data   (b) Standard antenna on the test tower
acquisition system

Fig. 5.24 Layout for antenna polarization characteristics measurement experiment

$$v_{rm} = \beta G_{rm} \begin{bmatrix} A_h \cdot \exp(\theta_{Ah}) \\ A_V \cdot \exp(\theta_{AV}) \end{bmatrix}^T \cdot \begin{bmatrix} S_h \cdot \exp(\theta_{sh}) \\ S_V \cdot \exp(\theta_{sV}) \end{bmatrix} \qquad (5.3.5)$$

Where θ_{Ah} and θ_{AV} are phase direction pattern of antenna co-polarization and cross polarization, respectively, and θ_{sh} and θ_{sV} are initial phase of the signal source transmission signal.

After vector operation is carried out and influence of amplitude factors is ignored, it is possible to obtain:

$$v_{rm} = A_h \cdot S_h \exp(\theta_{Ah} + \theta_{sh}) + A_V \cdot S_V \exp(\theta_{AV} + \theta_{sV}) \qquad (5.3.6)$$

To extract amplitude and phase of the signal, the above-mentioned equation is written in the form of complex number,

$$v_{rm} = A_h \cdot S_h \cdot \cos(\theta_{Ah} + \theta_{sh}) + j \cdot A_h \cdot S_h \cdot \sin(\theta_{Ah} + \theta_{sh})$$
$$+ A_V \cdot S_V \cos(\theta_{AV} + \theta_{sV}) + j \cdot A_V \cdot S_V \sin(\theta_{AV} + \theta_{sV}) \qquad (5.3.7)$$

Supposing that $\alpha = \theta_{Ah} + \theta_{sh}$, $\beta = \theta_{AV} + \theta_{sV}$, it is simplified:

$$v_{rm} = A_h \cdot S_h \cdot \cos(\alpha) + A_V \cdot S_V \cos(\beta) + j \cdot [A_V \cdot S_V \sin(\beta) + A_h \cdot S_h \cdot \sin(\alpha)] \qquad (5.3.8)$$

The extracted amplitude item is as follows:

$$|v_{rm}| = \sqrt{[A_h \cdot S_h \cdot \cos(\alpha) + A_V \cdot S_V \cos(\beta)]^2 + [A_V \cdot S_V \sin(\beta) + A_h \cdot S_h \cdot \sin(\alpha)]^2} \qquad (5.3.9)$$

The extracted phase item is as follows:

$$\xi = \operatorname{arct} g\left(\frac{A_V \cdot S_V \sin(\beta) + A_h \cdot S_h \cdot \sin(\alpha)}{A_h \cdot S_h \cdot \cos(\alpha) + A_V \cdot S_V \cos(\beta)}\right) \tag{5.3.10}$$

It can be seen from the above-mentioned equation that amplitude and phase of the receive signal modulate amplitude characteristics and phase characteristics, and measurement results obtained are amplitude $|v_{rm}|$ and relative phase ξ corresponding to transmit signal after superposing the antenna characteristics.

Meanwhile, intermediate frequency signal output of the receiver during actual measurement corresponds to two ways of signal after S mixed frequency of the signal source transmit signal is intersected with I, Q, and the intermediate frequency is f_m, the intermediate frequency signal, after being acquired, can be represented as follows:

$$I(t) = A_h \cdot S_h \cdot \cos(\alpha + 2\pi \cdot f_m \cdot PRI + \varphi_0) + A_V \cdot S_V \cos(\beta + 2\pi \cdot f_m \cdot PRI + \varphi_0) \tag{5.3.11}$$

$$Q(t) = A_V \cdot S_V \cdot \sin(\beta + 2\pi \cdot f_m \cdot PRI + \varphi_0) + A_h \cdot S_h \sin(\alpha + 2\pi \cdot f_m \cdot PRI + \varphi_0) \tag{5.3.12}$$

Where sampling period of the data acquisition system is PRI, φ is initial phase of the signal, S_h represents horizontal polarization component of the signal, and S_V represents vertical polarization component of the signal.

It can be seen from the above-mentioned equation that receive signal will superpose a phase item after being acquired, resulting in distortion of enable the measured phase direction pattern, causing sudden change of the phase. According to the above-mentioned equation, Fig. 5.25a shows sudden phase change result of the simulation caused by data acquisition, and jumping range is $\pm 180°$. Figure 5.25b gives actual cross polarization phase direction pattern, and it can be

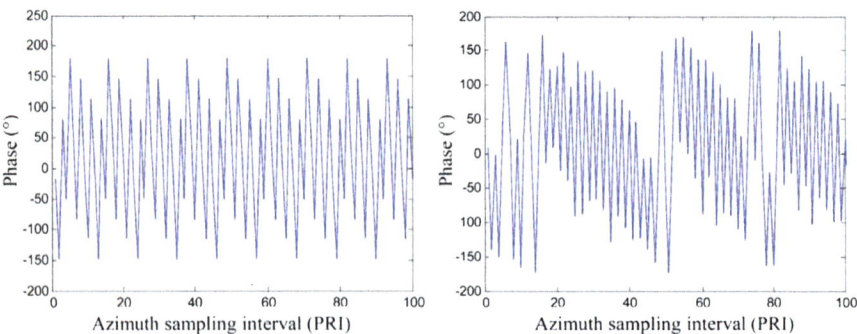

(a) Simulated phase characteristic   (b) Measurement phase direction pattern

Fig. 5.25 Sudden phase change resulted from data acquisition

seen that big phase jumping exists in 100 azimuth intervals in the main lobe, and it coincides with analysis conclusion.

5.3.2.3 Measurement Error of Non-ideal Factors and Correction

By measurement method set forth in Sect. 5.3.1, we carry out outfield measurement of a radar antenna, and processing result obtained is different from electromagnetic calculation result. Through analysis, it is known that its main cause is that outfield measurement conditions are different from electromagnetic conditions: electromagnetic calculation method is equivalent to simultaneous measurement of polarization state of the receive electric field at equivalent range r and with different (θ, ϕ) for single-frequency signal transmitted by the radar. This method cannot be realized during outfield actual operation, and experiment method in the measurement plan is to send a single frequency continuous wave signal to the radar from specific tower. The radar receiver does not work, but receives signal only. The radar antenna continuously scans receive signal recorded at different azimuths.

In addition, there are non-ideal factors in outfield measurement:

1. In the process of antenna scanning, the antenna's phase center does not coincide with rotating center of the antenna. The resulting change in antenna phase center can be represented:

$$\Delta\varphi = 2\pi \frac{r(1 - \cos \psi)}{\lambda} \qquad (5.3.13)$$

Where r is distance from projection resulted from deviation of rotating center to apex of the paraboloid, and is angle formed between the direct line from the focal point to the apex of the paraboloid and its symmetrical plane.

Therefore, phase change in co-polarization at about $0°$ azimuth angle in the measurement phase direction pattern shown in Fig. 5.27a is cosine-shaped, whose frequency is as follows:

$$f = \frac{r \sin \psi}{\lambda} \qquad (5.3.14)$$

2. Receive signal of the radar is received at different time, and absolute phase of transmit signal is different at different time. Because repeat frequency of the sampling card of $PRF = 410$ Hz and intermediate signal frequency is $f_0 = 1$ MHz do not meet sampling theorem, fuzzy number of the frequency is over 2000, and such factors as change in signal frequency and inaccuracy of the sampling clock will cause scattered frequency component in echo wave.

3. Co-polarization is not measured at the same time for cross polarization, when conducting two times of observation, the transmit signal frequency and θ, φ corresponding to pulse echo sampling are different totally.

The above-mentioned non-ideal factors will cause error in phase characteristics measurement and analysis of the antenna, so that there is certain gap between antenna phase characteristics measured and real antenna characteristic. Then, through modeling the above-mentioned certainty factors and comprehensively considering processing results of measurement data, influence of the above-mentioned factors on the antenna spatial polarization characteristics are corrected mainly by following steps:

First, analyze slow time domain in original measurement echo wave; then conduct frequency compensation to shift frequency of the receive signal to about zero frequency, so as to eliminate influence of the transmit signal frequency shift; lastly, to suppress influence of clutter frequency on characteristic measurement, it is required to conduct low pass filter of the orthogonal polarization echo signal. Figures 5.26 and 5.27 give results after partial measurement errors are corrected.

It can be seen from Fig. 5.26 that symmetrical center of the frequency spectrum of co-polarization and cross polarization is not nearby zero frequency, and spectrum center of co-polarization and cross polarization is at different position. This is

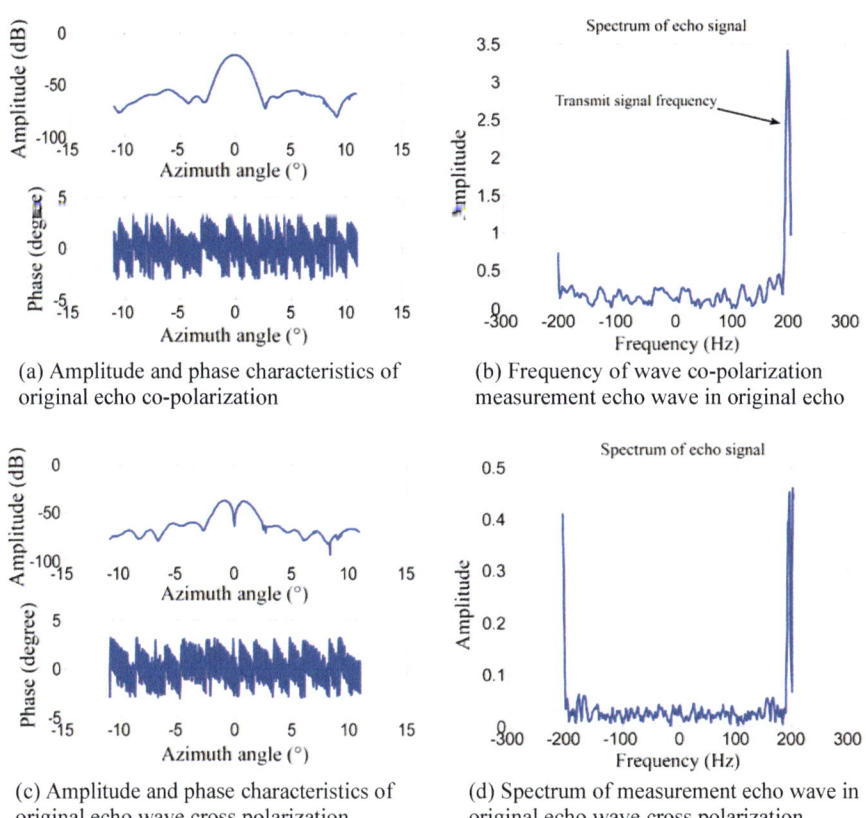

(a) Amplitude and phase characteristics of original echo co-polarization

(b) Frequency of wave co-polarization measurement echo wave in original echo

(c) Amplitude and phase characteristics of original echo wave cross polarization

(d) Spectrum of measurement echo wave in original echo wave cross polarization

Fig. 5.26 Processing result of measurement data

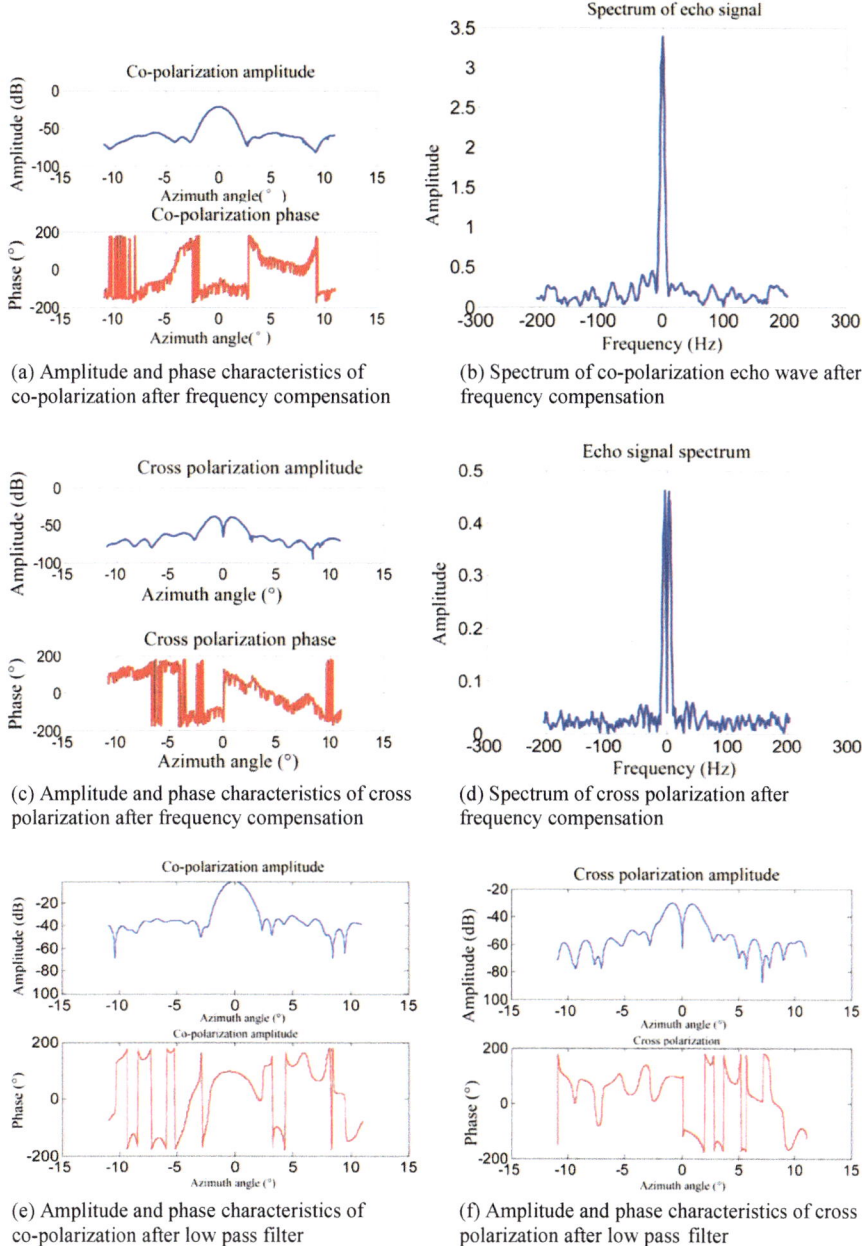

(a) Amplitude and phase characteristics of
co-polarization after frequency compensation

(b) Spectrum of co-polarization echo wave after
frequency compensation

(c) Amplitude and phase characteristics of cross
polarization after frequency compensation

(d) Spectrum of cross polarization after
frequency compensation

(e) Amplitude and phase characteristics of
co-polarization after low pass filter

(f) Amplitude and phase characteristics of cross
polarization after low pass filter

Fig. 5.27 Processing result of measurement data after frequency compensation and low pass filter

resulted from transmit signal frequency shift. For this reason, measurement result of antenna phase characteristics is interfered, so that phase characteristics of co-polarization and cross polarization show violent quick fluctuation.

As shown in Fig. 5.27, upon frequency compensation, the frequency center of receive signal is shifted to approx. zero frequency, so that quick fluctuation in phase characteristics of co-polarization and cross polarization is eliminated. In addition to zero frequency, some scattering frequency exists. To suppress influence of scattering frequency on antenna spatial polarization characteristics, the echo signal passes another low pass filter, and output amplitude characteristics and phase characteristics after filter are shown in Fig. 5.27. It can be seen that, after error correction, phase direction pattern of co-polarization and cross polarization is transited smoothly, and in the main lobe, co-polarization phase direction pattern shows certain even symmetry, and cross polarization direction pattern shows odd symmetry. This coincides with theoretical analysis results.

This section first analyzes amplitude and phase characteristics of co-polarization and cross polarization radiation field of a paraboloid dish antenna. The processing results show, due to restriction by non-ideal factors, actual outfield measurement result is significantly different from theoretical calculation, pointing out causes producing the antenna phase direction pattern measurement error. The errors resulted from non-ideal factors are compensated and corrected to better reflect real antenna polarization phase characteristics. The corrected result is well coincided with the theoretical value, proving correctness of error analysis and correction method.

Chapter 6
Matrix Measurement Method Based on Antenna Spatial Polarization Characteristics

Radar target polarization measurement is a basic problem in the field of radar polarization technology. How to acquire polarization information is undoubtedly a basic problem in the investigation of the radar detection technology. Accurate measurement of polarization scattering matrix ([S] Matrix) is the basis of application in the field of target classification and recognition. At present, there are mainly two kinds of polarization measurement radar system, one is the alternate polarization measurement system, and the other is the simultaneous polarization measurement system. Alternate polarization measurement system transmit a plurality of different polarized pulses and use two orthogonal polarized antennas to receive the signal simultaneously, so that the target scattering matrix can be estimated by the processing of two adjacent pulse echoes. Simultaneous polarization measurement system overcomes the inherent shortcomings of the system, and uses two orthogonal polarization channels simultaneously transmit and receive pulses with space diversity waveforms, and can measure polarization within single pulse. However, there is a common feature of these two kinds of polarization measurement methods, that is, the need for two polarization channels.

In this chapter, we propose two new methods for measuring polarization scattering matrix [156], which is named as "time domain measurement method" and "frequency domain measurement method" by using spatial polarization characteristics of antenna, and the theoretical analysis and computer simulation are presented. Although the two signal processing methods is different, but their essence is same, that is no longer pursuit of two ideal orthogonal polarization channels. On the contrary, these two methods make use of the impure polarization characteristic and the spatial variation properties. The methods break through the traditional thought of using the two orthogonal polarization channels in polarization measurement system, only one polarization channels is needed to realize target polarization characteristic measurement.

© National Defense Industry Press, Beijing and Springer Nature Singapore Pte Ltd. 2019   213
H. Dai et al., *Spatial Polarization Characteristics of Radar Antenna*,
https://doi.org/10.1007/978-981-10-8794-3_6

6.1   Time Domain Measurement of Polarization Scattering Matrix Based on Antenna Spatial Polarization Characteristics

Algorithm principle

Assuming the radar is a coherent radar system. The modulation signal $S(t)$ consists of N pulses with period T. Carrier frequency signal is $e^{j2\pi f_0 t}$ with frequency f_0. The coherent RF pulse train is transmitted through the antenna. The antenna scans in spatial domain. If space value is sampled every pulse repetition period T, the corresponding spatial scanning interval is $\Delta\varphi_s = \omega_s/f_r$. Assuming $m = 1$ corresponding $t = 0$, the form of the M-th transmitted vector pulse signal (2 × 1 vector) can be expressed as

$$x_m(t) = A \cdot e^{j2\pi f_0 t} \cdot \upsilon(t_m) \cdot G_{\text{t}m}(\boldsymbol{P}) \cdot \boldsymbol{h}_{\text{t}m}(\boldsymbol{P}) \qquad (6.1.1)$$

Among them, tm = t-(m-1)T, m = 1, 2, …, M, represents the delay between the start of the pulse. A represents the magnitude; normalized antenna voltage pattern $G_{\text{t}m}(\boldsymbol{P})$ and antenna polarization vector $h_{\text{t}m}(\boldsymbol{P})$ are functions of space angle $\boldsymbol{P}(\theta, \varphi)$, the following abbreviated $G_{\text{t}m}$ and $\boldsymbol{h}_{\text{t}m}$; $\upsilon(t)$ is pulse modulation function of the signal, which commonly used rectangular pulse, linear FM pulse or phase coded pulse form.

Assuming the target scattering matrix are constant within the coherent time (NT) and the signal bandwidth, denoted as $\boldsymbol{S} = \begin{bmatrix} S_{11} & S_{12} \\ S_{21} & S_{22} \end{bmatrix}$. Under the single static conditions linear target scattering with reciprocity [180], the scattering matrix is a symmetric matrix, that is $S_{12} = S_{21}$.

The radial velocity can be seen as constant in the coherence time, recorded as V_r. The Doppler frequency is $f_d = \frac{-2V_r f_0}{c}$, and C is the speed of light. After the target scattering, the mth vector echo signal is

$$\boldsymbol{y}_m(t) = B \cdot e^{j2\pi (f_0 + f_d)(t - \tau_m)} \cdot \upsilon(t_m - \tau_m) \cdot G_{\text{t}m} \cdot \boldsymbol{S}\boldsymbol{h}_{\text{t}m} \qquad (6.1.2)$$

Among them, B is the amplitude; $\tau_m = \tau + \frac{2V_r(m-1)T}{c}$ is the echo delay of the mth pulse; $\tau = \frac{2R}{c}$ is the echo delay of the first pulse. Generally $\frac{2V_r \tau_p}{c} \ll \frac{1}{\Delta f} \tau_p$ [τ_p is the pulse width, and Δf is the bandwidth of $\upsilon(t)$], therefore, pulse broadening or compression caused by Doppler effect can be ignored.

Echo signal received by radar antenna is

$$v_{\text{r}m}(t) = \beta \cdot e^{j2\pi (f_0 + f_d)(t - \tau_m)} \cdot \upsilon(t_m - \tau_m) \cdot G_{\text{r}m} \cdot G_{\text{t}m} \cdot \boldsymbol{h}_{\text{r}m}^T \boldsymbol{S}\boldsymbol{h}_{\text{t}m} \qquad (6.1.3)$$

In the formula, β is amplitude, which does not include the antenna polarization and pattern; $\beta = \frac{k_{RF}}{16\pi^2 R^4 L_R} \cdot \sqrt{\frac{P_t}{4\pi L_t}} \cdot G_m^2$, k_{RF} is RF amplifier coefficient. R is the distance between radar and target; P_t is the peak power; I_t is the transmitting loss; L_R is the receiving loss; G_{max} is the maximum gain of antenna; G_m and $\boldsymbol{h}_m$ are the normalized voltage gain and the polarization state vector of antenna.

Because the antenna has reciprocity, and in a very short time interval, $G_{rm} = G_{tm} = G_m, \boldsymbol{h}_{tm} = \boldsymbol{h}_{rm} = \boldsymbol{h}_m$ the formula can be written as

$$v_{rm}(t) = \beta \cdot e^{j2\pi (f_0 + f_d)(t - \tau_m)} \cdot v(t_m - \tau_m) \cdot G_m^2 \cdot \boldsymbol{h}_m^T \boldsymbol{S} \boldsymbol{h}_m \qquad (6.1.4)$$

The antenna is scanned in azimuth $[-\varphi_0/2, +\varphi_0/2]$; $\Delta\varphi_s = \omega_s/f_r = \omega_s T \omega_s$, ω_s is the scanning speed of the antenna; T is the pulse repetition period; the number of pulses (sample points) is $M = \varphi_0/\Delta\varphi_s$. Within the range of the scan, the signal received by the radar is pulse train

$$v_r(t) = \sum_{m=1}^{M} v_{rm}(t) \qquad (6.1.5)$$

First, the received signal is mixed, the radio frequency signal is changed into IF signal or zero IF signal

$$Z(t) = v_r(t) \cdot (q(t))^* = \sum_{m=1}^{M} \beta \cdot e^{j2\pi f_d(t - \tau_m)} \cdot e^{-j2\pi f_0 \tau_m - j\phi} \cdot v(t_m - \tau_m) \cdot G_m^2 \cdot \boldsymbol{h}_m^T \boldsymbol{S} \boldsymbol{h}_m$$

$$(6.1.6)$$

where $q(t) = e^{j(2\pi f_0 t + \phi)}$ is the mixing signal, and ϕ is the initial phase, $\Delta\phi_m = -2\pi f_0 \tau_m - \phi$, there is

$$Z(t) = \sum_{m=1}^{M} \beta \cdot e^{j\Delta\phi_m} \cdot e^{j2\pi f_d(t - \tau_m)} \cdot v(t_m - \tau_m) \cdot G_m^2 \cdot \boldsymbol{h}_m^T \boldsymbol{S} \boldsymbol{h}_m \qquad (6.1.7)$$

Assuming impulse response of the matching pulse is $h(t) = v^*(\tau_p - t)$, then the matching process is

$$R(t) = \int_{-\infty}^{+\infty} h(t - \lambda) Z(\lambda) d\lambda \qquad (6.1.8)$$

The result for the mth pulse matched filter is $R_m(t)$, then

$$R(t) = \sum_{m=1}^{M} R_m(t) \tag{6.1.9}$$

where

$$
\begin{aligned}
R_m(t) &= \int_{-\infty}^{+\infty} h(t-\lambda)Z_m(\lambda)\mathrm{d}\lambda \\
&= \beta \cdot G_m^2 \boldsymbol{h}_m^{\mathrm{T}} \boldsymbol{S}\boldsymbol{h}_m \cdot \mathrm{e}^{\mathrm{j}\Delta\phi_m} \cdot \mathrm{e}^{\mathrm{j}2\pi f_d(t-\tau_m-\tau_p)} \\
&\quad \int_{-\infty}^{+\infty} \upsilon^*(g) \cdot \upsilon(g+t-\tau_p-\tau_m) \cdot \mathrm{e}^{\mathrm{j}2\pi f_d g}\mathrm{d}g
\end{aligned}
\tag{6.1.10}
$$

Assuming transmitted signal is a rectangular pulse signal. There are M peak points corresponding time $t = (m-1)T + \tau_p + \tau_m$, and peak value is

$$
\begin{aligned}
R_m &= \beta \cdot G_m^2 \boldsymbol{h}_m^{\mathrm{T}} \boldsymbol{S}\boldsymbol{h}_m \cdot \mathrm{e}^{\mathrm{j}2\pi f_d(m-1)T} \cdot \mathrm{e}^{\mathrm{j}(-2\pi f_0 \tau_m - \phi)} \cdot \int_{-\infty}^{+\infty} |\upsilon(g)|^2 \cdot \mathrm{e}^{\mathrm{j}2\pi f_d g}\mathrm{d}g, \mathrm{m} \\
&= 1, 2, \ldots, \mathrm{M}
\end{aligned}
\tag{6.1.11}
$$

Generally there is $\frac{2R}{c} \gg \frac{2V_r t}{c}$, $\tau_m = \tau = \frac{2R}{c}$, the upper formula can be written as

$$K \tag{6.1.12}$$

Since $\beta \cdot \left[\int_{-\infty}^{+\infty} |\upsilon(g)|^2 \cdot \mathrm{e}^{\mathrm{j}2\pi f_d g}\mathrm{d}g \right] \cdot \mathrm{e}^{-\mathrm{j}(2\pi f_0 \tau + \phi)}$ is a constant, denoted as K, then

$$K = \beta \cdot \left[\int_{-\infty}^{+\infty} |\upsilon(g)|^2 \cdot \mathrm{e}^{\mathrm{j}2\pi f_d g}\mathrm{d}g \right] \cdot \mathrm{e}^{-\mathrm{j}(2\pi f_0 \tau + \phi)} \tag{6.1.13}$$

The formula (6.1.12) can be written as

$$R_m = K \cdot \mathrm{e}^{\mathrm{j}2\pi f_d(m-1)T} \cdot G_m^2 \boldsymbol{h}_m^{\mathrm{T}} \boldsymbol{S}\boldsymbol{h}_m \tag{6.1.14}$$

On the polarization basis $(\hat{\boldsymbol{h}}, \hat{\boldsymbol{v}})$, the electric field of the antenna at the mth scanning spot is

$$\boldsymbol{e}_m = \begin{bmatrix} e_{m\mathrm{H}} \\ e_{m\mathrm{V}} \end{bmatrix} = G_m \boldsymbol{h}_m = \begin{bmatrix} G_m h_{m\mathrm{H}} \\ G_m h_{m\mathrm{V}} \end{bmatrix} \tag{6.1.15}$$

The antenna pattern and the polarization characteristic are the functions of spatial domain. After mixed and matching filtered, the amplitude of each peak is a function of the target scattering matrix

$$R_m = K \cdot \mathrm{e}^{\mathrm{j}2\pi f_d(m-1)T} \cdot G_m^2(\boldsymbol{P}) \boldsymbol{h}_m(\boldsymbol{P})^{\mathrm{T}} \boldsymbol{S} \boldsymbol{h}_m(\boldsymbol{P}) \tag{6.1.16}$$

In   which, $K = \beta \cdot \left[\int_{-\infty}^{+\infty} |\upsilon(g)|^2 \cdot \mathrm{e}^{\mathrm{j}2\pi f_d g} \mathrm{d}g \right] \cdot \mathrm{e}^{-\mathrm{j}(2\pi f_0 \tau + \phi)}$ is   constant. Signal-noise-ratio expressed as $\mathrm{SNR} = \frac{|K|^2}{\sigma^2}$ reflects the target echo power, but does not contain the target scattering intensity and polarization.

$S = \begin{bmatrix} S_{11} & S_{12} \\ S_{12} & S_{22} \end{bmatrix}$ can be stretched into three-dimensional column vector $\boldsymbol{S} = \begin{bmatrix} S_{11} & 2S_{12} & s_{22} \end{bmatrix}^{\mathrm{T}}$, then there is

$$\boldsymbol{h}^{\mathrm{T}} \boldsymbol{S} \boldsymbol{h} = h_{\mathrm{H}}^2 S_{11} + 2h_{\mathrm{H}} h_{\mathrm{V}} S_{12} + h_{\mathrm{V}}^2 S_{22} \tag{6.1.17}$$

We discuss two conditions as following:

1. when the target is stationary, fd = 0, there is

$$\boldsymbol{R} = K \cdot \boldsymbol{GH} \cdot \boldsymbol{S} \tag{6.1.18}$$

where $\boldsymbol{R} = [R_1 R_2 R_M]^{\mathrm{T}}$, R1, R2 … … RM are amplitudes of each peak point.

$$\boldsymbol{S} = [S_{11} \, 2S_{12} \, S_{22}]^{\mathrm{T}}; \boldsymbol{G} = \begin{bmatrix} G_1^2 & 0 & \cdots & 0 \\ 0 & G_2^2 & \cdots & 0 \\ \vdots & \vdots & \ddots & \vdots \\ 0 & 0 & \cdots & G_M^2 \end{bmatrix}; \boldsymbol{H} = \begin{bmatrix} h_{\mathrm{H1}}^2 & h_{\mathrm{H1}} h_{\mathrm{V1}} & h_{\mathrm{V1}}^2 \\ h_{\mathrm{H2}}^2 & h_{\mathrm{H2}} h_{\mathrm{V2}} & h_{\mathrm{V2}}^2 \\ \vdots & \vdots & \vdots \\ h_{\mathrm{HM}}^2 & h_{\mathrm{HM}} h_{\mathrm{VM}} & h_{\mathrm{VM}}^2 \end{bmatrix}$$

The electric field matrix expressed as $\boldsymbol{E} = \boldsymbol{GH}$, then the formula (6.1.18) can be written as

$$\boldsymbol{R} = K \cdot \boldsymbol{ES} \tag{6.1.19}$$

2. when the target is moving, fd ≠ 0, the Doppler frequency should be taken into account, then there is

$$R = K \cdot P \cdot ES \tag{6.1.20}$$

where

$$P = \begin{bmatrix} 1 & 0 & \cdots & 0 \\ 0 & e^{j2\pi f_d T} & \cdots & 0 \\ \vdots & \vdots & \ddots & \vdots \\ 0 & 0 & \cdots & e^{j2\pi f_d (M-1)T} \end{bmatrix} \tag{6.1.21}$$

Whether the target is static or moving, it can be expressed as a formula $R = K \cdot P \cdot ES$: when the target is static, the coefficient matrix is $P = I_{M \times M}$; when the target is moving, the coefficient matrix P is expressed as the formula (6.1.21). Taking the noise into consideration, the upper formula can be written as

$$R = K \cdot P \cdot ES + n \tag{6.1.22}$$

In which, n is the complex Gauss white noise vector, $n \sim N(0, R_n)$. Where $R_n = \sigma^2 \cdot I_{M \times M}$, σ^2 is the observation noise variance.

Based on the Eq. (6.1.22), the scattering matrix of the target can be estimated by the least square method. Consider two situations: ① the radar does not know whether the target is static or moving in the process of measurement. Polarization scattering matrix of the target is estimated by formula $R = K \cdot ES + n$, so the result is unbiased estimation when the target is static, and the result is biased estimation when the target is moving. ② If the radar has the ability to measure Doppler frequency and get the target Doppler frequency estimation $\hat{f}$, "Doppler compensation" should be considered first, and then the target scattering matrix can be estimated based on the formula

$$R = K \cdot P' \cdot ES + n \tag{6.1.23}$$

In which

$$P' = \begin{bmatrix} 1 & 0 & \cdots & 0 \\ 0 & e^{j2\pi \hat{f} T} & \cdots & 0 \\ \vdots & \vdots & \ddots & \vdots \\ 0 & 0 & \cdots & e^{j2\pi \hat{f} (M-1)T} \end{bmatrix} \tag{6.1.24}$$

If the coefficient $K = \beta \cdot \left[\int_{-\infty}^{+\infty} |v(g)|^2 \cdot e^{j2\pi f_d g} dg \right] \cdot e^{-j(2\pi f_0 \tau + \phi)}$ is not known, the acquisition of the coefficient K is a difficult problem when the polarization scattering matrix of the target is estimated based on $\hat{S} = \frac{1}{K} \left(E^H E \right)^{-1} E^H R$. S can be estimated by $\left(E^H E \right)^{-1} E^H R$ first, and then S_{12}/S_{11}, S_{12}/S_{22} etc. are calculated. Even if the full polarization scattering matrix cannot be obtained, the relative components of target PSM is significant in distinguish false targets based on the antenna's spatial polarization characteristic. The simulation analysis is given in Sect 6.1.2.

In this section, the target polarization scattering matrix estimation method is discussed in the time domain. In the following Sect. 6.1.2, the target polarization scattering matrix is obtained by echo Fourier transforms. As the two kind of method to obtain the target polarization scattering matrix, the method discussed in this section is called "time domain method", and the method discussed in next section is called "frequency domain method".

6.1.1 Theoretical Analysis of the Algorithm Performance

After the target echo sequence is received by the radar, the signal is processed by the mixer and become IF signal or zero IF signal. The IF signal is processed by the matched filter and the peak value of the filtered signal is obtained as $\{R_m\}, m = 1, 2, \ldots, M$. Then echo peak value vector $\boldsymbol{R} = [R_1 R_2 R_M]^\mathrm{T}$ is established. According to Sect. 6.1.1, $\boldsymbol{R}$ is the function of element in the target scatter matrix S. The least square estimation of target scatter matrix S column vector can be obtained

$$\hat{S} = \frac{1}{K}(E^\mathrm{H}E)^{-1}E^\mathrm{H}R \tag{6.1.25}$$

When the target is stationary, or the target is moving motion but its Doppler frequency can be estimated accurately, that is $\tilde{f}_d = f_d - \hat{f}_d$, $\hat{S}$ obtained by formula (6.1.25) is unbiased estimation.

Estimation error of target polarization scattering matrix can be expressed

$$\tilde{S} = S - \hat{S} \tag{6.1.26}$$

Then $\tilde{S}$ is subject to the complex Gauss distribution, i.e., $S \sim N(0, R_{\tilde{s}})$. The variance of estimation error

$$R_{\tilde{s}} = \mathrm{var}[\tilde{S}] = \frac{\sigma^2}{K^2}(E^\mathrm{H}E)^{-1} \tag{6.1.27}$$

When the target is moving and the Doppler estimation error exists, the target polarization scattering matrix can be estimated still by the formula $\hat{S} = (1/K) \cdot (E^\mathrm{H}E)^{-1}E^\mathrm{H}R$, and the est $\tilde{S} \sim N(E\tilde{S}, R_{\tilde{s}})$ imation error $\tilde{S} = S - \hat{S}$ is still subject to the distribution of the complex Gauss, but it is a bias estimation

$$E\tilde{S} = E[\tilde{S}] = (E^H E)^{-1} E^H (P'' - I_{M \times M}) ES \tag{6.1.28}$$

Among them, the matrix P'' is

$$P'' = \begin{bmatrix} 1 & 0 & \cdots & 0 \\ 0 & e^{j2\pi \tilde{f}_d T} & \cdots & 0 \\ \vdots & \vdots & \ddots & \vdots \\ 0 & 0 & \cdots & e^{j2\pi \tilde{f}_d (M-1) T} \end{bmatrix} \tag{6.1.29}$$

The estimation error variance of $\hat{S}$ is still

$$R_{\tilde{S}} = \mathrm{var}[\tilde{S}] = \frac{\sigma^2}{K^2} (E^H E)^{-1} \tag{6.1.30}$$

Because the estimation error of $\tilde{S}$ subject complex Gauss distribution, $\tilde{S} \sim N(E\tilde{S}, R_{\tilde{S}})$, the probability density function is

$$f(\tilde{S}) = \frac{1}{(2\pi)^{3/2} |R_{\tilde{S}}|^{\frac{1}{2}}} \exp\left\{ -\frac{1}{2} (\tilde{S} - E\tilde{S})^H R_{\tilde{S}}^{-1} (\tilde{S} - E\tilde{S}) \right\} \tag{6.1.31}$$

The probability density contour $(\tilde{S} - E\tilde{S})^H R_{\tilde{S}}^{-1} (\tilde{S} - E\tilde{S})$ is determined by the ellipsoid [195]:

$$(\tilde{S} - E\tilde{S})^H R_{\tilde{S}}^{-1} (\tilde{S} - E\tilde{S}) = c^2 \tag{6.1.32}$$

The centers of these ellipsoids are $E\tilde{S}$, and the axis is $\pm c \sqrt{\lambda_i e_i}$, in which

$$R_{\tilde{S}} e_i = \lambda_i e_i, i = 1, 2, 3. \tag{6.1.33}$$

According to the theory of multivariate statistical analysis, there is

$$(\tilde{S} - E\tilde{S})^H R_{\tilde{S}}^{-1} (\tilde{S} - E\tilde{S}) \le d^2 = \chi_3^2(\alpha) \tag{6.1.34}$$

The probability of $\tilde{S}$ in the circle is $1 - \alpha$ (of which $\chi_3^2(\alpha)$ is the 100α percentile of 3 free degree χ^2 distribution). That is, the probability of $\tilde{S}$ falling in the ellipsoid of formulas (6.1.32) and (6.1.33) is $1 - \alpha$.

When the target is static, the probability density contour is ellipsoid with the center of O. If the probability of $\tilde{S}$ in the ellipsoid is required to be $P_0 = 1 - \alpha_0$, each axis of ellipsoid can be calculated by χ^2 distribution table, formulas (6.1.33) and (6.1.34). The long axis of the ellipsoid is R_0. The ellipsoid region is called the "admissible region". When the Target is moving, the center of ellipsoid is $E\tilde{S}$. If the distance of $E\tilde{S}$ deviation from O, which is the center of the probability density contour ellipsoid when the target is static, is small relative to R_0, the probability of $\tilde{S}$ in the permitted area is still very large. In the engineering application, $\tilde{S}$ is still acceptable.

When the target is moving and the longest ellipsoid axis R_1 satisfies $R_1 = R_0 - \|ES\|$, this ellipsoid is inscribed in the ellipsoid when the target is static. Assuming the Probability in the ellipsoid is P_1, there is $P_1 \leq P \leq P_0$. Therefore, if it is required that $P \geq 1 - \alpha_1$, the long axis of the ellipsoid should equal to $\sqrt{\chi_{3^2}(\alpha_1)} \cdot \sqrt{\lambda_i}, i = 1, 2, 3$ [which, λ_i is determined by the formula (6.1.33)]. Then $P \geq P_1 = 1 - \alpha_1$.

Thus, in the case of target motion, a sufficient condition for the probability $P \geq 1 - \alpha_1$ of estimation error $\tilde{S}$ in the admissible region is:

$$\|E\tilde{S}\| \leq \left(\sqrt{\chi_{3^2}(\alpha_0)} - \sqrt{\chi_{3^2}(\alpha_1)} \right) \lambda_{max} \qquad (6.1.35)$$

where λ_{max} is the longest axis of $\lambda_i, i = 1, 2, 3$ determined by the 6.1.33.

It can be seen from formula (6.1.28) and (6.1.35), as long as the accuracy of the estimation is high enough, $\|ES\|$ will be small. Even if the Doppler frequency cannot be measured accurately, vector estimation error $\tilde{S}$ is still acceptable, and it does not affect the algorithm performance.

6.1.2 Simulation Experiment and Analysis

At present, the "measurement error function" is the standard to judge the measured performance. It is defined as the difference between the real scattering matrix S and the measurement matrix $\hat{S}$, i.e. $e = S - \hat{S}$. The error e is a vector, and it only shows the absolute magnitude of the difference between the target scattering matrix S and the estimated matrix $\hat{S}$. It cannot represent the structure difference between the two victors. Therefore, the performance of the algorithm is measured by using the "measurement error" of the target polarization scattering matrix, "relative measurement error of each component" and the "similarity measure ξ between two vectors".

The relative measurement error of each component is used for analyzing the relative measurement accuracy of each component of the target scattering matrix. It can represent the difference between the measured value and the real value of the target PSM. The relative measurement error of each component is defined as:

$$e_{S_{11}} = \left| \frac{S_{11} - \hat{S}_{11}}{S_{11}} \right|, e_{S_{12}} = \left| \frac{S_{12} - \hat{S}_{12}}{S_{12}} \right|, e_{S_{22}} = \left| \frac{S_{22} - \hat{S}_{22}}{S_{22}} \right| \qquad (6.1.36)$$

In particular, if $S_{11} = 0$, $\dot{e}_{S_{11}}$ is defined as

$$e_{S_{11}} = \left| S_{11} - \hat{S}_{11} \right| \qquad (6.1.37)$$

If $S_{12} = 0$, $e_{S_{12}}$ is defined as

$$e_{S_{12}} = \left| S_{12} - \hat{S}_{12} \right| \qquad (6.1.38)$$

If $S_{22} = 0$, $e_{S_{22}}$ is defined as

$$e_{S_{22}} = \left| S_{22} - \hat{S}_{22} \right| \qquad (6.1.39)$$

As a general performance of the algorithm, the "similarity measure ξ" between the two vectors is a better representation of the similarity between the measurement matrix and the real matrix. If the measured value $\hat{S}$ and the true value S of the polarization scattering matrix are stretched to four-dimensional vector $\hat{X}$ and X, the absolute value of the cosine of the angle between $\hat{X}$ and X is

$$\xi = \frac{\left| X^H \hat{X} \right|}{\|X\| \cdot \|\hat{X}\|} \qquad (6.1.40)$$

where $\|\cdot\|$ is the Frobenius norm of vector. Because $\hat{X}$ and X are 4 dimensional complex vector, in the calculation of the cosine of the angle between the vector, they can be stretched to 8 dimensional vector and substituted into the formula (6.1.40). It is obvious that $0 \le \xi \le 1$, and there is a better measurement performance when the value ξ is more close to 1. Since the numerator and the denominator of ξ are correlative it is difficult to roll out exact analytical expressions for the mean and variance of ξ. When discussing the performance of the similarity measure ξ, it is obtained by numerical method. The performance of the algorithm is analyzed by computer simulation in the follow.

The radar system parameters are set as follows. Radar operates in C-band frequency, operating frequency f = 5 GHz, transmit power $P_t = 100$ W. The offset parabolic antenna is used which is discussed in Sect. 3.4.2.1, the aperture wavelength ratio $D/\lambda = 50$, the focus diameter ratio $F/D = 0.5$, the bias angle $\psi_0 = 44°$, the half power beam width $\varphi_{3dB} = 1.4°$. The antenna rotating speed is 6 Round/min, the maximum gain $G_t = 30$ dB, the pulse repetition frequency $f_r = 1$ KHz, the receiver bandwidth $B_n = 0.5$ MHz, the noise factor $F_n = 3$ dB, and the system loss $L_r = 10$ dB.

There are two targets. One is a matching target and the other is a standard metal ball target. The polarization scattering matrix of a matching target model is $S = \begin{bmatrix} 1 & 0.2 - 0.1j \\ 0.2 - 0.1j & 1 + 0.3j \end{bmatrix}$, which is measured in microwave anechoic chamber. Scattering matrix of the standard metal ball target is 2 order unit matrix. The distance between the target and radar is $R = 5$ km. For the matching target, according to the radar equation, the signal to noise ratio of each channel is estimated to be: 22.6 dB of HH channel, 22.8 dB of VV, and 16.1 dB of HV channel.

The radar is scanning between $\lfloor -\varphi_{3dB}/2, +\varphi_{3dB}/2 \rfloor$ in the horizontal plane. The number of simulation times is 300. Figure 6.1 illustrates the distribution of complex elements estimation the scattering matrix of matching target.

The polarization scattering matrix estimation is $\hat{S}$, and measurement error of each element is $\tilde{S}_{11} = \hat{S}_{11} - S_{11}$, $\tilde{S}_{12} = \hat{S}_{12} - S_{12}$ and $\tilde{S}_{22} = \hat{S}_{22} - S_{22}$. The simulation results are shown in Table 6.1.

Figure 6.2 illustrates the distribution of complex elements estimation the scattering matrix of standard metal ball. Table 6.2 shows the Statistical characteristics of measurement error.

From Figs. 6.1 and 6.2, Tables 6.1 and 6.2, when the co-polarization of the radar antenna is "vertical polarization", the accuracy of the vertical polarization components S_{22} of the target polarization scattering matrix is the highest. S_{12} is next. The accuracy of the Horizontal polarization component S_{11} is the lowest.

When the co-polarization of the radar antenna is horizontal polarization and other parameter of the radar are set as proceeding, the complex plane distribution of the PSM element estimation of the test target is shown in Fig. 6.3. Table 6.3 shows the statistical characteristics of the PSM measurement error.

Similarly, Fig. 6.4 illustrates the distribution of complex elements estimation the scattering matrix of standard metal ball. Table 6.4 shows the Statistical characteristics of measurement error.

From Figs. 6.3 and 6.4, Tables 6.3 and 6.4, we can see that the accuracy of the horizontal polarization components S11 of the target scattering matrix is the highest when the co-polarization of the radar antenna is "horizontal polarized". Next is S_{12}. The accuracy of the vertical polarization component S22 is the lowest.

The results of the analysis and a variety of simulation results show that this method of polarization scattering matrix estimation has higher accuracy when the

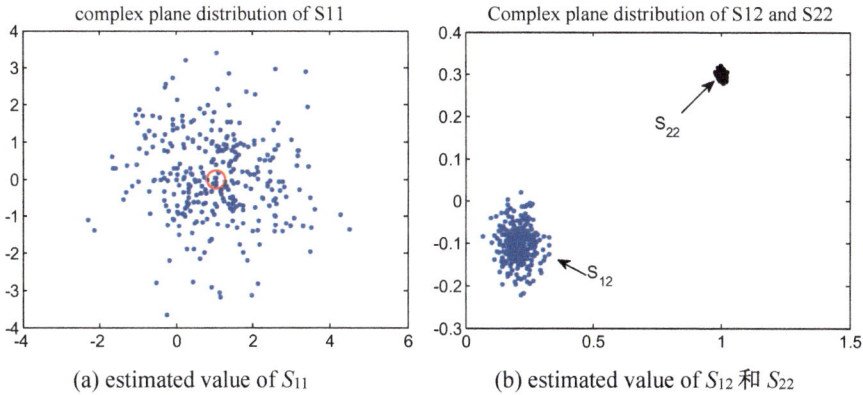

Fig. 6.1 The complex plane distribution of the PSM element estimation of the test target (the Co-polarization is the vertical polarization)

Table 6.1 Statistical characteristics of the PSM measurement error of test target (the Co-polarization is the vertical polarization)

Polarization scattering matrix estimation	Statistical characteristics of $\tilde{S}_{11}$	Statistical characteristics of $\tilde{S}_{12}$	Statistical characteristics of $\tilde{S}_{22}$
$\begin{bmatrix} 0.9263 - 0.0322\text{j} & 0.1986 - 0.0916\text{j} \\ 0.1986 - 0.0916\text{j} & 1.0000 + 0.3000\text{j} \end{bmatrix}$	Mean: $-0.0737 - 0.0322$	Mean: $-0.0014 + 0.0084\text{j}$	Mean: $(0.5588 + 2.8623\text{j}) \times 10^{-5}$
	Standard deviation: 1.724	Standard deviation: 0.0600	Standard deviation: 0.0098

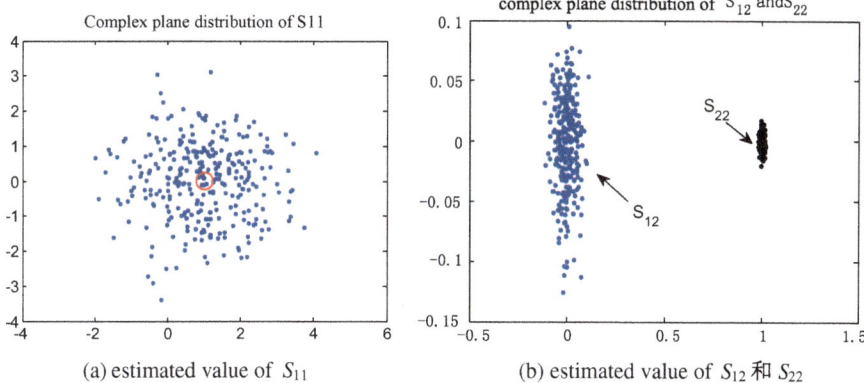

Fig. 6.2 Complex plane distribution of the PSM element estimation of the standard metal ball target (the Co-polarization is the vertical polarization)

estimated component is same to the radar polarization, and has a lowest accuracy when the estimated component is same to the cross-polarization of radar antenna. The accuracy of the middle component is intervenient. For example, when the co-polarization of the radar is horizontal polarization, the estimation accuracy of S11 component is the highest. And next is S12. The accuracy of S22 is the lowest. When the co-polarization of the radar is vertical polarization, the estimation accuracy of S22 component is the highest. And next is S12. The accuracy of S11 is the lowest. This result is easy to understand, because the co-polarization information is more abundant. So, in the use of antenna spatial polarization characteristics to estimate the target polarization scattering matrix, the estimation accuracy of component corresponding to main polarization of the antenna is higher than other components. The estimation accuracy of this method is independent of target polarization scattering matrix.

Applicability analysis of time domain method

The algorithm performance is discussed and simulated in the preceding section. The following section will be a brief analysis of the fault tolerance of the algorithm. Then, this section investigates the influence of main factors such as signal to noise ratio, the size of the observation range, Doppler frequency measurement accuracy and the degree of antenna spatial polarization characteristics and so on.

1. Fault tolerance analysis

For the two coordinate radar scanning in azimuth direction, the elevation data is not completely accurate. Therefore, it is necessary to analyze the influence of the angle measurement error on the performance of the algorithm.

The radar and test target in Fig. 3.1 are discussed in the following simulation. The main polarization of the C-band offset parabolic antenna is vertical. The antenna is scanned in the azimuth $\lfloor -\varphi_{3dB}/2, +\varphi_{3dB}/2 \rfloor$, and the antenna rotating

Table 6.2 Statistical characteristics of the PSM measurement error of standard metal ball (the Co-polarization is the vertical polarization)

Polarization scattering matrix estimation	Statistical characteristics of $\tilde{S}_{11}$	Statistical characteristics of $\tilde{S}_{12}$	Statistical characteristics of $\tilde{S}_{22}$
$\begin{bmatrix} 1.0531 + 0.0520j & -0.0067 + 0.0001j \\ -0.0067 + 0.0001j & 0.9996 + 0.0003j \end{bmatrix}$	Mean: $0.0531 + 0.05j$	Mean: $-0.0067 + 0.0001j$	Mean: $(-3.7618 + 3.4455j) \times 10^{-4}$
	Standard deviation: 1.6026	Standard deviation: 0.0549	Standard deviation: 0.0094

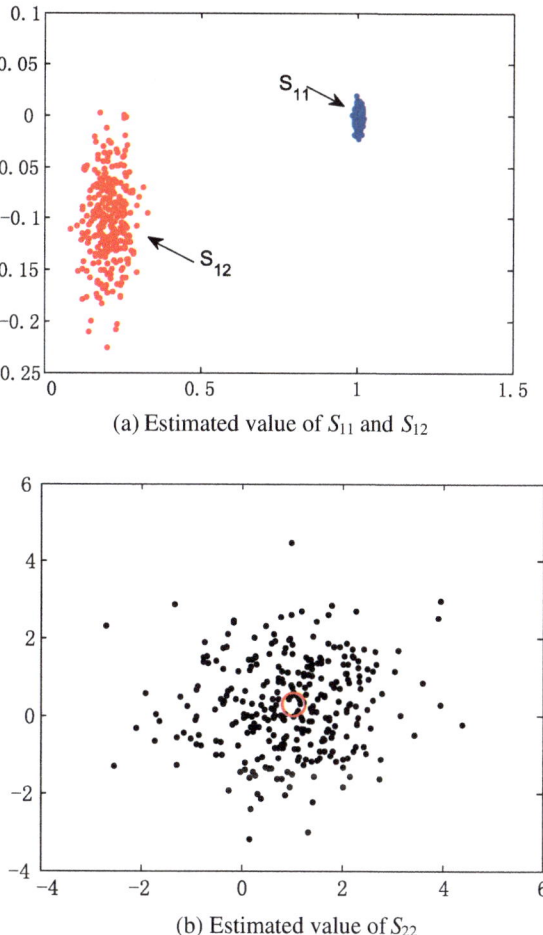

Fig. 6.3 The complex plane distribution of the PSM element estimation of the test target (the Co-polarization is the horizontal polarization)

(a) Estimated value of S_{11} and S_{12}

(b) Estimated value of S_{22}

speed is 6 round/min, and the radar pulse repetition frequency is fr = 1 kHz. Figure 6.5 shows estimation performance of the polarization scattering matrix S along with elevation error. Figure 6.5a shows the mean distribution of absolute value of the cosine of the angle in formula (6.1.40). Figure 6.5b–d estimation accuracy distribution of S11, S12 and S22 component in formula (6.1.36), where the horizontal axis is the signal-noise-ratio. Each curve represents different elevation measurement error. In each case, the Monte Carlo simulation times are 300.

It can be seen from Fig. 6.5 when the angle data obtained by the system have some measurement error, although the estimation performance of the target polarization scattering matrix can be slightly decreased, but the impact is not obvious. Also, various simulations show that the fault tolerance performance of this algorithm is good.

Table 6.3 Statistical characteristics of the PSM measurement error of test target (the Co-polarization is the horizontal polarization)

Polarization scattering matrix estimation $\hat{S}$	Statistical characteristics of $\tilde{S}_{11}$	Statistical characteristics of $\tilde{S}_{12}$	Statistical characteristics of $\tilde{S}_{22}$
$\begin{bmatrix} 1.0000 - 0.0003j & 0.1996 - 0.1001j \\ 0.1996 - 0.1001j & 1.0679 + 0.2946j \end{bmatrix}$	Mean: $(0.4585 + 3.4642j) \times 10^{-4}$ Standard deviation: 0.0054	Mean: $(-4.01 + 0.9805j) \times 10^{-4}$ Standard deviation: 0.0563	Mean: $0.0679 - 0.0054j$ Standard deviation: 1.6635

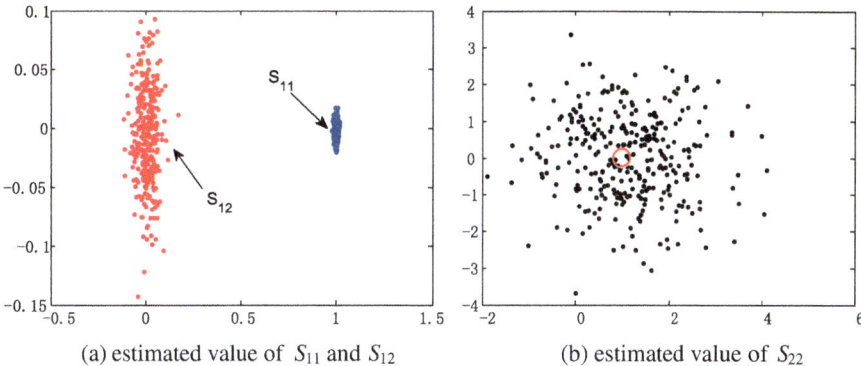

(a) estimated value of S_{11} and S_{12} (b) estimated value of S_{22}

Fig. 6.4 The complex plane distribution of the PSM element estimation of standard metal ball (the Co-polarization is the horizontal polarization)

At the same time, when the main polarization of the radar antenna is vertical polarized, the estimation accuracy of S22 component is better than that of S11 component. $|\xi|$ approaches to 1 and the estimation error of target polarization scattering matrix approaches to 0 while the SNR is higher. That is, the SNR higher, the estimation accuracy better. It can be seen from Fig. 6.5b–d that the estimation accuracy of S22 component is still good when the SNR is low. But the estimation accuracy of S12 and S11 will be good only if the SNR is high.

When the main polarization of the radar antenna is horizontal polarization, the similar simulations are carried. Results show that the fault tolerance performance of this algorithm is still good. The estimation accuracy of S11 component is better than that of S22 component. The estimation accuracy of S11 is still good when the SNR is low. But the estimation accuracy of S12 and S22 will be good only if the SNR is high.

2. Influence of observation range

The observation range ratio is $a = \varphi_0/\varphi_{3dB}$, that is the ratio of the observed range φ_0 and the half power beamwidth φ_{3dB}. The simulation scene in Fig. 6.1 is still used to investigate the relationship between the performance of the algorithm and the observation range. Figure 6.6 shows the estimation performance of the target polarization scattering matrix when $a = 1, 1.5, 2, 2.5$, in which each curve represents a different observation range. Figure 6.6a is distribution of mean value of the angle cosine absolute value. Figure 6.6b–d is the relative accuracy distribution of S11, S12 and S22 respectively.

It can be seen from Fig. 6.6, when the SNR is same, $|\xi|$ becomes large and relative error of the target polarization scattering matrix becomes small with increase of the observation range. When the observation range is changed from $[-\varphi_{3dB}/2, +\varphi_{3dB}/2]$ to $[-0.75\varphi_{3dB}, +0.75\varphi_{3dB}]$ (that is $a = 1 \rightarrow 1.5$), the performance of the algorithm improves obviously. When the observation range is changed from $[-0.75\varphi_{3dB}, +0.75\varphi_{3dB}]$ to $\lfloor-\varphi_{3dB}, +\varphi_{3dB}\rfloor$ (that is $a = 1.5 \rightarrow 2$),

Table 6.4 Statistical characteristics of the PSM measurement error of standard metal ball (the Co-polarization is the horizontal polarization)

Polarization scattering matrix estimation $\hat{\mathbf{S}}$	Statistical characteristics of $\tilde{S}_{11}$	Statistical characteristics of $\tilde{S}_{12}$	Statistical characteristics of $\tilde{S}_{22}$
$\begin{bmatrix} 1.0000 - 0.0005j & -0.0005 + 0.0055j \\ -0.0005 + 0.0055j & 1.0527 - 0.0231j \end{bmatrix}$	Mean: $(0.4816 - 4.6580) \times 10^{-4}$ Standard deviation: 0.0095	Mean: $-0.0005 + 0.0055j$ Standard deviation: 0.0590	Mean: $0.0527 - 0.0231j$ Standard deviation: 1.6470

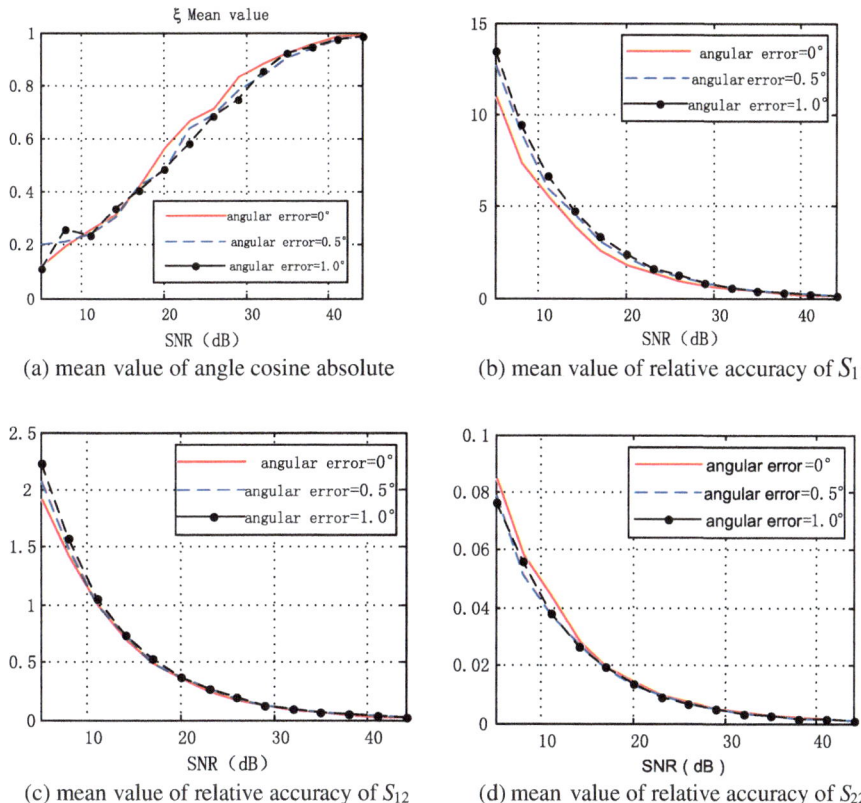

Fig. 6.5 Estimation performance of S under different angle measurement accuracy

the performance of the algorithm improves. When the observation range is changed from $\lfloor -\varphi_{3dB}, +\varphi_{3dB} \rfloor$ to $\lfloor -1.25\varphi_{3dB}, +1.25\varphi_{3dB} \rfloor$ (that is $a = 2 \rightarrow 2.5$), the performance of the algorithm improves a little, but it is no as obvious as the former two conditions.

The nature of the observation range expansion is to fully use the antenna spatial information, but it does not mean that the max observation range is optimal. The signal to noise ratio will gradually decrease if the target leaves from the beam center. It is more obvious when the target is far from the beam center, such as the target is in the side lobe. So the performance of the algorithm improves obvious when $a = 1 \rightarrow 1.5$, but it is no so obvious when $a = 2 \rightarrow 2.5$. On the one hand, observation range expansion can make full use of the antenna's spatial information. On the one hand, signal to noise ratio is reduced via the increase of the observation range and distance from the beam center. Generally speaking, two times or 2.5 times the half power beam width is a good choice for observation range φ_0 (i.e., $\varphi_0 = 2\varphi_{3dB}$ or $\varphi_0 = 2.5\varphi_{3dB}$).

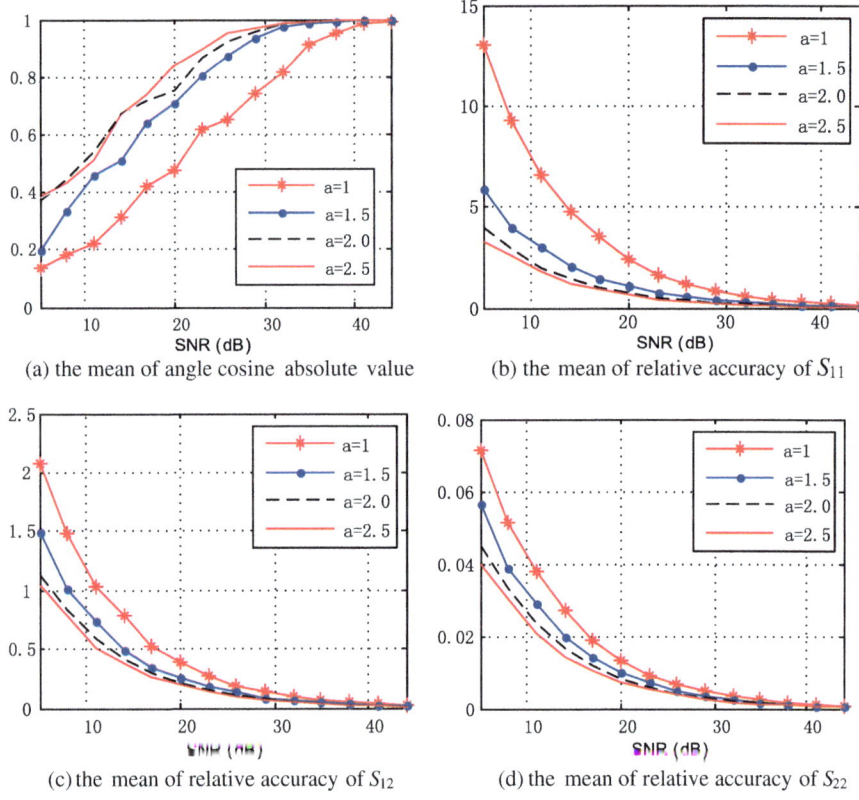

(a) the mean of angle cosine absolute value

(b) the mean of relative accuracy of S_{11}

(c) the mean of relative accuracy of S_{12}

(d) the mean of relative accuracy of S_{22}

Fig. 6.6 Estimation performance of S under different observation range

3. Influence of Doppler shift

According to the analysis of Sect. 6.1.2, the estimation of the scattering matrix of the target is biased in the presence of the Doppler frequency measurement error. The previous discussion is based on the assumption of the Doppler frequency measurement error $f_d = 0$. The performance of target polarization scattering matrix under different Doppler frequency measurement accuracy is discussed, where four kinds of spatial polarization characteristic antenna in Sect. 3.6 is used and the antenna is scanning between $\lfloor -\varphi_{3dB}/2, \ +\varphi_{3dB}/2 \rfloor$.

The test target in preceding section is used all the same. Figure 6.7 shows mean value of absolute value $|\xi|$ via the f_d, where ξ is the cosine of the angle between the true value S and estimated value $\hat{S}$ of the target polarization scattering matrix, and f_d is the Doppler frequency measurement error. Figure 6.7a is the antenna model 1, and Kpolar = 14. Figure 6.7b is the antenna model 2 and Kpolar = 22. Figure 6.7c

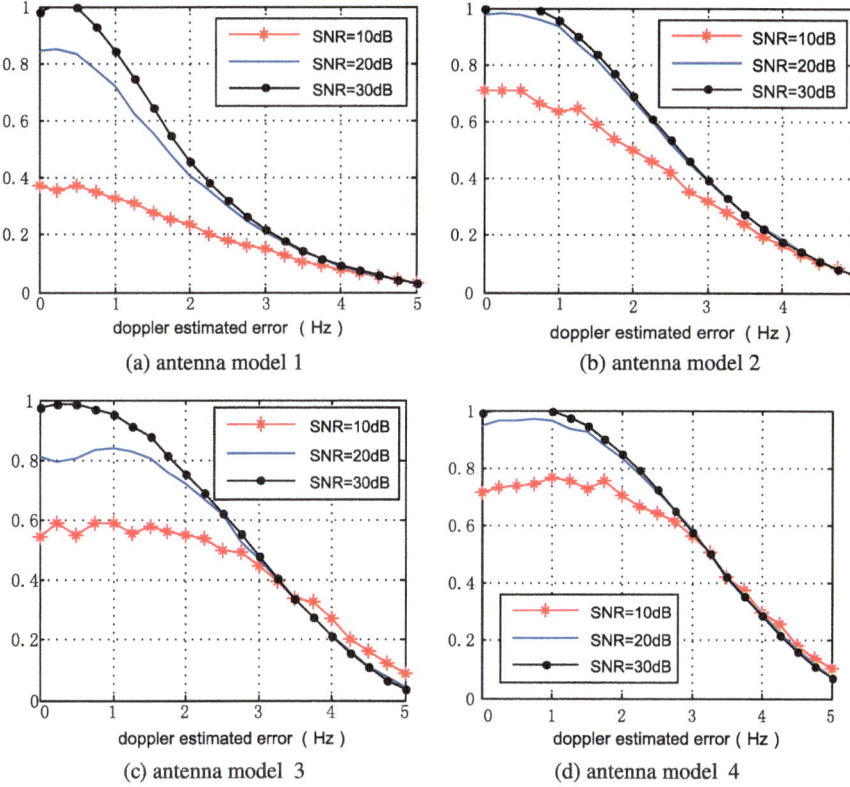

Fig. 6.7 Estimation performance of S under different Doppler frequency measurement error

is the antenna model 3 and Kpolar = 14. Figure 6.7d is the antenna model 4 and Kpolar = 22.

According to Fig. 6.7, when the measurement error of the Doppler frequency increases, the performance of the algorithm decreases dramatically. Even though it can improve SNR and observation range, the performance of the algorithm is still bad. This is the limitation of the algorithm.

4. Influence of the antenna spatial polarization characteristic

The algorithm performance is discussed when the polarization characteristics of the antenna are changing. Assuming the radar antenna main polarization is horizontal polarization, and belong to model 1. The antenna half power beamwidth is $\varphi_{3dB} = 1.4°$. The antenna is scanning between $[-\varphi_{3dB}, +\varphi_{3dB}]$ in azimuth plane. The test target is still used. In the case of different values of Kpolar, which is the spatial change rate of antenna polarization angle, Fig. 6.8 shows the curve of estimation performance of PSM varies with the SNR. Figure 6.8a is the curve of mean absolute value of angle cosine. Figure 6.8b–d show mean estimation accuracy of S11, S12 and S22 respectively. Horizontal axis is SNR. Each curve represents different values of Kpolar. The number of simulation in each case was 300.

From Fig. 6.8a, it can be seen that the greater the value of Kpolar, the antenna spatial polarization change in the same scanning range is more obvious, and the polarization information is more abundant, and the performance of the algorithm improve obviously. Therefore, it is an effective way to improve the performance of the algorithm to use antenna with obvious spatial polarization characteristic in the condition that the main radar performance cannot be affected.

For the horizontal polarization antenna, the estimation accuracy of S11 is the highest. From Fig. 6.8b–d, when the value of Kpolar is greater, the estimation accuracy of S22 can improve obviously. The estimation accuracy of S12 improves too, but no so obviously. The estimation accuracy of S11 changes very little. According to Fig. 6.8 and other various simulations, if the antenna with obvious spatial polarization characteristic is used, the measurement accuracy of the component whose performance is bad originally can improve obviously, and the measurement accuracy of the component whose performance is high originally changes very little.

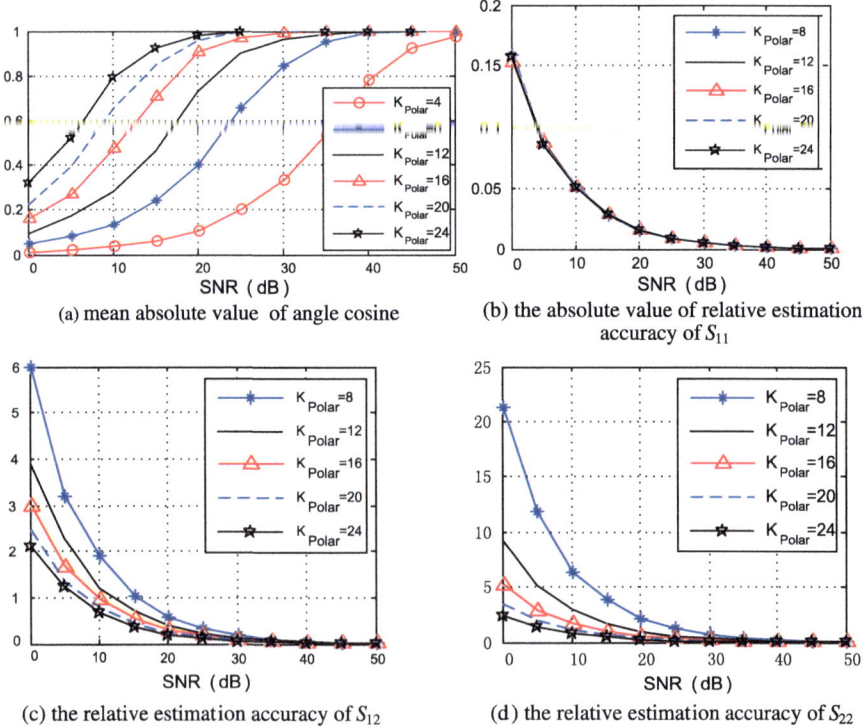

(a) mean absolute value of angle cosine

(b) the absolute value of relative estimation accuracy of S_{11}

(c) the relative estimation accuracy of S_{12}

(d) the relative estimation accuracy of S_{22}

Fig. 6.8 The relationship between estimation performance of PSM and change rate of antenna spatial polarization (model 1)

When the spatial polarization characteristics of radar antenna meet the "model 4", and the spatial variation rate Kpolar of antenna polarization angle variety, the estimation performance of the target PSM is shown in Fig. 6.9.

By Fig. 6.9, when the main polarization of radar antenna is vertical polarization, the estimation accuracy of S11 is the lowest, and the estimation accuracy of S22 is the highest. When the antenna polarization characteristic in spatial variation is improved, the estimation accuracy of S11 improves obviously, and the estimation accuracy of S22 component improves very little.

6.2 Frequency Domain Measurement of Target Polarization Scattering Matrix Based on Antenna Spatial Polarization Characteristics

Algorithm Principle

According to polarization theory, in the horizontal and vertical polarization basis as $(\hat{h}, \hat{v})$, the received voltage of the antenna for the target can be expressed as

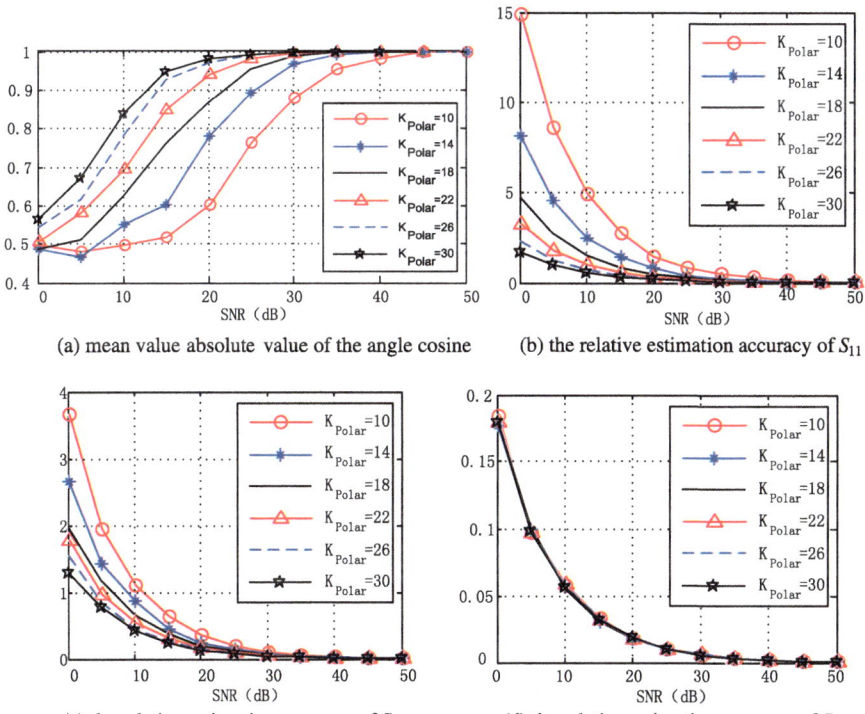

(a) mean value absolute value of the angle cosine

(b) the relative estimation accuracy of S_{11}

(c) the relative estimation accuracy of S_{12}

(d) the relative estimation accuracy of S_{22}

Fig. 6.9 The relationship between estimation performance of PSM and change rate of antenna spatial polarization (model 1)

$$v_r(\varphi) = \beta_1 \cdot G_t(\varphi)G_r(\varphi) \cdot \boldsymbol{h}_r^T(\varphi)\boldsymbol{S}\boldsymbol{h}_t(\varphi) \tag{6.2.1}$$

where β_1 is amplitude, it is the function of the elements in radar equation, including the maximum gain for the transmitting antenna, distance, transmit power, transmission loss, transmission loss etc. The polarization scattering matrix of the target is S. The polarization vector of the transmitting antenna and receiving antenna are $\boldsymbol{h}_t = [h_{tH}, h_{tV}]^T$ and $\boldsymbol{h}_r = [h_{rH}, h_{rV}]^T$ separately, G_t and G_r as the antenna pattern of the transmitting antenna and receiving antenna. The polarization vector and gain of the antenna are the function of the spatial angular coordinates. Suppose radar antenna scanning circularly in the azimuth direction, and note the azimuth angle as φ, and then every other variable can be noted as $\boldsymbol{h}_t(\varphi)$, $\boldsymbol{h}_r(\varphi)$, $G_t(\varphi)$, $G_r(\varphi)$.

When the radar transmitting antenna is used for receiving, it can be learned by the reciprocity principle that the receiving polarization vector satisfys $\boldsymbol{h}_r(\varphi) = \boldsymbol{h}_t(\varphi)$ and $G_t(\varphi) = G_r(\varphi)$, put them into above formula and we can obtained

$$v_r(\varphi) = \beta \cdot G_t^2(\varphi) \cdot \boldsymbol{h}_t^T(\varphi)\boldsymbol{S}\boldsymbol{h}_t(\varphi) \tag{6.2.2}$$

where β is signal amplitude, that is determined together by the radar receiver processing gain as well as the values of the different elements in the radar equation excluding the radar scattering cross section, but irrelevant to the radar polarization and target scattering matrix.

For simplicity, received voltage signal can be normalized with the amplitude, i.e. let $\beta = 1$ in the Eq. (6.2.2), and the received voltage of the antenna for the target can be expressed as

$$
\begin{aligned}
v_r(\varphi) &= G_t^2(\varphi) \cdot \boldsymbol{h}_t^T(\varphi)\boldsymbol{S}\boldsymbol{h}_t(\varphi) \\
&= F_H^2(\varphi) \cdot S_{11} + 2F_H(\varphi)F_V(\varphi) \cdot S_{12} + F_V^2(\varphi) \cdot S_{22}
\end{aligned} \tag{6.2.3}
$$

The spatial spectrum can be obtained by the Fourier transform for the radar received target echo as following

$$V_r(f_\varphi) = \int_{-\varphi_0/2}^{+\varphi_0/2} v_r(\varphi)e^{-j2\pi f_\varphi \varphi}d\varphi \tag{6.2.4}$$

where f_φ is the spatial frequency, $f_\varphi = 1/\Delta\varphi_s$, and φ_0 is the observation window width.

Put Eq. (6.2.3) into the above equation and we get

$$V_r(f_\varphi) = \int_{-\varphi_0/2}^{\varphi_0/2} \times \exp(-j2\pi f_\varphi \varphi) \, d\varphi$$

$$= S_{11} \cdot \int_{-\varphi_0/2}^{\varphi_0/2} F_H^2(\varphi) \times \exp(-j2\pi f_\varphi \varphi) \, d\varphi + 2S_{12} \cdot \int_{-\varphi_0/2}^{\varphi_0/2} F_H(\varphi) F_V(\varphi) \times \exp(-j2\pi f_\varphi \varphi) \, d\varphi$$

$$+ S_{22} \cdot \int_{-\varphi_0/2}^{\varphi_0/2} F_V^2(\varphi) \times \exp(-j2\pi f_\varphi \varphi) \, d\varphi$$

$$= k_{11}(f_\varphi) \cdot S_{11} + k_{12}(f_\varphi) \cdot 2S_{12} + k_{22}(f_\varphi) \cdot S_{22}$$

$$(6.2.5)$$

where

$$k_{11}(f_\varphi) = \int_{-\varphi_0/2}^{\varphi_0/2} F_H^2(\varphi) \times \exp(-j2\pi f_\varphi \varphi) \, d\varphi;$$

$$k_{12}(f_\varphi) = \int_{-\varphi_0/2}^{\varphi_0/2} F_H(\varphi) F_V(\varphi) \times \exp(-j2\pi f_\varphi \varphi) \, d\varphi;$$

$$k_{22}(f_\varphi) = \int_{-\varphi_0/2}^{\varphi_0/2} F_V^2(\varphi) \times \exp(-j2\pi f_\varphi \varphi) \, d\varphi$$

It can be seen from Eq. (6.2.5), that the spectrum $V_r(f_\varphi)$ of the received voltage of the radar antenna is a function of the elements in the target polarization scattering matrix, and each coefficient has the corresponding physical meaning. $k_{11}(f_\varphi)$ is the spectrum of the power pattern of the antenna horizontal polarization component, $k_{12}(f_\varphi)$ is the coupling spectrum of the power pattern by the vertical polarization radiation pattern and the horizontal polarization radiation pattern, $k_{22}(f_\varphi)$ is the spectrum of the power pattern of the antenna vertical polarization component. In particular, when $f_\varphi = 0$, $k_{11}(0)$ denote receiving power of the antenna horizontal polarization field at the spatial area $[-\varphi_0/2, +\varphi_0/2]$, $k_{12}(0)$ is the receiving power of the coupling field of the antenna horizontal polarization and the vertical polarization at the area $[-\varphi_0/2, +\varphi_0/2]$, $k_{22}(0)$ is the receiving power of the antenna vertical polarization field at the spatial area $[-\varphi_0/2, +\varphi_0/2]$.

In engineering, antenna polarization impurity is usually characterized by "main polarization radiation pattern $F_m(\varphi)$" and "cross polarization radiation pattern $F_c(\varphi)$". For example, when the main polarization of the radar antenna is horizontal polarization, the antenna's spatial electric field vector can be expressed as

$$\begin{bmatrix} F_{\mathrm{m}}(\varphi) \\ F_{\mathrm{c}}(\varphi) \end{bmatrix} = G(\varphi) \cdot \boldsymbol{h}(\varphi) = \begin{bmatrix} G(\varphi) \cdot h_{\mathrm{H}}(\varphi) \\ G(\varphi) \cdot h_{\mathrm{V}}(\varphi) \end{bmatrix} = \begin{bmatrix} F_{\mathrm{H}}(\varphi) \\ F_{\mathrm{V}}(\varphi) \end{bmatrix} \qquad (6.2.6)$$

Similarly, when the main polarization of the antenna is vertical polarization, the antenna's spatial electric field vector can be expressed as

$$\begin{bmatrix} F_{\mathrm{m}}(\varphi) \\ F_{\mathrm{c}}(\varphi) \end{bmatrix} = G(\varphi) \cdot \boldsymbol{h}(\varphi) = \begin{bmatrix} G(\varphi) \cdot h_{\mathrm{V}}(\varphi) \\ G(\varphi) \cdot h_{\mathrm{H}}(\varphi) \end{bmatrix} = \begin{bmatrix} F_{\mathrm{V}}(\varphi) \\ F_{\mathrm{H}}(\varphi) \end{bmatrix} \qquad (6.2.7)$$

According to the antenna principle, the main polarization and cross polarization radiation pattern of the antenna can be measured or calculated, which make it possible to measure the target scattering matrix by using spatial polarization characteristics of antenna.

Suppose a radar transmit pulse signal, and the pulse repetition frequency is f_{r}, the scan speed of radar antenna is ωs. The sampling interval of the radar to the target is $\Delta\varphi_{\mathrm{s}}$, and then $\Delta\varphi_{\mathrm{s}} = \omega_{\mathrm{s}}/f_{\mathrm{r}}$. We can get received voltage sequence $\{v_{\mathrm{r}}(\varphi_n)\}$ after sampling the target echo signal $v_{\mathrm{r}}(\varphi)$, where $\varphi_n = -\varphi_0/2 + n\Delta\varphi_{\mathrm{s}}$, $n = 1, 2, \ldots, N_{\mathrm{s}} = [\varphi_0/\Delta\varphi_{\mathrm{s}}]$. $[\cdot]$ denote integral operator. The spatial spectrum can be obtained by the discrete Fourier transform for the received voltage sequence as following

$$\hat{V}_{\mathrm{r}}(f_\varphi) = \sum_{n=1}^{N_{\mathrm{s}}} v_{\mathrm{r}}(\varphi_n) \mathrm{e}^{-\mathrm{j}2\pi f_\varphi \varphi_n} \Delta\varphi_{\mathrm{s}} \qquad (6.2.8)$$

Where $V_{\mathrm{r}}(f_\varphi)$ is discrete approximation formula for $V_{\mathrm{r}}(f_\varphi)$. The two equation can be considered as the equivalent when the sampling interval $\Delta\varphi_{\mathrm{s}}$ is enough small.

From the structure of $V_{\mathrm{r}}(f_\varphi)$ in the previous discussion, it can be learn $V_{\mathrm{r}}(f_\varphi)$ is the joint characterization quantum of each elements of target polarization scattering matrix, and each coefficient is a function of the main polarization and cross polarization radiation pattern of antenna. Therefore, we can get target polarization scattering matrix by selecting some key spatial frequency points of interest, calculating received echo voltage spectrum $V_{\mathrm{r}}(f_\varphi)$ and the corresponding coefficients of the different elements in the scattering matrix which are both corresponded to the frequency, constructing the linear equations and taking simultaneous solution.

Suppose the frequency points of interest as $f_\varphi = f_{\varphi_1}, f_{\varphi_2}, \ldots, f_{\varphi_N}$, linear equations can be constructed as follows:

$$\boldsymbol{V}_{\mathrm{r}} = \boldsymbol{KS} \qquad (6.2.9)$$

In the equation, the target polarization scattering matrix can be recorded as the form of column vector as follows.

$$S = [S_{11}\ 2S_{12}\ S_{22}]^{\mathrm{T}} \tag{6.2.10}$$

$$V_{\mathrm{r}} = \left[V_{\mathrm{r}}(f_{\varphi J})\ V_{\mathrm{r}}(f_{\varphi_2})\ \cdots\ V_{\mathrm{r}}(f_{\varphi_N})\right]^{\mathrm{T}} \tag{6.2.11}$$

$$k(f_{\varphi_1}),\ k(f_{\varphi 2}),\ \ldots,\ k(f_{\varphi N}) \tag{6.2.12}$$

Here, K is name as "spectrum coefficient matrix". $k(f_{\varphi_1})$, $k(f_{\varphi 2})$, ..., $k(f_{\varphi N})$, ..., denotes row vector of matrix K, which are called "spectrum coefficient row vector".

Thus, the problem of [S] matrix estimation is transformed into: Find three linear independent spectrum coefficient row vector which can be denoted K_1, K_2, K_3, and constitute the 3×3 reversible matrix $K_{3\times3}$, so that the solution of the equation is deduced as following

$$S = K_{3\times3}^{-1} V_{\mathrm{r}(3\times1)} \tag{6.2.13}$$

This is a problem of constructing and solving the consistent linear equations, but it has essential difference. We are trying to find an answer: whether a reversible matrix $K_{3\times3}$ exists? If there is, is it the only one? If not the only, then the solutions obtained by the various reversible matrices $K_{3\times3}$ are consistent? In this case, how to treat this situation? If linear equations $V_{\mathrm{r}} = KS$ can be regarded as insoluble simply or mine their physical meaning and make use of them? For specific problems, there are several possible cases:

(1) When the rank of the spectrum coefficient row vector corresponding to each frequency point of interest is $\mathrm{rank}(K_{\mathrm{row}}) \geq 3$, there exist consequentially three typical spectrum coefficient row vector K_1, K_2, K_3 which are linear independent, and compose 3×3 non-singular matrix $K_{3\times3}$, and then we can get the polarization scattering matrix of the target from equation $S = K_{3\times3}^{-1} V_{\mathrm{r}(3\times1)}$. If the matrix $K_{3\times3}$ exist and is unique, there is a unique solution to the equations $V_{\mathrm{r}} = KS$.

(2) When $\mathrm{rank}(K_{\mathrm{row}}) \geq 3$, there may be another case: the constructing method of the matrix $K_{3\times3}$ is not unique, neither the solution educed from equation $S = K_{3\times3}^{-1} V_{\mathrm{r}(3\times1)}$ is unique. This corresponds to the case of the insoluble equations $V_{\mathrm{r}} = KS$, in which case, if it is considered that the polarization scattering matrix of the target can not be solved? The answer is no. The case that the radar and target are in the complex electromagnetic environment, for example, the existence of the objective factors such as noise, clutter, passive or active jamming etc., will lead to the fact that the actual received voltage sequence deviates from the true value.

When constructing spectrum coefficient matrix K with different typical frequency point, there may lead to the inconsistent case of the solved target scattering matrix, but it does not indicate that the algorithm is insoluble.

There are two most commonly used methods: method one is to select a set of the most three representative (has strongest physical meaning) spectrum coefficient row vector, construct the reversible matrix $K_{3\times3}$, take its solution $S = K_{3\times3}^{-1}V_{r(3\times1)}$ as the estimation of the target polarization scattering matrix. Method two is to select multi-group of spectral coefficient row vector, to obtain the multiple estimations of the target scattering matrix S, and to get the final estimation according to certain rules (such as "to get the average value").

(3) When the rank of the spectrum coefficient matrix K, composed of the spectrum coefficient row vector corresponding to each frequency point of interest, is rank$(K) < 3$, it indicates that the maximum number of linear independent vectors of all row vectors and the column vectors of the matrix K is less than 3, the number of linear independent equations is less than the number of variables in this case, which correspond to the case of infinite number of solutions for the equation $V_r = KS$. However, in the actual problem of solving the target polarization scattering matrix, it does not indicate that we can obtain an infinite number of solutions at this time. Instead, it is not possible to obtain the full PSM of the target, but it is possible to solve one or two elements in the target scattering matrix.

Because of the wide variety of radar antennas, the directional characteristics and polarization characteristics of different antennas are very different, which lead to much difference in the selection of key frequency points and the structure of the linear equations. In this section, a "general used" equations are constructed without involving the specific antenna patterns. Through discussing the solution of linear equations, the basic principle of solving the target polarization scattering matrix by using the spatial polarization characteristic is discussed, and the feasibility and limitations of the algorithm are pointed out.

6.2.1   Theoretical Analysis of the Performance for the Algorithm

After the target echo sequence is received by the radar, the observation sequence is intercepted with the azimuth window φ_0, and the DFFT is used for the sequence in the window at the frequency points $f_\varphi = f_{\varphi1}, f_{\varphi2}, f_{\varphi3}$, construct the spectral coefficient matrix K and then the PSM of target is solved. The flow chart is shown in Fig. 6.10.

Because the radar echo signal is mixed with the noise and clutter information besides the target echo signal, the actual received target echo of the radar is denoted

Fig. 6.10 The estimation flow chart of the target PSM

as $\hat{v}_r(\varphi_n)$ at the scanning angle. ε_n is receiver noise, and the influence of cluster and jamming are not considered temporarily, we have

$$\hat{v}_r(\varphi_n) = v_r(\varphi_n) + \varepsilon_n \tag{6.2.14}$$

Suppose the receiver bandwidth is B_n, noise temperature is F_n, and we have $\varepsilon_n \sim N(0, \sigma_\varepsilon^2)$, i.e., the noise obeys complex Gaussian distribution. Noise variance or average power is $\sigma_\varepsilon^2 = k_B T_0 B_n F_n$, where $k_B = 1.38 \times 10^{-23} J/K$, which is the Boltzmann constant. $T_0 = 290$ K is the standard temperature.

After obtaining the spatial echo sequence $\{\hat{v}_r(\varphi_n)\}$ of the target, the observation sequence is intercepted with φ_0 as the spatial window, and then the Fourier transform is used for the sequence in the window at the typical frequency points, i.e., calculate the equation as follows

$$\hat{v}_r(f_\varphi) = \sum_{n=1}^{N_s} \hat{v}_r(\varphi_n) e^{-j2\pi f_\varphi \varphi_n} \Delta \varphi_s = V_r(f_\varphi) + \sum_{n=1}^{N_s} \varepsilon_n e^{-j2\pi f_\varphi \varphi_n} \Delta \varphi_s \tag{6.2.15}$$

In the above formula, when $\Delta \varphi_s$ is enough small, and $\hat{v}_r(f_\varphi) \approx V_r(f_\varphi)$, then the above formula can be written as

$$\delta_f = \hat{v}_r(f_\varphi) - V_r(f_\varphi) = \sum_{n=1}^{N_s} \varepsilon_n e^{-j2\pi f_\varphi \varphi_n} \Delta \varphi_s \tag{6.2.16}$$

where δ_f is the error when the spatial spectrum of the target is calculated according to the target's spatial echo sequence.

It can be seen from the above equation, δ_f is linear weighted sum of the radar observation noise ε_n at various spatial angle. In common cases, ε_n is independent each other, therefore we have

$$\delta_f \sim N(0, \sigma_\delta^2), \sigma_\delta^2 = \sigma_\varepsilon^2 \varphi_0 \Delta \varphi_s \tag{6.2.17}$$

When $\Delta \varphi_s$ is enough small, the cross-correlation between the two frequency points at $f_\varphi = f_{\varphi p}$ and $f_\varphi = f_{\varphi q}$ can be calculated as

$$\left\langle \delta_{fp} \delta_{fq}^* \right\rangle = \sigma_\varepsilon^2 \sum_{n=1}^{N_s} e^{-j2\pi (f_p - f_q)\varphi_n} \Delta \varphi_s^2 = \sigma_\varepsilon^2 \Delta \varphi_s \int_{-\varphi_0/2}^{\varphi_0/2} e^{-j2\pi (f_p - f_q)\varphi} d\varphi \tag{6.2.18}$$

$$= \sigma_\varepsilon^2 \text{sa}\{\pi(f_p - f_q)\varphi_0\}$$

By the formula (6.2.18), the correlation of the corresponding error at the frequency points $f_\varphi = f_{\varphi_1}, f_{\varphi_2}, f_{\varphi_3}$ is related to the distance between the frequency points. When the distance between f_p and f_q is close, δ_{f_1} δ_{f_2} δ_{f_3} are the normal variables which are related to each other; When the distance between f_p and f_q is far away, that satisfy $\left|sa\{\pi(f_p - f_q)\varphi_0\}\right| \ll 1$, the variables as, δ_{f_1} δ_{f_2} and δ_{f_3} can be regarded as independent and identically distributed normal variables.

Suppose the target PSM vector measurement value, the real value and the measurement error are recorded as follows: $\hat{S} = [\hat{S}_{11} \ \hat{S}_{12} \ \hat{S}_{22}]^T$, $S = [S_{11} \ 2S_{12} \ S_{22}]^T$ and $\Delta S = [\Delta S_{11} \ \ \Delta 2S_{12} \ \ \Delta S_{22}]^T$. Denote the spectrum vector measurement value, the real value and the measurement error as $\hat{v}_r = [\hat{v}_r(f_1) \ \hat{v}_r(f_2) \ \hat{v}_r(f_3)]^T$ $V_r = [V_r(f_1) \ V_r(f_2) \ V_r(f_3)]^T$, and $\Delta V_r = [\delta_{f1} \ \delta_{f2} \ \delta_{f3}]^T$. It can be learned from previous analysis that $\Delta V_r \sim N(0, \sigma_\delta^2 I_3)$. According to formula (5.3.13), we can get

$$\hat{S} = K^{-1}\hat{V}_r \tag{6.2.19}$$

Therefore, the measurement error of the target polarization scattering matrix can be expressed as:

$$\Delta S = \hat{S} - S = K^{-1}(\hat{V}_r - V_r) = K^{-1}\Delta V_r \tag{6.2.20}$$

From the formula (6.2.20), ΔS is a linear transformation for ΔV_r, and ΔS is still a normal random vector, denoted as $\Delta S \sim N(\mu_{\Delta S}, \Gamma_{\Delta S})$, which can be easily educed that $\mu_{\Delta S} = K^{-1}\mu_{\Delta V_r} = 0$, $\Gamma_{\Delta S} = K^{-1}R_{\Delta V_r}(K^{-1})^H$, where $\mu_{\Delta V_r}$ and $R_{\Delta V_r}$ are the mean value and the variance of vector ΔV_r separately, then we get

$$\Delta S \sim N(0, \sigma_\delta^2(K^H K)^{-1}) \tag{6.2.21}$$

The formula shows that the estimation error of the three elements of the target PSM is subject to zero mean normal distribution, and the estimated variance is a function of σ_δ^2 and the spectral coefficient matrix K.

It can be seen from Eqs. (6.2.17) to (6.2.21), that it is an effective way to improve the accuracy of target polarization measurement by improving the receiver sensitivity and reducing the spatial sampling interval $\Delta\varphi_s$, which can be achieved by increasing the radar pulse repetition rate or reducing the antenna mechanical scanning speed. In a more accurate sense, the former sentence should be: "Improving the signal to noise ratio and reducing the spatial sampling interval" is an effective way to improve the accuracy of target polarization measurement. However, the measurement accuracy of target PSM element is closely related to the antenna pattern, antenna spatial polarization characteristic and the width φ_0 of spatial observation window. For different types of radar antenna and different types of targets, the performance of the algorithm cannot lump together. Due to limited space, the performance of the algorithm will be simulated and analyzed with typical radar antenna as an example in the following.

Simulation analysis of the performance for the algorithm

Radar system parameters are set as follows: radar work in C band, the working frequency is f = 5 GHz. An offset parabolic antenna with vertical polarization and the matching test target as outlined above are still used as discussed from previous sect. 3.4.2.1. The half power beam antenna width is $\varphi_{3dB} = 1.4°$, antenna rotate speed is 6 rpm, radar pulse repetition frequency is 1 kHz. When the antenna scans separately at the range of $[-\varphi_{3dB}/2, +\varphi_{3dB}/2]$, $[-3\varphi_{3dB}/4, +3\varphi_{3dB}/4]$, $[-\varphi_{3dB}, +\varphi_{3dB}]$ in the horizontal azimuth direction, The changing curve of the 1st-order or 2nd-order moment of the vector angle cosine absolute value $|\xi|$ between the target PSM estimated value $\hat{S}$ and the true value versus the signal to noise ratio SNR is shown in Fig. 6.11, where every curve denote the case that observation range rate $a = \varphi_0/\varphi_{3dB}$ take different typical values, and the times of Monte Carlo simulation in various cases is 300 times.

It can be seen from Fig. 6.11, the performance of the algorithm can be greatly improved by improving the system's signal to noise ratio and increasing the observation range. Figure 6.12 shows the changing curve of the mean value and the variance of the angle cosine absolute value between the target PSM estimated value and the true value versus the signal to noise ratio SNR under different conditions of elevation angle measurement accuracy. The other parameters are set to be constant, Fig. 6.13 shows the estimation performance of the target PSM under different conditions of elevation angle measurement error, when the main polarization of the radar antenna is horizontally polarized.

From Figs. 6.12 and 6.13 shows, when the elevation data obtained is in the presence of measurement error of slightly, although the target PSM estimation performance will decline slightly, but the change was not obvious; at the same time, the condition number of spectrum coefficient matrix K obtained is cond($\mathbf{K}$)= 102.5. Simulation results under various conditions show that the algorithm has better fault tolerance.

Suppose antenna satisfies the spatial polarization characteristic which is shown in the "model one". The half power beam antenna width is $\varphi_{3dB} = 1.4°$. The antenna scans at the range of $\lfloor -\varphi_{3dB}, +\varphi_{3dB} \rfloor$ in the azimuth direction. The target

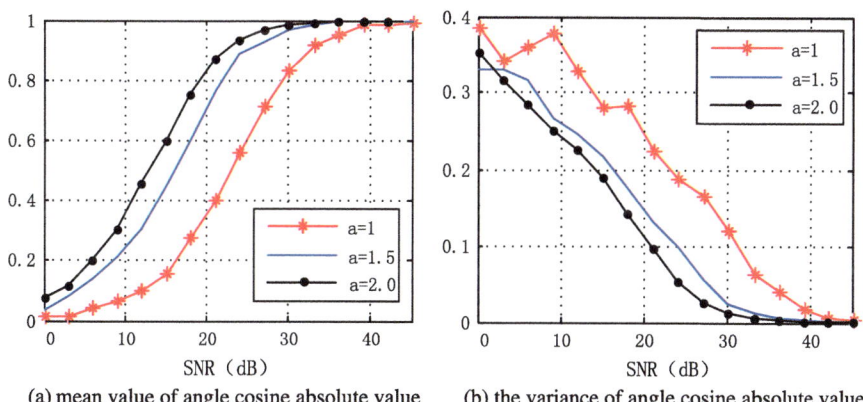

(a) mean value of angle cosine absolute value   (b) the variance of angle cosine absolute value

Fig. 6.11 Estimation performance of S under different conditions of the observation range

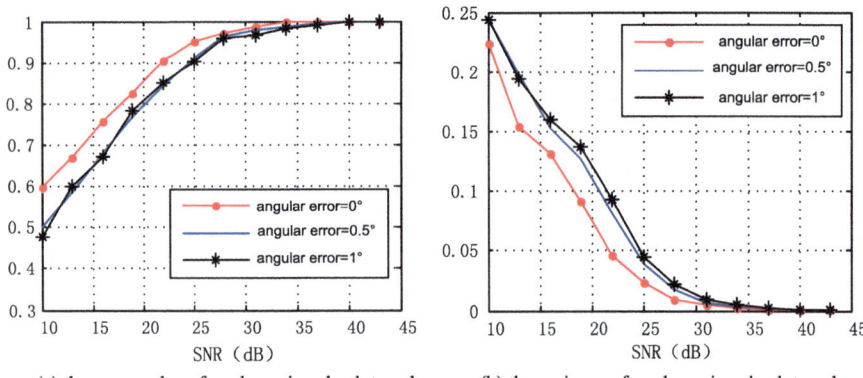

(a) the mean value of angle cosine absolute value   (b) the variance of angle cosine absolute value

Fig. 6.12 Estimation performance of S under different conditions of the elevation angle measurement error (the main polarization of the antenna is vertical polarization)

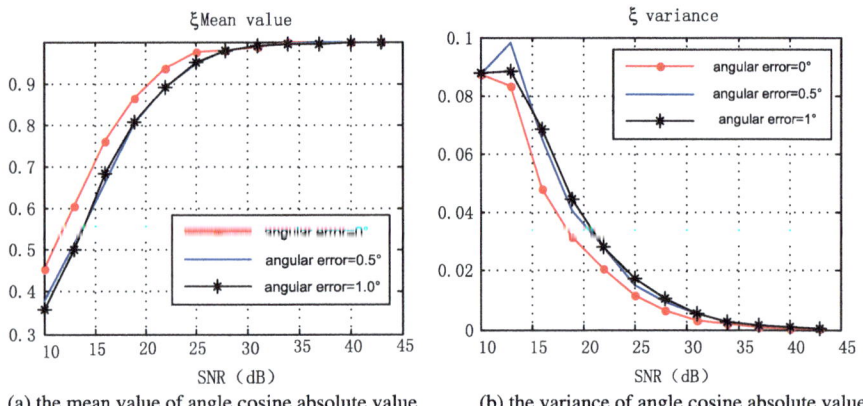

(a) the mean value of angle cosine absolute value   (b) the variance of angle cosine absolute value

Fig. 6.13 Estimation performance of S under different conditions of the elevation angle measurement error (the main polarization of the antenna is horizontal polarization)

is still the matching test target as outlined above. When the spatial polarization change rate of the antenna takes different typical value, the changing curve of the target PSM estimation performance versus SNR is shown in Fig. 6.14. Figure 6.14a shows the changing curve of the mean value of the angle cosine absolute value versus the signal to noise ratio SNR. Figures 6.14a–d are separately the changing curve of relative estimation accuracy of the target PSM's component of S11, S12 and S22 versus SNR, where every curve denote the case that the spatial polarization change rate Kpolar of the antenna take different values, and the times of simulation in various cases is 500 times.

The other parameters are constant, and Fig. 6.15 shows the performance of the target polarization scattering matrix when the main polarization of the radar antenna

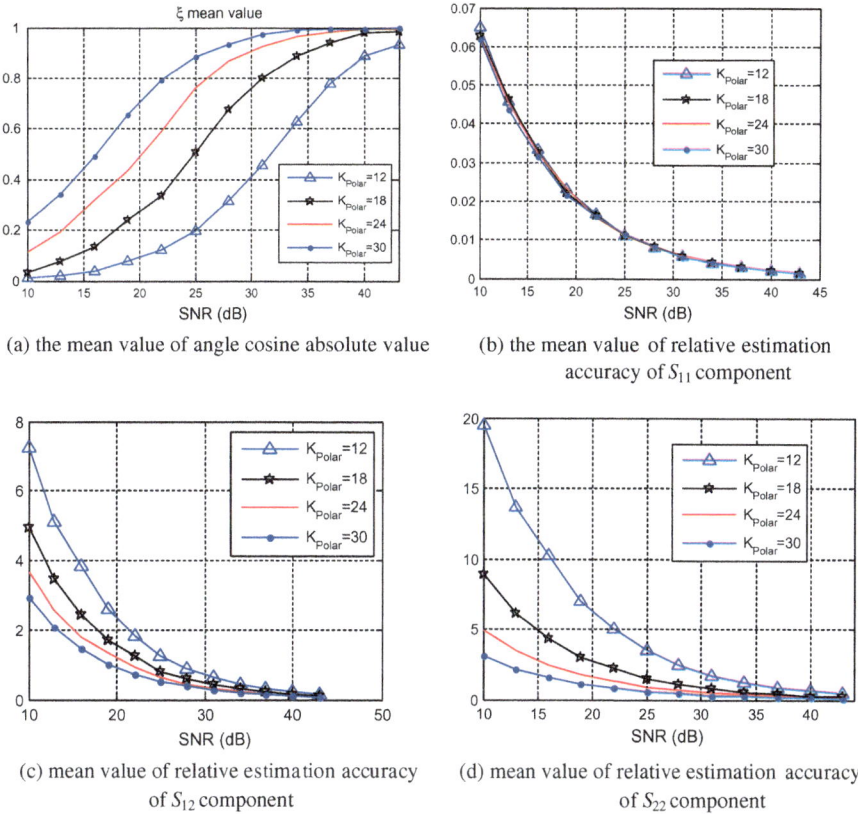

(a) the mean value of angle cosine absolute value

(b) the mean value of relative estimation accuracy of S_{11} component

(c) mean value of relative estimation accuracy of S_{12} component

(d) mean value of relative estimation accuracy of S_{22} component

Fig. 6.14 The relationship between the target PSM estimation performance and the spatial polarization change rate of the antenna (model 1)

is vertical polarization under distinctness degree conditions of the antenna spatial polarization characteristic.

Figures 6.14 and 6.15 shows that as Kpolar value becomes larger, the polarization variation of radar antenna is more obvious, and more polarization information can be used. As a result, the estimation performance of the algorithm will be better in the same scanning space. Furthermore, the effect of the distinctness degree of the antenna polarization characteristic changing in airspace on the estimation of the component with poor accuracy of the measurement is more significant.

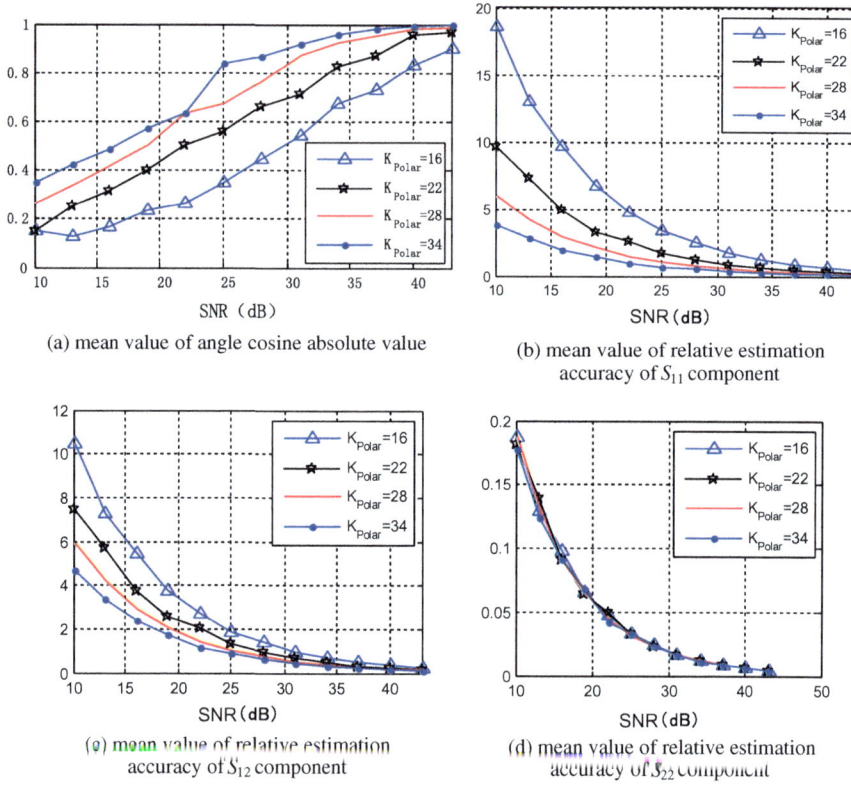

(a) mean value of angle cosine absolute value

(b) mean value of relative estimation accuracy of S_{11} component

(c) mean value of relative estimation accuracy of S_{12} component

(d) mean value of relative estimation accuracy of S_{22} component

Fig. 6.15 The relationship between the target PSM estimation performance and the spatial polarization change rate of the antenna (model 4)

Using the "frequency domain measurement method" discussed in this section, we can get the same conclusion as the "time domain measurement method" in Sect. 6.1 compared with the proposed estimation accuracy. However, as a new method to analyze the problem from different perspectives and to find a way to solve the problem through multiple approaches, this section of the discussion is very meaningful.

Chapter 7
Blanketing Jamming Countermeasure Method Based on Spatial Polarization Characteristics of Antenna

In increasing complicated and adverse electromagnetic field in battlefield, survival ability and working performance of individual electronic information systems including modern radar, communication, navigation and reconnaissance systems faces increasingly stern challenges, so it is required to enhance anti-jamming ability, improve signal receiving quality and dig useful information to the highest extent. In existing conditions, most jamming and anti-jamming measures for ordinary radars are carried out in time domain, frequency domain and space domain. For example, range gate pull-out resistance, repeat frequency agility, pulse width identification, frequency agility, side lobe blanking, side lobe cancellation and etc., do not use polarization information sufficiently. With respect to this problem, we discuss jamming polarization estimation and jamming suppression filter by use of antenna's spatial polarization characteristics, and theoretical method for false target jamming identification and experiment results.

Based on theoretical framework formed in previous five chapters, Sect. 7.2 in this chapter studies algorithm and principle based on echo wave signal polarization decomposition and polarization parameter estimation of the antenna spatial polarization characteristics [157, 158], points out influence of AGC circuit on algorithm performance, and gives corresponding compensation algorithm; Sect. 7.3 first discusses design method for virtual spatial polarization receive technology [159], and studies influence of polarization filter on output signal amplitude and phase characteristics, and proposes a kind of spatial null-phase-shift jamming suppression polarization filter (SNPS-ISPF). By use of this technology, it cannot only suppress jamming effectively and improve signal to interference ratio (SIR), but also can ensure coherence of output signal, eliminate influence of polarization filter on signal amplitude and phase characteristics, realizing compatibility of coherence processing with polarization information processing. Section 7.4 studies adverse influence of elevation angle measurement error on polarization estimation and anti-jamming performance, proposes a virtual spatial multi-channel parallel polarization filter processing method, so as to ensure soundness and effectiveness of jamming suppression performance in blind estimation of elevation angle; Sect. 7.5 analyzes

© National Defense Industry Press, Beijing and Springer Nature Singapore Pte Ltd. 2019 247
H. Dai et al., *Spatial Polarization Characteristics of Radar Antenna*,
https://doi.org/10.1007/978-981-10-8794-3_7

polarization filter performance from four prospects, i.e., antenna polarization characteristics, channel amplitude and phase error, channel noise influence and polarization measurement algorithm, and gives demonstration of effectiveness of polarization filter in polarization valuation error conditions. The study indicates that it is important to optimize output and processing of polarization channel during evaluation and verification of polarization filter effectiveness, and accuracy of the polarization estimator does not directly restrict filtering effectiveness of the whole polarization filter. This is because orthogonal polarization double-channel output data and time- frequency-domain measurement polarization has included polarization error, and polarization filter vector is based on polarization estimation error, but this error is compensated during polarization filter, without influence on effectiveness of polarization filter. This conclusion is important to improve ordinary radar signal processing method, and increase polarization measurement and anti-jamming ability.

Active false target jamming has traditional deception jamming effect. and with rapid development of microwave technology, digital radio frequency memory technology (DRFM) and microelectronic technology, modern active jamming system has developed to the extent that it can autonomously produce false target which is highly approximated to target echo with respect to its energy, waveform and phase modulation and other aspects [168–174]; Therefore, a great deal of lifelike targets has blanketing jamming effect. Although the defense party also can take some corresponding countermeasures at hardware and signal processing layers, e.g., sending orthogonal or approximately orthogonal random phase jamming signal, and using polarization scattering characteristic difference for polarization discrimination, and other methods, but these methods are prone to restriction by radar signal form and polarization measurement ability, and have significant barrier for engineering application. For most modern missile defense radar systems, due to restriction by equipment and technology, their anti-jamming ability at hardware layer and signal processing layer is limited very much. False target jamming that cannot be discriminated by the radar signal processing will form spot trace, and enter the radar data processor, and even form several stable flight path, so that the radar processor is saturated to achieve the purpose of shield the real target. These highly deceptive false targets are huge threat to modern radar defense system undoubtedly.

Existing polarization discrimination algorithm mainly utilizes difference between active false target and radar target with respect to their polarization characteristics [107–111], and part-time or simultaneous polarization measurement radar to transmit polarization modulation signal to measure polarization scattering matrix of real target; Meanwhile, orthogonal polarization receiving of the jamming signal is conducted to measure equivalent scattering matrix of the jamming signal. But, in practice, due to restriction by various factors, existing radar sensors cannot accurately obtain complete and accurate polarization scattering information of the target or jamming signal, i.e. polarization measurement information is deviated from estimation, with big error. Therefore, discrimination of target and jamming through such natures as reciprocity and irregularity of polarization scattering matrix is limited, and single-polarization radar does not have polarization measurement and polarization anti-jamming ability. Section 7.5 in this chapter ingeniously utilizes difference between polarization characteristics of mono-pulse radar amplitude and —difference beam, and studies a processing method for estimating echo polarization and false target jamming polarization, designs sound characteristic parameters for real and false target discrimination based on polarization similarity, gives false target jamming discrimination method and processing flow. This method does not require scattering matrix characteristics of real and false targets for judgment, and reduces calculation and processing difficulty, and is easier to achieve its purpose.

7.1 Polarization Decomposition and Parameter Estimation Method of Return Signal

7.1.1 Radar Observation Equation Building

Offset feed type parabolic antenna is one of most common radar antennas. By taking typical offset feed type parabolic antenna as example, the following figure gives a offset feed type parabolic antenna diagram in the vertical tangent plane, where origin of the coordinate system $OXYZ$ is at apex of the parabolic plane, origin of the coordinate system $O_rX_rY_rZ_r$ is at center of the reflection plane, and Y axis of individual coordinate system is not given in the figure. Polarization characteristics of the antenna is analyzed and calculated in the coordinate system $O_rX_rY_rZ_r$: supposing that angular coordinate of electric wave propagation direction of the antenna in the coordinate system $O_rX_rY_rZ_r$ is (θ, φ), where θ is angle between electric wave propagation direction and Z_r axis, φ is angle between projection in the $X_rO_rY_r$ plane and X_r axis in electric wave propagation direction, and polarization base is named as $\hat{\theta}$ and $\hat{\varphi}$, respectively. Polarization o the electric wave can be represented at this polarization base (Fig. 7.1):

$$E = E_1\hat{\theta} + E_2\hat{\varphi} \qquad (7.1.1)$$

Supposing that the antenna conducts mechanical scanning in azimuth direction, the radar target is unchanged at coordinate θ in $O_rX_rY_rZ_r$, and φ linearity varies; when the target is placed within specified beam wide, number of pulses transmitted by the radar is as follows:

$$N = PRF \cdot BW/f \qquad (7.1.2)$$

where PRF is pulse repeat frequency expressed in Hz, BW is specified beam width of the antenna expressed in degree, f is antenna mechanical scanning rate expressed

Fig. 7.1 Offset feed type paraboloid diagram

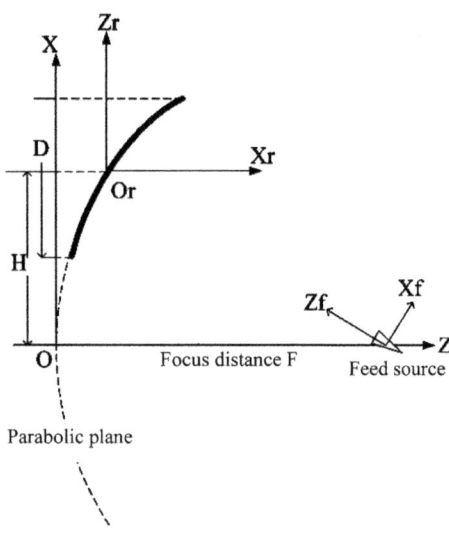

in degree/second, and for typical mechanical scanning radar, N value is usually above ten and below twenty.

At horizontal and vertical polarization base, supposing that co-polarization component of the antenna is horizontal polarization g_H, cross polarization component is vertical polarization g_V, through voltage sequence received by spatial scanning of the radar antenna, a radar observation equation [159] is built as follows:

$$\begin{bmatrix} v_1 \\ v_2 \\ \vdots \\ v_N \end{bmatrix} = \begin{bmatrix} g_H(\varphi_1) & g_V(\varphi_1) \\ g_H(\varphi_2) & g_V(\varphi_2) \\ \vdots & \vdots \\ g_H(\varphi_N) & g_V(\varphi_N) \end{bmatrix} \begin{bmatrix} J_H \\ J_V \end{bmatrix} + N \tag{7.1.3}$$

where, v_N is complex voltage measured and obtained by Nth *pulse echo receiver*, $g_H(\varphi_N)$ and $g_V(\varphi_N)$ are co-polarization component and cross polarization component of the radiation field in (θ, φ_n) direction, respectively, J_H and J_V is H component and V component of Jones vector of the target echo, respectively. The above-mentioned equation can be simplified as follows:

$$V = AJ + N \tag{7.1.4}$$

where, A is called observation matrix; supposing $N \sim (0, \sigma^2 I_N)$, I_N is Nth order unit matrix. The above-mentioned observation equation is linear.

7.1.2 Polarization Decomposition and Parameter Estimation

It can be seen from the Eq. (7.1.4) that voltage signal of the antenna received within a scanning period is not only function of the antenna's spatial polarization characteristics, but also function of time t. Therefore, voltage sequence obtained by antenna scanning is used to estimate orthogonal polarization channel echoing process, which is actually is the incoming wave polarization state estimation process. Polarization of the antenna for any two times of observation forms a group of complete bases for the polarization space, so time sequence is subject to polarization decomposition through any two observation values obtained during scanning, so as to design corresponding processing method, and estimate and obtain echo of the orthogonal polarization channel. The following gives specific mathematical derivation.

7.1.2.1 Jamming Signal Model

Supposing that peak gain of the jammer antenna is G_J, its polarization method can be approx. considered constant during scanning of the radar antenna. Under radar polarization base (H, V), it is written as $\boldsymbol{J}_{HV} = \begin{bmatrix} J_H \\ J_V \end{bmatrix}$, and $\|\boldsymbol{J}_{HV}\|^2 = 1$. The jammer's transmit power is P_J, and jamming signal waveform is $j(t)$. For jamming that is entered in main lobe direction, jamming signal received by the radar will be modulated by spatial polarization of the antenna. At specific azimuth angle φ_k, jamming signal received by the radar is as follows:

$$V_{r,J}(\varphi_k,t) = \sqrt{\frac{P_J G_J G \lambda^2}{(4\pi)^2 r_J^2} \cdot \frac{B_r}{B_J}} \cdot \mathrm{e}^{-j\frac{2\pi r_J}{\lambda}} \cdot \boldsymbol{g}^T(\varphi_k) \cdot \boldsymbol{J}_{HV} \cdot j(\varphi_k,t)$$
$$= K_J \cdot [g_H(\varphi_k)J_H + g_V(\varphi_k)J_V] \cdot j(\varphi_k,t) \tag{7.1.5}$$

where, r_J is jammer distance, $K_J = \sqrt{\frac{P_J G_J G \lambda^2}{(4\pi)^2 r_J^2} \cdot \frac{B_r}{B_J}} \cdot \mathrm{e}^{-j\frac{2\pi r_J}{\lambda}}$, $j(\varphi_k,t)$ is jamming signal at this azimuth angle, B_r is band width of radar receiver and B_J is frequency band width of the noise jamming signal.

Generally, active blanking jamming covers the whole search distance, occupies several distance elements. The search distance is classified into M distance resolution elements, which are named as τ_m, $m = 1, \ldots, M$. Distance resolution element containing jamming is written as τ_J, and jamming signal received by the radar at azimuth angle φ_k is written as $V_{r,J}(\varphi_k,t)$, $\varphi_k = \varphi_{-K}, \ldots, \varphi_K$. Meanwhile, supposing that jamming signal is in the same distance element, it varies randomly around the average value $j(\tau_J)$ at different azimuth angle, and its variable quantity is $\Delta j(\theta_k,\tau_J)$, and is subject to zero average value Gauss distribution, as well as its variance is δ_J^2. Thus, receive voltage in jamming distance element τ_J and at different azimuth angle can be written in the form of linear equation set as follows:

$$\begin{cases} V_J(\varphi_{-K}, \tau_J) = K_J \cdot [g_H(\varphi_{-K})J_H + g_V(\varphi_{-K})J_V] \cdot j(\tau_J) + \Delta j(\varphi_{-K}, \tau_J) \\ \cdots \\ V_J(\varphi_0, \tau_J) = K_J \cdot [g_H(\varphi_0)J_H + g_V(\varphi_0)J_V] \cdot j(\tau_J) + \Delta j(\varphi_0, \tau_J) \\ \cdots \\ V_J(\varphi_K, \tau_J) = K_J \cdot [g_H(\varphi_K)J_H + g_V(\varphi_K)J_V] \cdot j(\tau_J) + \Delta j(\varphi_K, \tau_J) \end{cases} \quad (7.1.6)$$

The above-mentioned equation is written in the form of matrix as follows:

$$\boldsymbol{V}_J(\varphi, \tau_J) = K_J \cdot \boldsymbol{G}(\varphi) \cdot \boldsymbol{J} \cdot j(\tau_J) + \Delta \boldsymbol{j}(\varphi, \tau_J) \quad (7.1.7)$$

where, $\boldsymbol{V}_J(\varphi, \tau_J) = \begin{bmatrix} V_J(\varphi_{-K}, \tau_J) \\ \cdots \\ V_J(\varphi_0, \tau_J) \\ \cdots \\ V_J(\varphi_K, \tau_J) \end{bmatrix}$, $\boldsymbol{G}(\varphi) = \begin{bmatrix} g_H(\varphi_{-K}) & g_V(\varphi_{-K}) \\ \cdots \\ g_H(\varphi_0) & g_V(\varphi_0) \\ \cdots \\ g_H(\varphi_K) & g_V(\varphi_K) \end{bmatrix}$,

$\Delta \boldsymbol{j}(\theta, \tau_J) = \begin{bmatrix} \Delta j(\theta_{-K}, \tau_J) \\ \cdots \\ \Delta j(\theta_0, \tau_J) \\ \cdots \\ \Delta j(\theta_K, \tau_J) \end{bmatrix}$, and $\boldsymbol{I}$ is unit array.

Supposing $\boldsymbol{x}(\tau_J) = \begin{bmatrix} x_H(\tau_J) \\ x_V(\tau_J) \end{bmatrix} = K_J \cdot j(\tau_J) \cdot \boldsymbol{J}$, the above-mentioned equation can be simplified as follows:

$$\boldsymbol{V}_J(\varphi, \tau_J) = \boldsymbol{G}(\varphi) \cdot \boldsymbol{x}(\tau_J) + \Delta \boldsymbol{j}(\varphi, \tau_J) \quad (7.1.8)$$

7.1.2.2   Target Signal Model

Supposing that radar transmit power is P_t, transmit narrow band pulse signal is $e(t)$, and pulse width is τ_p, there is a target in the antenna scanning scope, the distance is r_T, and corresponding round-trip delay $\tau_0 = \frac{2r_T}{c}$. Supposing that target polarization scattering matrix maintains unchanged, it is written as $\boldsymbol{S} = \begin{bmatrix} s_{11} & s_{12} \\ s_{21} & s_{22} \end{bmatrix}$ at polarization basis (H, V), and in single static conditions, it meets reciprocity ($s_{12} = s_{21}$). Thus, target echo signal received by the radar at azimuth angle φ_k is as follows:

$$v_T(\varphi_k, t) = \sqrt{\frac{P_t G^2 \lambda}{(4\pi)^3 r_T^4}} \cdot e^{-\frac{4\pi r_T}{\lambda}} \cdot \boldsymbol{g}^T(\varphi_k) \cdot \boldsymbol{S} \cdot \boldsymbol{g}(\varphi_k) \cdot e(t - \tau_0)$$

$$= K_T \cdot \boldsymbol{g}^T(\varphi_k) \cdot \boldsymbol{E}_a(\varphi_k) \cdot e(t - \tau_0) \tag{7.1.9}$$

where, $\boldsymbol{E}_a(\varphi_k) = \begin{bmatrix} E_{a,H}(\varphi_k) \\ E_{a,V}(\varphi_k) \end{bmatrix} = \boldsymbol{S} \cdot \boldsymbol{g}(\varphi_k)$ is polarization vector of the target

scattering echo, $\varphi_k = \varphi_{-K}, \ldots, \varphi_K$, $K_T = \sqrt{\frac{P_t G^2 \lambda}{(4\pi)^3 r_T^4}} \cdot e^{-\frac{4\pi r_T}{\lambda}}$.

For narrow band radar, distance resolution element occupied by target echo is written as τ_T, echo signal received by the target distance element is written as $v_T(\varphi_k, \tau_T)$ In $\varphi_{-k} \sim \varphi_k$ azimuth angle, polarization vector of the target scattering echo is written as follows:

$$\boldsymbol{E}_a(\varphi) = \begin{bmatrix} E_{a,H}(\varphi_{-K}) & \cdots & E_{a,H}(\varphi_0) & \cdots & E_{a,H}(\varphi_K) \\ E_{a,V}(\varphi_{-K}) & \cdots & E_{a,V}(\varphi_0) & \cdots & E_{a,V}(\varphi_K) \end{bmatrix} = \boldsymbol{S} \cdot \boldsymbol{G}^T(\varphi) \tag{7.1.10}$$

For target distance element, target echo signal received by the radar at individual azimuth angle can be written in the form of matrix as follows:

$$v_T(\varphi, \tau_T) = K_T \cdot D\{\boldsymbol{G}(\varphi) \cdot \boldsymbol{E}_a(\varphi)\} \cdot e(\tau_T) \tag{7.1.11}$$

where $D\{\cdot\}$ represents that main diagonal element of the matrix is taken for operation.

7.1.2.3   Radar Receive Signal Model

Considering that receive signal is influenced by channel noise, signal received by the radar in jamming distance element is as follows:

$$\boldsymbol{r}(\varphi, \tau_m) = V_J(\varphi, \tau_m) + \boldsymbol{n}(\varphi, \tau_m)$$

$$= \boldsymbol{G}(\varphi) \cdot \boldsymbol{x}(\tau_m) + \Delta \boldsymbol{j}(\varphi, \tau_m) + \boldsymbol{n}(\varphi, \tau_m), \quad \tau_m = \tau_J \tag{7.1.12}$$

While signal received by the radar in target distance element is superposition of jamming signal and target echo signal, which is represented as follows:

$$\boldsymbol{r}(\varphi, \tau_m) = V_J(\varphi, \tau_m) + V_T(\varphi, \tau_m) + \boldsymbol{n}(\varphi, \tau_m)$$

$$= \boldsymbol{G}(\varphi) \cdot \boldsymbol{x}(\tau_m) + \Delta \boldsymbol{j}(\varphi, \tau_m) + \boldsymbol{s}(\varphi, \tau_m) + \boldsymbol{n}(\varphi, \tau_m) \quad \tau_m = \tau_T \tag{7.1.13}$$

where, $n(\varphi, \tau_m) = \begin{bmatrix} n(\varphi_{-K}, \tau_m) \\ \cdots \\ n(\varphi_0, \tau_m) \\ \cdots \\ n(\varphi_K, \tau_m) \end{bmatrix}$ is channel noise vector,

$$s(\varphi, \tau_m) = V_T(\varphi, \tau_m).$$

7.1.2.4 Polarization Estimation and Virtual Polarization Decomposition Algorithm

It can be known from the above-mentioned analysis that when the radar antenna beam scans a space area, there are $2K + 1$ azimuth angle scattering samples, which are $\varphi_{-K}, \ldots, \varphi_0, \ldots, \varphi_K$, respectively. Receive voltage signal of the radar at each azimuth angle is $r(\varphi_k, t)$, which is divided into M distance resolution elements, and $2K + 1$ voltages for the same distance element are formed into vector $r(\varphi, \tau_m)$. Signal polarization decomposition is to utilize known antenna spatial polarization characteristics $G(\varphi)$, and voltage vector $r(\varphi, \tau_m)$ of each distance element should be decomposed by least square algorithm, and it is estimated to obtain two ways of orthogonal polarization signal $v(\tau_m)$, which is represented by following equation:

$$v(\tau_m) = arg \quad \min_{v(\tau_m)} \|r(\theta, \tau_m) - G(\varphi) \cdot v(\tau_m)\|^2 \qquad (7.1.14)$$

The above-mentioned equation can be expanded to obtain:

$$v(\tau_m) = \left[G^H(\varphi)G(\varphi) \right]^{-1} \cdot G^H(\varphi) \cdot r(\varphi, \tau_m) \qquad (7.1.15)$$

where, $v(\tau_m) = \begin{bmatrix} v_H(\tau_m) \\ v_V(\tau_m) \end{bmatrix}$, $\tau_m = \tau_1, \ldots, \tau_M$.

In the above-mentioned equation, $G(\varphi)$ is specifically written as:

$$G = \begin{bmatrix} g_H(\varphi_1) & g_V(\theta_1) \\ \vdots & \vdots \\ g_H(\theta_N) & g_V(\theta_N) \end{bmatrix} \qquad (7.1.16)$$

Then,

$$G^H = \begin{bmatrix} g_H(\varphi_1)^* & \cdots & g_H(\varphi_N)^* \\ g_V(\varphi_1)^* & \cdots & g_V(\varphi_N)^* \end{bmatrix} \quad (7.1.17)$$

Thus,

$$[G^H G]^{-1} G^H = \begin{bmatrix} \sum\limits_{n=1}^{N} g_H^2(\varphi_n) & \sum\limits_{n=1}^{N} g_H(\varphi_n) g_V(\varphi_n) \\ \sum\limits_{n=1}^{N} g_V(\varphi_n) g_H(\varphi_n) & \sum\limits_{n=1}^{N} g_V^2(\varphi_n) \end{bmatrix}^{-1}$$
$$\times \begin{bmatrix} g_H(\varphi_1)^* & \cdots & g_H(\varphi_N)^* \\ g_V(\varphi_1)^* & \cdots & g_V(\varphi_N)^* \end{bmatrix} \quad (7.1.18)$$

For the reason that $g_H(\varphi_n)$ is orthogonal to $g_V(\varphi_n)$ in radiation field spherical coordinate system, the above-mentioned equation can be simplified as follows:

$$[G^H G]^{-1} G^H = \begin{bmatrix} \sum\limits_{n=1}^{N} g_H^2(\varphi_n) & 0 \\ 0 & \sum\limits_{n=1}^{N} g_V^2(\varphi_n) \end{bmatrix}^{-1} \begin{bmatrix} g_H(\varphi_1)^* & \cdots & g_H(\varphi_N)^* \\ g_V(\varphi_1)^* & \cdots & g_V(\varphi_N)^* \end{bmatrix}$$
$$= \begin{bmatrix} \dfrac{g_H(\varphi_1)^*}{\sum_{n=1}^{N} g_H^2(\varphi_n)} & \dfrac{g_H(\varphi_2)^*}{\sum_{n=1}^{N} g_H^2(\varphi_n)} & \cdots & \dfrac{g_H(\varphi_N)^*}{\sum_{n=1}^{N} g_H^2(\varphi_n)} \\ \dfrac{g_V(\varphi_1)^*}{\sum_{n=1}^{N} g_V^2(\varphi_n)} & \dfrac{g_V(\varphi_2)^*}{\sum_{n=1}^{N} g_V^2(\varphi_n)} & \cdots & \dfrac{g_V(\varphi_N)^*}{\sum_{n=1}^{N} g_V^2(\varphi_n)} \end{bmatrix}^{2 \times N}$$
$$(7.1.19)$$

The Eq. (7.1.15) can be further rewritten as follows:

$$
\begin{aligned}
v(\tau_m) &= \left[G^H(\varphi)G(\varphi)\right]^{-1} \cdot G^H(\varphi) \cdot r(\varphi, \tau_m) \\
&= \left[G^H(\varphi)G(\varphi)\right]^{-1} \cdot G^H(\varphi) \cdot \left[v_J(\varphi_1, \tau_J) \quad \cdots \quad v_J(\varphi_{n-m}, \tau_J) \quad \cdots \quad v_J(\varphi_N, \tau_J)\right]^T \\
&= \begin{bmatrix} \dfrac{g_H(\varphi_1)^*}{\sum_{n=1}^N g_H^2(\varphi_n)} & \dfrac{g_H(\varphi_2)^*}{\sum_{n=1}^N g_H^2(\varphi_n)} & \cdots & \dfrac{g_H(\varphi_N)^*}{\sum_{n=1}^N g_H^2(\varphi_n)} \\[2mm] \dfrac{g_V(\varphi_1)^*}{\sum_{n=1}^N g_V^2(\varphi_n)} & \dfrac{g_V(\varphi_2)^*}{\sum_{n=1}^N g_V^2(\varphi_n)} & \cdots & \dfrac{g_V(\varphi_N)^*}{\sum_{n=1}^N g_V^2(\varphi_n)} \end{bmatrix}^{2\times N}
\begin{bmatrix} K_J \cdot [g_H(\varphi_1)J_H + g_V(\varphi_1)J_V] \\ K_J \cdot [g_H(\varphi_2)J_H + g_V(\varphi_2)J_V] \\ \vdots \\ K_J \cdot [g_H(\varphi_N)J_H + g_V(\varphi_N)J_V] \end{bmatrix}^{N\times M} \\
&= \begin{bmatrix} \dfrac{g_H(\varphi_1)^*}{\sum_{n=1}^N g_H^2(\varphi_n)} & \dfrac{g_H(\varphi_2)^*}{\sum_{n=1}^N g_H^2(\varphi_n)} & \cdots & \dfrac{g_H(\varphi_N)^*}{\sum_{n=1}^N g_H^2(\varphi_n)} \\[2mm] \dfrac{g_V(\varphi_1)^*}{\sum_{n=1}^N g_V^2(\varphi_n)} & \dfrac{g_V(\varphi_2)^*}{\sum_{n=1}^N g_V^2(\varphi_n)} & \cdots & \dfrac{g_V(\varphi_N)^*}{\sum_{n=1}^N g_V^2(\varphi_n)} \end{bmatrix}^{2\times N}
K_J \begin{bmatrix} g_H(\varphi_1) & g_V(\varphi_1) \\ g_H(\varphi_2) & g_V(\varphi_2) \\ \vdots & \vdots \\ g_H(\varphi_N) & g_V(\varphi_N) \end{bmatrix}^{N\times 2} \\
&\quad \times \begin{bmatrix} J_H(\tau_1) & J_H(\tau_1) & \cdots & J_H(\tau_M) \\ J_V(\tau_1) & J_V(\tau_2) & \cdots & J_V(\tau_M) \end{bmatrix}^{2\times M} \\
&= K_J \begin{bmatrix} J_H(\tau_1) & J_H(\tau_1) & \cdots & J_H(\tau_M) \\ J_V(\tau_1) & J_V(\tau_2) & \cdots & J_V(\tau_M) \end{bmatrix}^{2\times M}
\end{aligned}
$$

$$(7.1.20)$$

After polarization decomposition of the above-mentioned signal, receive signal of each distance element is decomposed into two ways of orthogonal polarization signal. In condition where no receiver noise is considered, receive signal for jamming distance element only includes jamming signal, and it can be known from parameter estimation theory that decomposition result of the Eq. (7.1.15) is the least square estimation of $x(\tau_J)$, i.e. $\hat{x}_{LS}(\tau_J) = v(\tau_J)$, which is subject to energy normalization to obtain polarization state estimate value $\hat{J} = \dfrac{\hat{x}_{LS}(\tau_J)}{\|\hat{x}_{LS}(\tau_J)\|}$ of the jamming signal.

It can be known from nature of the least square algorithm that jamming polarization estimation meets following nature: Estimation result is unbiased estimation of the jamming polarization, i.e.:

$$E\{\hat{x}(\tau_J)\} = x(\tau_J) \tag{7.1.21}$$

Variance of the estimation result is as follows:

$$V\{\hat{x}(\tau_J)\} = \left[G^H(\varphi)G(\varphi)\right]^{-1} \cdot \left(\delta_n^2 + \delta_J^2\right) \tag{7.1.22}$$

For target distance element τ_T, the receive signal is superposition of jamming signal and target signal, and the polarization decomposition result is biased estimation of jamming polarization state. The Eq. (7.1.13) is substituted into the Eq. (7.1.15) to obtain $E\{v(\tau_T)\} = x(\tau_T) + \left[G^H(\varphi)G(\varphi)\right]^{-1} \cdot G^H(\varphi) \cdot s(\varphi, \tau_T)$. This shows orthogonal polarization signal obtained for target distance element is deviated from the jamming polarization state, and the deviation is related to antenna spatial polarization state $G(\varphi)$ and target polarization scattering matrix $[S]$. Thus, two ways of orthogonal polarization signal obtained by polarization decomposition has different characteristics in different distance element, which lays a basis for design and application of subsequent polarization anti-jamming technologies.

To analyze influence of signal/noise ratio on jamming polarization estimation, it can be known from the Eq. (7.1.15) that the least square estimation of the jamming polarization J is as follows:

$$\hat{J} = \left(G^H G\right)^{-1} G^H V = J + \left(G^H G\right)^{-1} G^H N = J + WN \tag{7.1.23}$$

Element $\hat{J}_H$ and $\hat{J}_V$ of $\hat{J}$ is echo of orthogonal polarization channel H and V. Dual orthogonal polarization channel echo $\hat{J}$ is composed of two parts, of which one part is incoming wave J, and the other part is noise WN. WN is subject to normal distribution, and its average value is zero, with variance of $\left(G^H G\right)^{-1} \sigma^2$. Antenna polarization for any two times of observation forms a group of complete base for polarization space, so through observation values obtain for any two times during scanning, it is possible to obtain echo of the orthogonal polarization channel. Considering influence of the receiver noise on estimation result, if observation value obtained for two times are used only, noise level of the orthogonal polarization channel, especially noise level of V channel echo, will be higher. By several times of observation, it is possible to decrease $\left(G^H G\right)^{-1} \sigma^2$, reduce noise level, and improve estimation accuracy. This is actually equivalent to coherent accumulation carried out by several times of observation for improving signal/noise ratio.

The following figure gives estimation results of incoming polarization (Monte Carlo number is 500) when SNR is 30 dB, real value of incoming polarization is $[1 \quad j]$, variation range of azimuth scanning angle φ is $[-5°, 5°]$, and interval is $0.1°$. SNR is defined as ratio of signal energy to noise energy when maximum radiation direction of the antenna is aligned with incoming wave. Maximum radiation direction cross polarization is zero, i.e. $g_V(0) = 0$, so

$$\text{SNR} = 10 \log\left(\frac{|g_H(0)J_H|^2}{\sigma^2}\right) \tag{7.1.24}$$

It can be seen from Fig. 7.2 that estimation error of J_V is much higher than J_H, i.e. two ways of orthogonal polarization channel echo have different noise level, and difference in their noise level is bigger. Equivalent noise power difference between cross polarization and co-polarization channels is about 30 dB. Usually, study on virtual polarization reception shows that the noise level remains unchanged when supposing that noise level of orthogonal channels is equal to virtualize polarization reception. But, virtual polarization technology based on

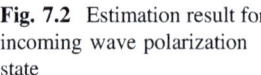

Fig. 7.2 Estimation result for incoming wave polarization state

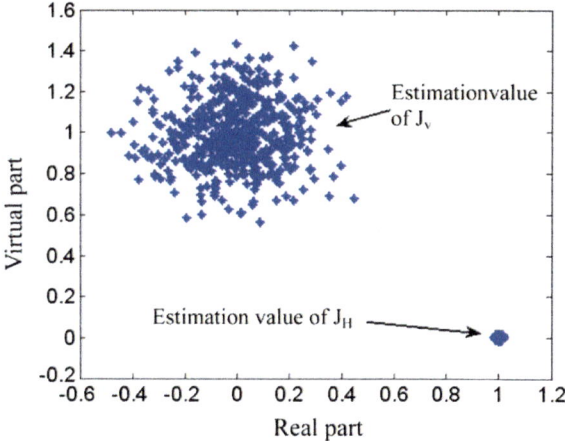

antenna spatial polarization characteristics does not meet this supposition, and noise level difference between H and V channel is 30 dB.

7.1.2.5 Correction of Radar Observation Equation Under Influence of AGC Circuit

During tracking of the radar, an automatic gain control (AGC) circuit must be provided. To ensure automatic direction tracking of the target, it is required that intensity of angular error signal output by the receiver is only related to angle θ by which the target is deviated from the antenna axis (error angle), and not related to such factors as distance from the target and reflection area (i.e. error signal is normalized). If no AGC is provided in the receiver, such requirement is not satisfied because in actual operation, even if θ remains unchanged, intensity of receiver's input signal also varies with change in distance from the target (and target reflection area), i.e., input signal is weak and the output signal is also weak if the target is at long distance; and if the target is at short distance, the input signal is strong and the output signal is also strong.

For this, AGC must be used to enable average value of the target echo to remain unchanged, i.e. when the echo increases, the receiver's gain is reduced correspondingly to enable the echo amplitude to remain unchanged, then amplitude of error signal voltage is equal at same error angle θ, so it is not related to distance of the target, being able to ensure correct tracking of the target.

Considering automatic gain control, echo signal will get weak after the target is deviated from the antenna's main lobe center. To keep inspection of the target signal, AGC may work, so as to enable sequence of voltage received by the antenna during spatial scanning is not necessarily subject to linear variation; In addition, when jamming signal fluctuates violently and jamming power is big, automatic gain control circuit is actuated necessarily to enable the anti-jamming circuit of the receiver to be saturated, so that when the main lobe or side lobe, and even back lobe radiates the jammer, it is still possible to receive strong jamming signal, and at this moment, the observation Eq. (5.2.3) can be rewritten by following expression:

$$
\begin{bmatrix} v_1 \\ v_2 \\ \vdots \\ v_N \end{bmatrix} = \begin{bmatrix} k_1 g_H(\varphi_1)J_H + k_1 g_V(\varphi_1)J_V \\ k_2 g_H(\varphi_2)J_H + k_2 g_V(\varphi_2)J_V \\ \vdots \\ k_n g_H(\varphi_N)J_H + k_n g_V(\varphi_N)J_V \end{bmatrix}
$$

$$
= \begin{bmatrix} k_1 g_H(\varphi_1) & k_1 g_V(\varphi_1) \\ k_2 g_H(\varphi_2) & k_2 g_V(\varphi_2) \\ \vdots & \vdots \\ k_n g_H(\varphi_N) & k_n g_V(\varphi_N) \end{bmatrix} \cdot \begin{bmatrix} J_H \\ J_V \end{bmatrix} \tag{7.1.25}
$$

$$
= \begin{bmatrix} k_1 & 0 & \cdots & 0 \\ 0 & k_2 & \cdots & 0 \\ 0 & 0 & \ddots & 0 \\ 0 & 0 & \cdots & k_N \end{bmatrix} \cdot \begin{bmatrix} g_H(\varphi_1) & g_V(\varphi_1) \\ g_H(\varphi_2) & g_V(\varphi_1) \\ \vdots & \vdots \\ g_H(\varphi_N) & g_V(\varphi_N) \end{bmatrix} \cdot \begin{bmatrix} J_H \\ J_V \end{bmatrix}
$$

It can be seen from the above-mentioned equation that adjustment coefficient of AGC can be written in the form of diagonal matrix $\begin{bmatrix} k_1 & 0 & \cdots & 0 \\ 0 & k_2 & \cdots & 0 \\ 0 & 0 & \ddots & 0 \\ 0 & 0 & \cdots & k_N \end{bmatrix}$, when AGC control coefficient at individual observation angle can be obtained accurately or approximately, it is possible to compensate sequence of voltage received by the AGC loop, which can be represented as follows:

$$
\begin{bmatrix} k_1 & 0 & \cdots & 0 \\ 0 & k_2 & \cdots & 0 \\ 0 & 0 & \ddots & 0 \\ 0 & 0 & \cdots & k_N \end{bmatrix}^{-1} \begin{bmatrix} v_1 \\ v_2 \\ \vdots \\ v_N \end{bmatrix} = \begin{bmatrix} g_H(\varphi_1) & g_V(\varphi_1) \\ g_H(\varphi_2) & g_V(\varphi_1) \\ \vdots & \vdots \\ g_H(\varphi_N) & g_V(\varphi_N) \end{bmatrix} \cdot \begin{bmatrix} J_H \\ J_V \end{bmatrix} \tag{7.1.26}
$$

Therefore, polarization estimation of the signal can be solved as follows:

$$
\begin{bmatrix} J_H \\ J_V \end{bmatrix} = \left(\begin{bmatrix} g_H(\varphi_1) & g_V(\varphi_1) \\ g_H(\varphi_2) & g_V(\varphi_1) \\ \vdots & \vdots \\ g_H(\varphi_N) & g_V(\varphi_N) \end{bmatrix}^H \begin{bmatrix} g_H(\varphi_1) & g_V(\varphi_1) \\ g_H(\varphi_2) & g_V(\varphi_1) \\ \vdots & \vdots \\ g_H(\varphi_N) & g_V(\varphi_N) \end{bmatrix} \right)^{-1} \begin{bmatrix} g_H(\varphi_1) & g_V(\varphi_1) \\ g_H(\varphi_2) & g_V(\varphi_1) \\ \vdots & \vdots \\ g_H(\varphi_N) & g_V(\varphi_N) \end{bmatrix}
$$

$$
\times \begin{bmatrix} k_1 & 0 & \cdots & 0 \\ 0 & k_2 & \cdots & 0 \\ 0 & 0 & \ddots & 0 \\ 0 & 0 & \cdots & k_N \end{bmatrix}^{-1} \begin{bmatrix} v_1 \\ v_2 \\ \vdots \\ v_N \end{bmatrix}
$$

$$(7.1.27)$$

Supposing $V_r = \begin{bmatrix} v_1 & v_2 & \cdots & v_N \end{bmatrix}^T$, polarization decomposition under influence of automatic gain control circuit and estimated correction expression can be simplified as follows:

$$
\begin{bmatrix} J_H \\ J_V \end{bmatrix} = (G^H G)^{-1} G \cdot [AGC]^{-1} V_r \tag{7.1.28}
$$

This section takes account of modulation influence of the AGC circuit on echo signal which will results in distortion of the radar observation equation, leading to failure of the polarization decomposition and estimation method. By correction method given in this section, it is possible to compensate AGC's influence, so as to ensure effectiveness of polarization estimation method. The following will focus on processing method of active jamming suppression.

7.2 Spatial Null Phase-Shift Interference Suppression Polarization Filter Design

7.2.1 Spatial Virtual Polarization Filtering Technology

A. J. Poelman first proposes "Virtual polarization" concept [11–13] which has importance in subsequent design and research variable polarization radar, especially corresponding digital signal processing. The radar receive system polarization state can be changed in receive channel by means of "Virtual polarization", in addition to in antenna and feed line. The so-called "Virtual polarization" is polarization state in which the radar receive antenna and feed line are not changed actually, but the orthogonal dual polarization receive channel signal is subject to amplitude phase weighing, so as to reach same effect obtained by changing receive antenna polarization state. By virtual polarization, it is possible to electric wave received by the antenna (including target echo, clutter and jamming) is subject to any variable polarization processing, so as to achieve match (receive expected signal at full power) or orthogonal (fully eliminate jamming signal) condition. By use of digital signal processing, it is possible to easily and flexibly realize relevant operation required for virtual polarization.

Spatial virtual polarization filter technology given in this book is divided into two steps, first estimate echo of orthogonal polarization channel by the method given in Sect. 7.1 and by use of voltage sequence obtained by antenna scanning, and then realizes variable polarization reception of incoming wave by use of virtual polarization. The processing flow is shown in Fig. 7.3.

Supposing that virtual receive polarization vector of the radar is h, receive voltage of the radar receiver under receive polarization is as follows:

$$v = h^T \hat{J} = h^T J + N_v \tag{7.2.1}$$

Obviously,

$$E[v] = h^T J, \quad \text{var}[v] = G^H (G^H G)^{-1} h \sigma^2 \tag{7.2.2}$$

Voltage received by virtual polarization is composed of two parts, i.e. $h^T \hat{J}$ and N_v. N_v can be taken as corresponding receiver noise during h polarization reception, with variance of $h^H (G^H G)^{-1} h \sigma^2$. It can be seen that spatial polarization characteristics of the antenna is able to enable the radar to have certain virtual variable polarization capacity, but when receiving incoming wave by different polarization, equivalent noise level of the receiver is sensitive to the receive polarization.

Through mathematical derivation, it is possible to obtain following conclusion:

1. Supposing $K = (G^H G)^{-1}$, K is 2×2 Hermit matrix, the diagonal line element is a real number, its characteristic vector is x_1 and x_2, respectively, and corresponding characteristic value is λ_1 and λ_2 (supposing $\lambda_1 \leq \lambda_2$). If receive polarization of the antenna is unit energy, then:

$$\lambda_1 \sigma^2 \leq h^H (G^H G)^{-1} h \sigma^2 \leq \lambda_2 \sigma^2 \tag{7.2.3}$$

That's to say, when the receiver receives incoming wave by different virtual polarization, the receiver's noise power will have maximum and minimum values.

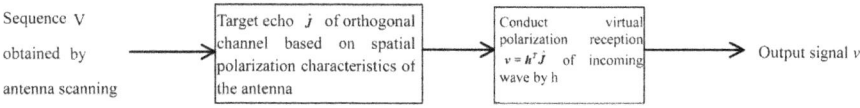

Fig. 7.3 Framework diagram for filter by spatial virtual technology

When taking characteristic vector of matrix K as virtual receive polarization of the antenna, noise power of the receiver reaches maximum/minimum value.

2. By decomposing G with respect to singular value,

$$G = U \Sigma V^H \tag{7.2.4}$$

It is possible to prove:

$$K = V \left(\Sigma^2 \right)^{-1} V^H \tag{7.2.5}$$

i.e., characteristic value of square matrix K is reciprocal for square of the first two singular values of the matrix A, and the characteristic matrix is V, $x_1 = V(1)$, $x_2 = V(2)$, where $V(1)$ and $V(2)$ is first and second column of the matrix V, respectively. When any polarization $\cos \gamma x_1 + \sin \gamma e^{j\phi} x_2$ is received, equivalent noise power is $\left(\lambda_1 \cos^2 \gamma + \lambda_2 \sin^2 \gamma \right) \sigma^2$.

3. When meeting $G_1^H G_2 = 0$ (G_1 and G_2 are listed in first column and second column of the matrix G), it can be derived to obtain $\lambda_1 = \left(G J_1^H G_1 \right)^{-1}$, $\lambda_2 = \left(G_2^H G_2 \right)^{-1}$, $x_1 = [1, 0]^T$, $x_2 = [0, 1]^T$.

If jamming polarization state is known and the noise level does not vary with receive polarization, polarization filter rule can be selected to minimize jamming energy P_I, i.e. select h to enable:

$$h_{opt} = \arg \min_h h^T J \tag{7.2.6}$$

Virtual variable polarization based on spatial polarization of the antenna can decrease jamming energy entering the receiver, but it is also possible to increase equivalent noise level, whose use scope is applicable to conditions with noise energy far less than jamming energy. According analysis in this section, noise level is sensitive to receive polarization, so it is also possible to propose another guideline, even if sum of jamming and noise energy P_{IN} is minimum, so that h is selected to obtain following equation:

$$\begin{aligned} h_{opt} &= \arg \min_h P_{IN} \\ &= \arg \min_h \left(\left| h^T J \right|^2 + h^H \left(A^H A \right)^{-1} h \sigma^2 \right) \end{aligned} \tag{7.2.7}$$

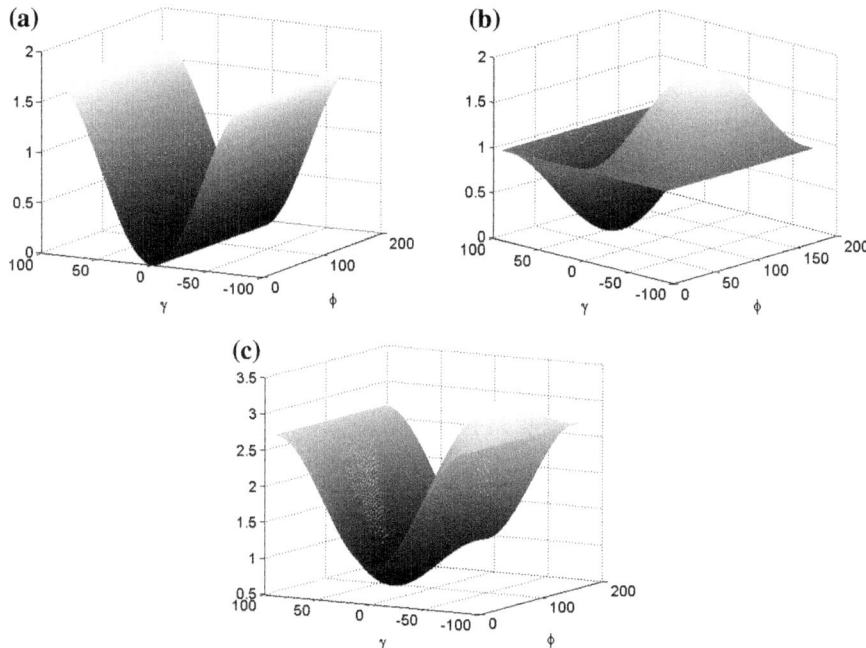

Fig. 7.4 Relationship for equivalent noise power, jamming energy and P_{IN} versus receive polarization (characterized by phase descriptor)

Improvement factor based on jamming suppression method of virtual polarization is defined as $\frac{P_{IN}|_{h=[1,0]}}{P_{IN}|_h}$. Supposing that polarization of jamming incoming wave is $[1,j]$, when INR = 15 dB, Fig. 7.4 gives relationship for equivalent noise power P_N, jamming power P_I, sum of jamming plus noise energy P_{IN} versus virtual receive polarization (characterized by phase descriptor).

When INR = 30 dB, optimum receive polarization obtained under P_I and P_{IN} minimization guideline is $\frac{1}{\sqrt{2}}[1,j]^T$, and improvement factor is 15.64 dB. When INR = 15 dB, optimum receive polarization obtained under P_{IN} minimization guideline is $[0.90, 0.43j]^T$, and improvement factor obtained is 2.65 dB, as well as improvement factor obtained under P_I minimization guideline is 0.65 dB. Figure 7.5 gives improvement factor versus INR curve under two guidelines. It can be seen that the lower the INR, the bigger the difference in improvement factor between two guidelines is. Theoretically, when INR is low, improvement obtained under P_I and P_{IN} minimization guideline is small; when INR is big, it has excellent anti-jamming effect. This condition is met under real countermeasure conditions.

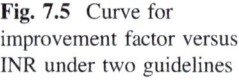

Fig. 7.5 Curve for improvement factor versus INR under two guidelines

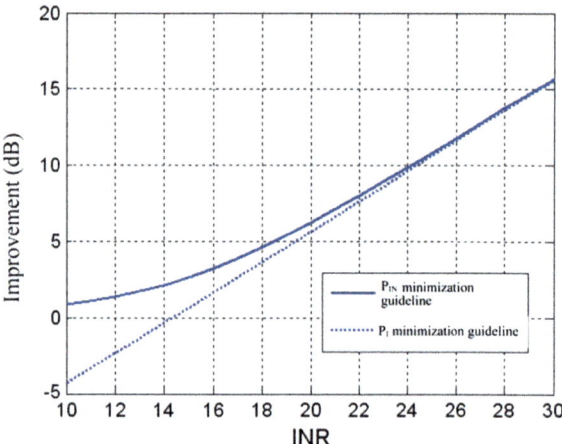

7.2.2 Spatial Null Phase-Shift Interference Suppression Polarization Filter

Modern radar mainly adopts coherent system, and phase of echo signal plays important role during signal processing. In the process of anti-jamming by use of virtual polarization filter technology, polarization state of jamming signal may be fluctuated to some extent, which enables parameters of polarization filter to be varied, meanwhile polarization filter will bring certain polarization loss in echo signal, resulting in amplitude deviation and phase distortion. To eliminate this influence and enable the polarization filter not to influence coherent processing of the radar, as well as ensure its compatibility and improve its polarization anti-jamming ability, a kind of spatial null phase shift suppression polarization filter is designed here. The following gives specific design thought.

Echo signal received by the radar does not only include jamming signal, but also contains target signal. After jamming suppression polarization filter of the echo signal, the output signal obtained is equivalent to scalar sum of the jamming signal and target signal filter.

$$
\begin{aligned}
V_O(t) &= \left(\boldsymbol{v}(\tau_m) + \boldsymbol{E}_S(t_m) \right)^T \cdot \boldsymbol{H} \\
&= \boldsymbol{v}(\tau_m)^T \cdot \boldsymbol{H} + \boldsymbol{E}_S(t_m)^T \cdot \boldsymbol{H}
\end{aligned}
\tag{7.2.8}
$$

ISPF filter designed according to Sect. 7.2.1 filters jamming component in the input signal, and the jamming component can be suppressed.

$$
V_O(t) = \boldsymbol{v}(\tau_m)^T \cdot \boldsymbol{H} \approx 0
\tag{7.2.9}
$$

When jamming component in the input signal, or the residual signal component occupied is very weak and can be ignored, and when considering target signal in the input signal, method and processing steps given in Sect. 7.2.2 can estimate equivalent orthogonal polarization echo of resolution elements occupied by the target signal, which can be represented as follows:

$$E_S(t) = E_s \begin{bmatrix} \cos \varepsilon_s & i_H \\ \sin \varepsilon_s \exp j(\delta_s) & i_V \end{bmatrix} \exp j(\omega_s t + \theta_s) \quad (7.2.10)$$

where ε_s is estimated polarization angle of the target echo, δ_s is estimated polarization angle phase difference of the target echo, θ_s is initial phase of the target signal, ω_s is angular frequency of the target signal.

Polarization filter of target echo component in the input signal can be represented as follows:

$$
\begin{aligned}
E_S(t)^{\mathrm{T}} \cdot H &= E_s \exp j(\omega_s t + \theta_s) \begin{bmatrix} \cos \varepsilon_s & i_H \\ \sin \varepsilon_s \exp j(\delta_s) & i_V \end{bmatrix}^{\mathrm{T}} \begin{bmatrix} \cos \varepsilon_r & i_H \\ \sin \varepsilon_r \exp(j\delta_r) & i_V \end{bmatrix} \\
&\approx E_s \exp[j(\omega_s t + \theta_s)][\cos \varepsilon_s \cdot \cos \varepsilon_r + \sin \varepsilon_s \exp j(\delta_s) \cdot \sin \varepsilon_r \exp(j\delta_r)] \\
&= E_s \exp[j(\omega_s t + \theta_s)] \sqrt{\begin{array}{c} (\cos \varepsilon_s \cdot \cos \varepsilon_r + \sin \varepsilon_s \sin \varepsilon_r \cos(\delta_s + \delta_r))^2 \\ + [\sin \varepsilon_s \sin \varepsilon_r \sin(\delta_s + \delta_r)]^2 \end{array}} \\
&\quad \times \tan^{-1}\left(\frac{\sin \varepsilon_s \sin \varepsilon_r \sin(\delta_s + \delta_r)}{\cos \varepsilon_s \cdot \cos \varepsilon_r + \sin \varepsilon_s \sin \varepsilon_r \cos(\delta_s + \delta_r)} \right)
\end{aligned}
$$

$$(7.2.11)$$

It can be seen from the above-mentioned equation that polarization filter introduces amplitude and phase deviation. This deviation mainly depends on polarization state and virtual polarization filter vector of the target signal. The following considers three kinds of special conditions for analysis:

When co-polarization of the radar antenna is vertical polarization, target echo is dominated by vertical polarization component, i.e. target echo approaches vertical polarization, then $\varepsilon_s = 90°$; thus, the Eq. (7.2.11) can be further rewritten as follows:

$$E_S(t)^T \cdot H \approx E_s \exp[j(\omega_s t + \theta_s)][\exp j(\delta_s) \cdot \sin \varepsilon_r \exp(j\delta_r)] \quad (7.2.12)$$

When co-polarization of the radar antenna is horizontal polarization, target echo is dominated by horizontal polarization component, i.e. target echo approaches horizontal polarization, then $\varepsilon_s = 0°$; thus, the above-mentioned equation can be further rewritten as follows:

$$E_S(t)^T \cdot H \approx E_s \exp[j(\omega_s t + \theta_s)] \cdot \cos \varepsilon_r \quad (7.2.13)$$

When the echo approaches 45° line polarization, then $\varepsilon_s = 45°$

$$E_S(t)^{\mathrm{T}} \cdot \boldsymbol{H} \approx \frac{\sqrt{2}}{2} E_s \exp[\,j(\omega_s t + \theta_s)][\cos \varepsilon_r + \sin \varepsilon_r \exp j(\delta_s + \delta_r)] \qquad (7.2.14)$$

The Eqs. (7.2.12)–(7.2.14) indicates polarization of the target signal is different, target signal amplitude distortion and phase deviation introduced by ISPF filter output is also different.

When polarization component of the echo signal is dominated by vertical polarization, use of polarization filter method set forth in Refs. [161,162] can obtain excellent jamming suppression effect. However, polarization of echo signal has uncertainty in the most radar, and is characterized by several polarization states, which mainly depends on transmit polarization of the radar and variable polarization effect of the target. For example, most air surveillance radars adopt horizontal polarization; thus, use of method in Ref. [161] will cause error, which does not only result in amplitude distortion of the target signal, but also influences coherent processing of the radar. Therefore, null phase shift polarization filter given in the Reference is not complete, and may be failed. The following builds a unified expression for spatial zero phase polarization filter, and gives detailed derivation.

It can be known from the Eq. (7.2.14) that amplitude characteristics of any polarization filter output signal from the ISPF filter are as follows:

$$
\begin{aligned}
\left| E_s(t)^{\mathrm{T}} \cdot \boldsymbol{H} \right|_{AMP} &= \left| E_s \exp[j(\omega_s t + \theta_s)] \right| \\
&\quad \sqrt{(\cos \varepsilon_s \cdot \cos \varepsilon_r + \sin \varepsilon_s \sin \varepsilon_r \cos(\delta_s + \delta_r))^2 + [\sin \varepsilon_s \sin \varepsilon_r \sin(\delta_s + \delta_r)]^2} \\
&= |E_s| \sqrt{\frac{\cos 2\varepsilon_s + 1}{2} \cdot \frac{\cos 2\varepsilon_r + 1}{2} + \frac{1 - \cos 2\varepsilon_s}{2} \cdot \frac{1 - \cos 2\varepsilon_r}{2} + \frac{1}{2} \sin 2\varepsilon_r \sin 2\varepsilon_s \cos(\delta_s + \delta_r)} \\
&= \frac{\sqrt{2}}{2} |E_s| \sqrt{\cos 2\varepsilon_s \cdot \cos 2\varepsilon_r + 1 + \sin 2\varepsilon_r \sin 2\varepsilon_s \cos(\delta_s + \delta_r)}
\end{aligned}
$$

$$(7.2.15)$$

Fig. 7.6 Spherical triangle diagram

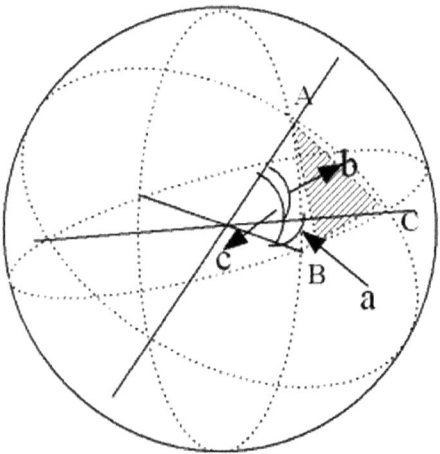

According to cosine law of spherical triangle edges (Fig. 7.6), it can be known:

$$\cos 2\varepsilon_s \cdot \cos 2\varepsilon_r + \sin 2\varepsilon_r \sin 2\varepsilon_s \cos(\delta_s + \delta_r) = \cos a \tag{7.2.16}$$

The Eq. (7.2.15) can be written as:

$$\left| E_S(t)^{\mathrm{T}} \cdot H \right| = \frac{\sqrt{2}}{2} |E_s| \sqrt{1 + \cos a} \tag{7.2.17}$$

The above-mentioned equation indicates that amplitude of the target signal output after polarization filter is $\frac{\sqrt{2}}{2}\sqrt{1 + \cos a}$ the original amplitude. An amplitude recovery filter can be connected in series to recover amplitude characteristics of the output signal. So, filter function for amplitude recovery can be represented as follows:

$$h_A = \sqrt{\frac{2}{1 + \cos a}} \tag{7.2.18}$$

Considering influence of ISPF on phase characteristics after filter of any polarization wave, which is shown as following equation:

$$\begin{aligned} \beta &= \arg[\cos \varepsilon_s \cdot \cos \varepsilon_r + \sin \varepsilon_s \sin \varepsilon_r \cdot \exp(\delta_s + \delta_r)] \\ &= \tan^{-1}\left(\frac{\sin \varepsilon_s \sin \varepsilon_r \sin(\delta_s + \delta_r)}{\cos \varepsilon_s \cdot \cos \varepsilon_r + \sin \varepsilon_s \sin \varepsilon_r \cos(\delta_s + \delta_r)}\right) \end{aligned} \tag{7.2.19}$$

A phase item β is introduced; a phase compensation function is designed to eliminate this phase item, which is represented as follows:

$$h_p = [\cos \varepsilon_s \cdot \cos \varepsilon_r + \sin \varepsilon_s \sin \varepsilon_r \cdot \exp(\delta_s + \delta_r)]^{-1} \tag{7.2.20}$$

It can be known from the Eqs. (7.2.18), (7.2.20) that, after ISPF processing of any polarization wave, the filters that are used for amplitude and phase recovery are written in a unified manner:

$$H_h = h_A \cdot h_P = \frac{\sqrt{\frac{2}{1 + \cos a}}}{\cos \varepsilon_s \cdot \cos \varepsilon_r + \sin \varepsilon_s \sin \varepsilon_r \cdot \exp(\delta_s + \delta_r)} \tag{7.2.21}$$

This filter is connected with ISPF in cascaded manner to form a Spatial Null Phase Shift—Interference Suppression Polarization Filter, abbreviated as SNPS-ISPF. Its diagram is shown in Fig. 7.7.

To sum up, SNPS-ISPF processing is divided into four steps: first, based on least square method, voltage sequence obtained by antenna scanning is decomposed to obtain echo of orthogonal polarization channel; second, extract polarization parameters of echo signal by time domain of frequency domain method; then, ISPF

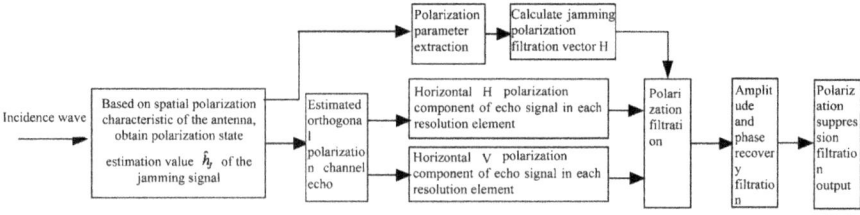

Fig. 7.7 Schematic diagram for spatial null shift interference suppression polarization filter

filter of the orthogonal polarization channel echo obtained at step one is carried out by virtual polarization receive method; lastly, the signal obtained after filter output is subject to amplitude phase recovery filter processing.

7.2.3 Experiment Analysis

To verify effectiveness of SNPS-ISPF, narrow band white noise interference is first taken to carry out simulation test to verify effectiveness and anti-jamming performance of the algorithm; and then, measurement data of the radar is taken to further verify effectiveness of the algorithm.

7.2.3.1 Simulation Experiment Analysis

The simulation experiment result is shown in Fig. 7.8. The simulation produces 4 sections of pulse data, and polarization state, frequency, amplitude and phase of each pulse are different. The first pulse is horizontal polarization (H), the second pulse vertical polarization (V), the third pulse 45° line polarization, and the fourth pulse 135° line polarization. The jamming signal is left-hand circular polarization (LCL), and the jamming signal is superposed to the first and fourth pulse. Figure 7.8a gives amplitude characteristics of echo signal before and after being interfered. After estimating polarization parameter of four pulse signal and orthogonal polarization channel signal, and ISPF and SNPS-ISPF are used for filter, amplitude characteristics of output signal is shown in Fig. 7.8b. It can be seen that after ISPF filter, the jamming signal is suppressed, but amplitude loss of the target signal also exists, so the target signal gets weak. However, after SNPS-ISPF filter, amplitude characteristics of the target signal remains consistent with that shown when it is not interfered, SNPS-ISPF increase amplitude and phase recovery of the target signal polarization, which compensates the polarization filter loss effectively. Meanwhile, it can be known from Fig. 7.8c, d that phase of ISPF filter output target signal changes, and SNPS-ISPF prevents phase shift introduced by polarization filter, ensuring normal working of the radar under coherent system.

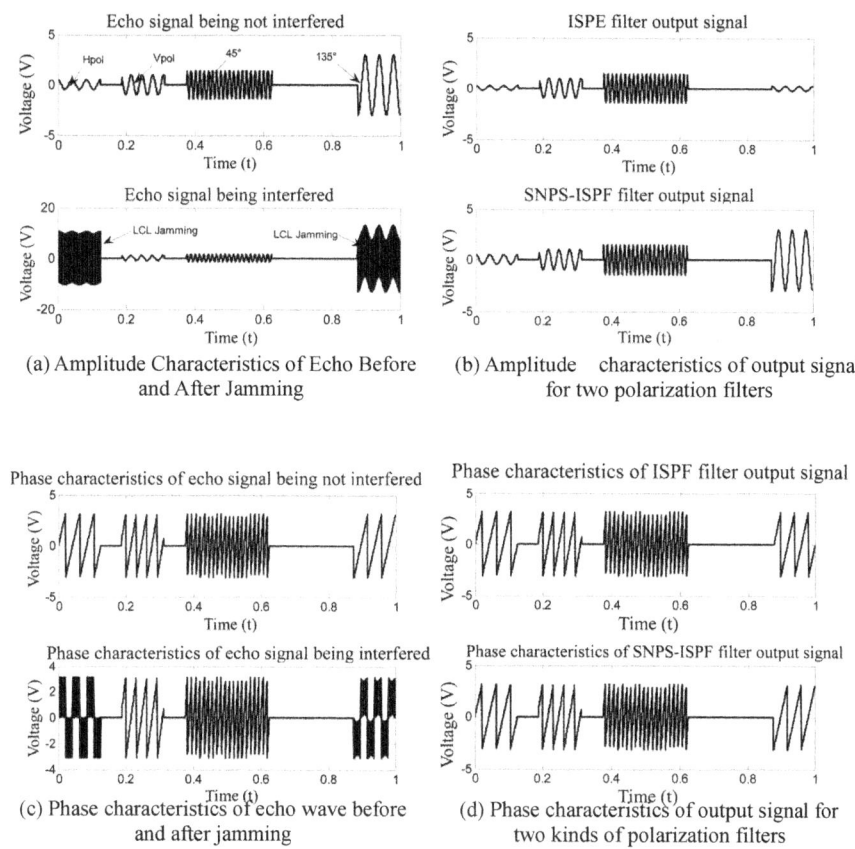

(a) Amplitude Characteristics of Echo Before and After Jamming

(b) Amplitude characteristics of output signal for two polarization filters

(c) Phase characteristics of echo wave before and after jamming

(d) Phase characteristics of output signal for two kinds of polarization filters

Fig. 7.8 Performance comparison between two kinds of polarization filters

7.2.3.2 Experiment Data Analysis

To verify effectiveness of the method proposed herein, outfield experiment data is collected and analyzed. During experiment, the jamming equipment is fixed to non-metallic test tower, and transmits 45° line polarization narrow band noise jamming. The radar antenna is in vertical polarization state, and works in receive condition only, as well as conducts omnidirectional scanning in azimuth direction at rate of 3 revolutions/minute. The signal source transmits vertical polarization signal as target echo signal, which is superposed to certain resolution element of the jamming signal. By adjusting the jamming signal power to a value bigger than target signal power, it is possible to estimate a Signal to Interference Ratio (SIR) of -35.9543 dB before polarization filter. Figure 7.9a, b give polarization parameter distribution estimated by use of echo wave sequence obtained by antenna scanning, and orthogonal polarization channel echo. It can be seen that estimated polarization ratio of interference signal is 18, polarization phase angle is 145°, and there is significant estimation error. This is because two ways of orthogonal polarization channel echo have different gain (V polarization channel is 30 dB stronger than H polarization channel). This error reflects intrinsic property under modulation of the antenna polarization characteristics, and is compensated during polarization filter. Meanwhile, it can also be seen that resolution element where the target is placed is different from interfered polarization. This will result in a sudden change in polarization valuation and decomposition, indicating effectiveness of polarization decomposition and valuation method. Figure 7.9c, d give amplitude and phase characteristics before and after ISPF and SNPS-ISPF filter, respectively. It can be seen that, before it is not processed, target signal is flooded in jamming signal, and it is impossible to effectively detect the target. Following ISPF filter, Signal to Interference Ratio (SIR) reaches 17.2077 dB, and jamming signal of individual distance elements can be effectively suppressed. After SNPS-ISPF filter, signal

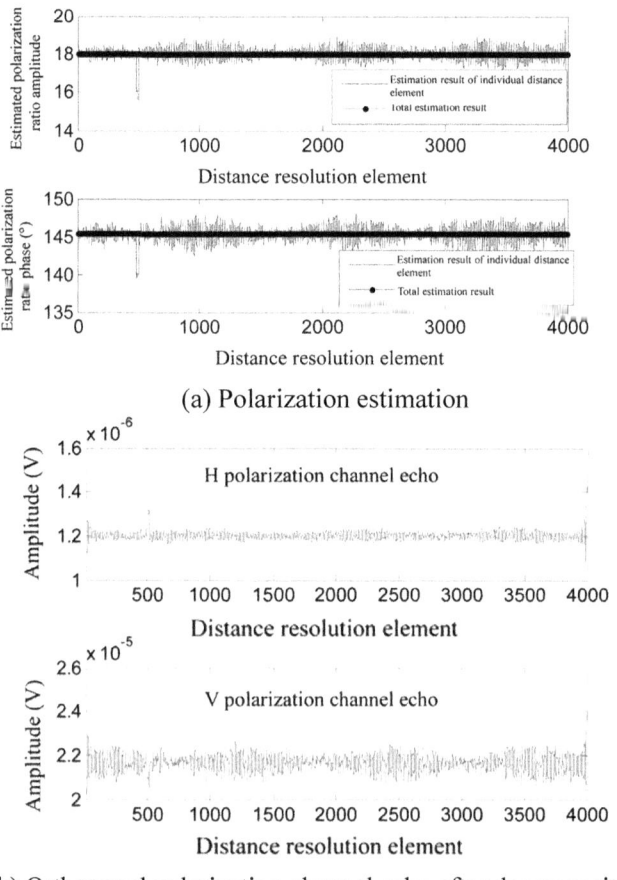

(a) Polarization estimation

(b) Orthogonal polarization channel echo after decomposition

Fig. 7.9 Measurement data verification of spatial null phase shift polarization filter

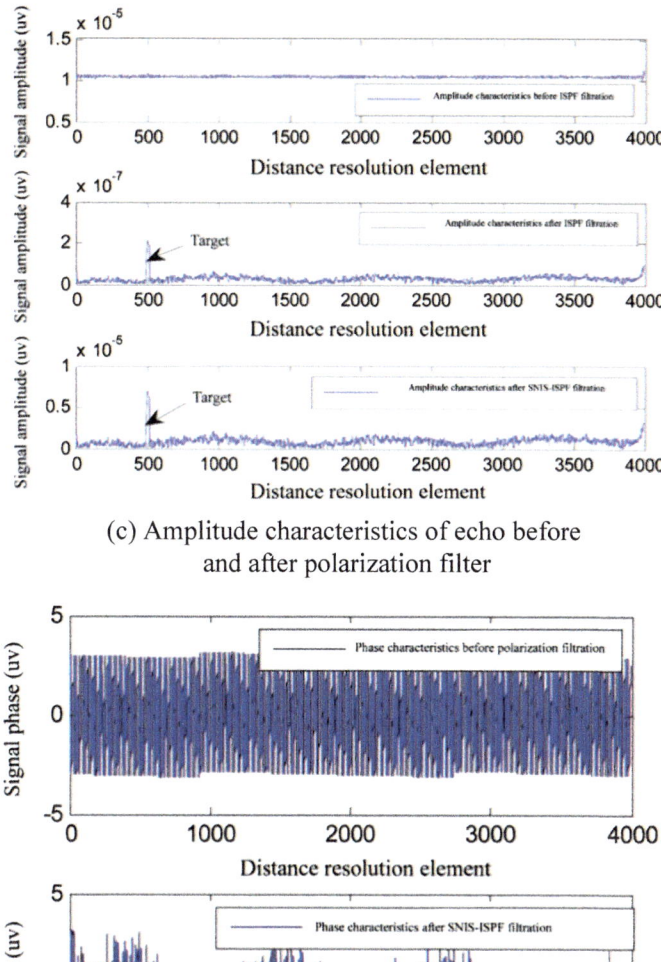

(c) Amplitude characteristics of echo before
and after polarization filter

(d) Phase characteristics of echo before
and after polarization filter

Fig. 7.9 (continued)

amplitude phase characteristics are effectively compensated. Based on the ISPF, it is improved by about 14 dB, proving effectiveness of the method proposed.

This section gives mathematical principles for obtaining jamming signal polarization by use of antenna spatial polarization characteristics, and analyzes unified expression for influence of ISPF filter on amplitude and phase characteristics of the target signal, which can be used to indicate polarization filter effect of target echo in individual polarization state, and gives expression for standard null phase shift spatial instantaneous polarization filter (SNPS-ISPF). Through simulation and measurement data, correctness and effectiveness of algorithm analysis are verified. This provides theoretical and practical basis for developing potential to process polarization information of the single polarization radar.

7.3 Spatial Virtual Multi-channel Parallel Polarization Filter in Elevation Blind Estimation Condition

Through the above-mentioned analysis and processing, it can be known that when target or interference azimuth and elevation can be measured accurately, orthogonal polarization decomposition, polarization estimation and polarization filter suppression of the receive signal can be realized by use of one-dimension (azimuth dimension) spatial polarization characteristics during antenna scanning; however, most warning radars are two-coordinate mechanical scanning mechanism radar in practical application, i.e. beam scanning is achieved by use of the whole antenna system or part of mechanical movement, and use of fan beam is one of common scanning method for circumferential scanning. The antenna's beam at horizontal plane is very narrow, and the azimuth resolution can be up to less than 1 degree and be measured at azimuth angle of the target accurately, while it is very wide at vertical plane, the scanning is rough, and elevation angle of the target cannot be measured accurately.

To accurately obtain spatial polarization characteristics of the parabolic antenna at azimuth dimension and elevation dimension. It is required to adopt GRASP9 developed by TICRA. Spatial polarization characteristics of the parabolic antenna obtained are shown in Fig. 7.10, where focal distance $F = 1.25$ m, antenna aperture $D = 1$ m, and working frequency of the antenna is 12 GHz. According definition of feed source during electromagnetic calculation, co-polarization component E_θ of this antenna is vertical polarization, i.e., $E_\theta = E_V$. Cross polarization component E_φ is horizontal polarization, i.e., $E_\varphi = E_H$. It can be seen that antenna co-polarization and cross polarization radiation field meet symmetrical requirements, but polarization characteristics in azimuth direction is different from that in elevation direction.

It can be seen from Fig. 7.11a that amplitude and phase of co-polarization radiation field at fixed elevation angle θ and at different azimuth angle φ is evenly symmetrical to $\varphi = 0$, i.e., $E_V(\varphi) = E_V(-\varphi)$. Cross component of the antenna at

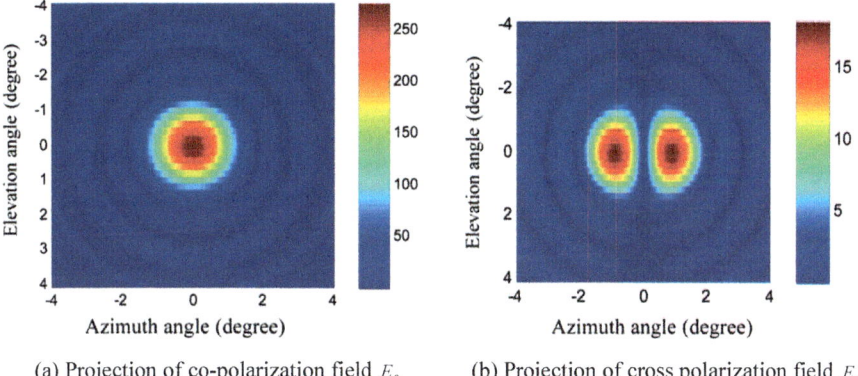

(a) Projection of co-polarization field E_θ (b) Projection of cross polarization field E_φ

Fig. 7.10 3D projection of spatial polarization characteristics of a parabolic reflection antenna

main lobe center position is weakest, and change in cross polarization component is obviously and monotonously increased within range from the center to the first zero point, and antenna polarization characteristics vary obviously and regularly. Figure 7.11a also gives azimuth polarization direction pattern in conditions of being deviated from the normal line direction by 0°, 0.2°, 0.6° and 1.2° in elevation direction, which is subject to the above-mentioned law; but as the deviation angle increases, the radiation gain declines obviously.

Figure 7.11b gives elevation direction pattern at fixed azimuth. It can be seen that its co-polarization and cross polarization also meet symmetry, where cross polarization amplitude is evenly symmetrical to $\varphi = 0$, and phase is jumped by 180° at position of $\varphi = 0$, showing odd symmetry, i.e. $E_H(\varphi) = -E_H(-\varphi)$. And, its cross polarization structure is similar to that of co-polarization direction pattern. Because co-polarization component shows same changing trend as the cross polarization component at main lobe, polarization characteristics of the antenna does not change obviously in elevation direction.

Due to difference in antenna polarization characteristics at different elevation angles, polarization estimation of the receive signal sequence by use of the method given in Sect. 7.2 will produce estimation error. At this moment, estimation results can be represented as follows:

$$v'(\tau_m) = \left[G^H(\varphi, \theta + \Delta\theta) G(\varphi, \theta + \Delta\theta) \right]^{-1} \cdot G^H(\varphi, \theta + \Delta\theta) \cdot r(\varphi, \theta, \tau_m) \quad (7.3.1)$$

where, $\Delta\theta$ is elevation angle measurement error, $G(\varphi, \theta + \Delta\theta)$ represents spatial polarization characteristics in azimuth direction under elevation angle error $\Delta\theta$.

If real polarization of receive signal is represented as $s = \left[\cos\gamma \quad \sin\gamma e^{j\phi} \right]^T$, Stocks vector is represented as follows:

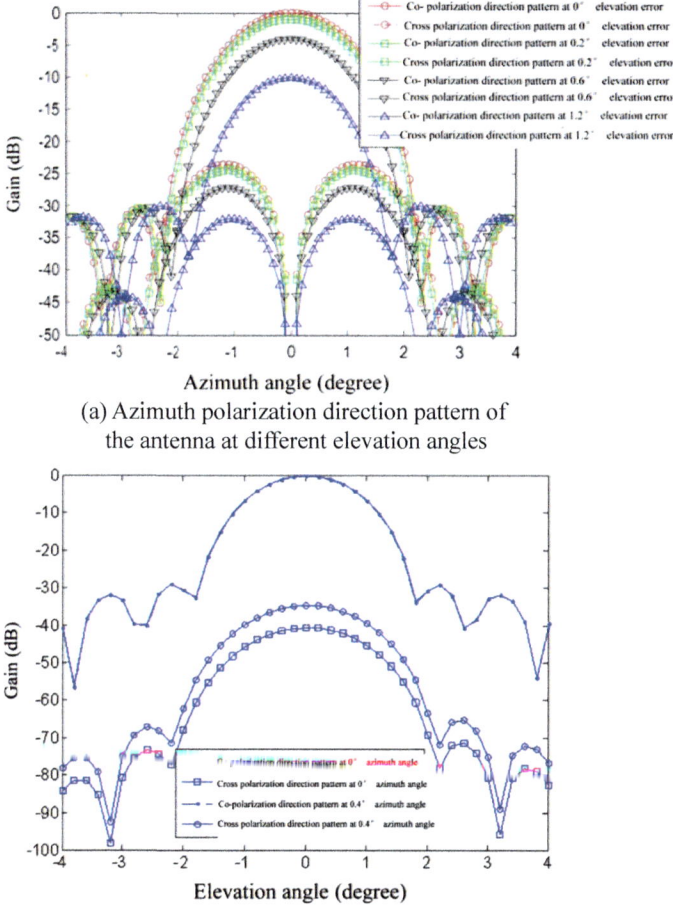

(a) Azimuth polarization direction pattern of
the antenna at different elevation angles

(b) Elevation polarization direction pattern of
the antenna at different azimuth angles

Fig. 7.11 2D distribution of spatial polarization characteristics of an offset parabolic reflection antenna

$$\boldsymbol{J}_s = R(\boldsymbol{s} \otimes \boldsymbol{s}^*) = R \begin{bmatrix} \cos \gamma \\ \sin \gamma \mathrm{e}^{j\phi} \end{bmatrix} \otimes \begin{bmatrix} \cos \gamma \\ \sin \gamma \mathrm{e}^{-j\phi} \end{bmatrix} = R \begin{bmatrix} \cos^2 \gamma \\ \cos \gamma \sin \gamma \mathrm{e}^{-j\phi} \\ \sin \gamma \cos \gamma \mathrm{e}^{j\phi} \\ \sin^2 \gamma \end{bmatrix} \quad (7.3.2)$$

where $\otimes$ represents Kronecker product, $R = \begin{bmatrix} 1 & 0 & 0 & 1 \\ 1 & 0 & 0 & -1 \\ 0 & 1 & 1 & 0 \\ 0 & j & -j & 0 \end{bmatrix}$.

Therefore,

$$J_s = R(s \otimes s^*) = \begin{bmatrix} 1 \\ \cos^2 \gamma - \sin^2 \gamma \\ \cos \gamma \sin \gamma e^{-j\phi} + \sin \gamma \cos \gamma e^{j\phi} \\ \cos \gamma \sin \gamma e^{-j\phi} j - \sin \gamma \cos \gamma e^{j\phi} j \end{bmatrix} = \begin{bmatrix} 1 \\ \cos^2 \gamma - \sin^2 \gamma \\ 2 \sin \gamma \cos \gamma \cos \phi \\ 2 \sin \gamma \cos \gamma \sin \phi \end{bmatrix}$$

$$(7.3.3)$$

The above-mentioned equation can be simplified as $J_s = [1 \quad g_s]$, where
$$g_s = \begin{bmatrix} \cos^2 \gamma - \sin^2 \gamma \\ 2 \sin \gamma \cos \gamma \cos \phi \\ 2 \sin \gamma \cos \gamma \sin \phi \end{bmatrix}.$$

Supposing that polarization estimation result can be represented as $\hat{s} = \begin{bmatrix} \cos \tilde{\gamma} & \sin \tilde{\gamma} e^{j\tilde{\phi}} \end{bmatrix}^T$ when there is elevation angle error. The Stokes vector can be

represented as $\tilde{J}_s = R(\tilde{s} \otimes \tilde{s}^*) = [1 \quad \tilde{g}_s]$, where $R = \begin{bmatrix} 1 & 0 & 0 & 1 \\ 1 & 0 & 0 & -1 \\ 0 & 1 & 1 & 0 \\ 0 & j & -j & 0 \end{bmatrix}$.

Where polarization filter vector $H_r = \begin{bmatrix} \cos \gamma_r \\ \sin \gamma_r e^{j\phi_r} \end{bmatrix}$, and polarization filter vector is orthogonal vector of the polarization estimated value, then

$$\begin{cases} \gamma_r = \gamma + \frac{\pi}{2} \\ \phi_r = -\phi \end{cases} \qquad (7.3.4)$$

That's to say, $H_r = \tilde{s}_\perp = \begin{bmatrix} -\sin \tilde{\gamma} & \cos \tilde{\gamma} e^{-j\tilde{\phi}} \end{bmatrix}^T$. Stokes vector can be represented as $\tilde{J}_\perp = R(\tilde{s}_\perp \otimes \tilde{s}_\perp^*) = [1 \quad \tilde{g}_{s\perp}^T]^T$, where $\tilde{g}_{s\perp} = \Lambda_{12}\tilde{g}_s$, $\Lambda_{12} = \text{diag}\{-1,-1,1\}$. Due to limitation by polarization estimation accuracy, $H_r = \hat{s}_\perp$ is not usually orthogonal to real interference polarization s, so residual interference power after polarization filter is calculated as follows [163]:

$$P_r = |\tilde{s}_\perp{}^T s|^2 = \frac{1}{2}\tilde{J}_\perp{}^T U_4 g_s = \frac{1}{2}(1 + \tilde{g}_{s\perp}{}^T \Lambda_3 g_s) \qquad (7.3.5)$$

where $U_4 = \text{diag}\{1,1,1,-1\}$, $\Lambda_3 = \text{diag}\{1,1,-1\}$.

Supposing that vector angle between signal polarization g_s and estimated polarization $\tilde{g}_s$ is ϑ, the above-mentioned equation can be simplified as follows:

$$P_r = \frac{1}{2}(1 - \tilde{g}_s{}^T \tilde{g}_s) = \frac{1}{2}(1 - \cos \vartheta) \qquad (7.3.6)$$

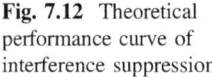

Fig. 7.12 Theoretical performance curve of interference suppression

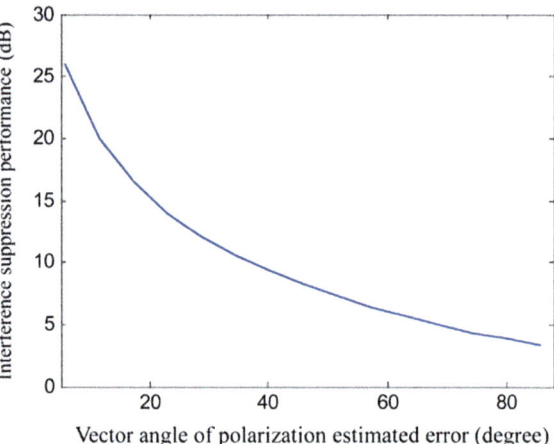

It can be seen that residual interference power after polarization (Fig. 7.12) filter depends on estimated error of interference polarization, and interference polarization depends on antenna polarization characteristics, so elevation angle estimation error will influence performance of polarization filter.

7.3.1 Spatial Virtual Multi-channel Parallel Polarization Filter

By the help of Poelman's multi-notch polarization filter [12] thought, a spatial virtual multi-channel polarization filter is designed, whose basic thought is to use antenna characteristic data at several elevation angle to decompose receive signal to obtain restructured multi-way orthogonal polarization signal output, then carry out interference suppression polarization filter, and lastly take "Logic product" of the filtered signal as final result to improve filter efficiency.

Supposing that elevation angle resolution is δ_θ, radar elevation angle value is θ_t, and antenna elevation beam width is θ_e, it is possible to adopt antenna characteristics $G(\varphi, \theta_t \pm \delta_\theta), G(\varphi, \theta_t \pm 2\delta_\theta), \ldots, G(\varphi, \theta_t \pm n\delta_\theta)$ at several elevation angle because the elevation measurement error is unknown (where $n \leq \left\lfloor \frac{\theta_e}{\delta_\theta} \right\rfloor$) to carry out polarization estimation of the echo sequence, so as to obtain $2n$ ways of polarization echo output $[v_1(\tau_m) \quad v_2(\tau_m) \quad \cdots \quad v_n(\tau_m)]^{2 \times 1}$; with respect to $2n$ ways of output signal, it is possible to build n polarization filters $H_{r1}, H_{r2}, \ldots H_{rn}$, whose polarization filter vector is different. After filter, following logic operation for n ways of output is carried out:

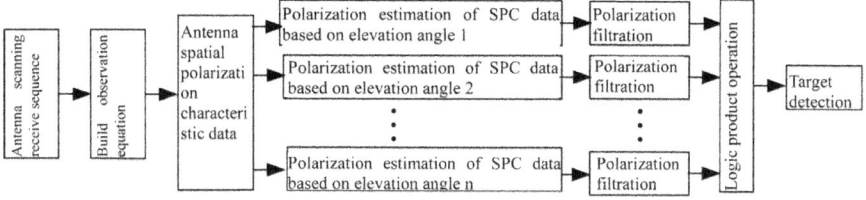

Fig. 7.13 Block diagram for spatial virtual multichannel polarization filter

$$O_f(\tau_m) = \min_{k=1}^{n}\{\boldsymbol{H}_{rk}^T \cdot \boldsymbol{v}_k(\tau_m)\} \qquad (7.3.7)$$

Figure 7.13 gives a block diagram for spatial virtual multichannel polarization filter.

7.3.2   Simulation Experiment and Result Analysis

Input signal to interference ratio (SIR_i) is defined as the ratio of energy of target echo signal received by the radar at azimuth angle θ_0 to that of interference signal, i.e.,

$$SIR_i = \frac{\left|E\left\{v_{r,T}(\varphi_0, t)^2\right\}\right|}{\left|E\left\{v_{r,J}(\varphi_0, t)^2\right\}\right|} \qquad (7.3.8)$$

The output signal to interference ratio (SIR_o) is defined as the ratio of energy of target distance signal to that of interference distance element signal after orthogonal polarization decomposition, polarization estimation and spatial virtual polarization filter, i.e.,

$$SIR_o = \frac{\left|E\left\{o(\tau_T)^2\right\}\right|}{\left|E\left\{o(\tau_J)^2\right\}\right|} \qquad (7.3.9)$$

To quantitatively describe anti-jamming performance of spatial virtual polarization filter, the improvement factor is defined as follows:

$$EIF = 10\lg\frac{SIR_o}{SIR_i} \qquad (7.3.10)$$

In experiment, azimuth scanning window width is taken as $-3°$ to $3°$, and there are 79 sampling points in total. Supposing that the jammer adopts left-hand circular

Fig. 7.14 Narrow band
blanketing jamming spectrum

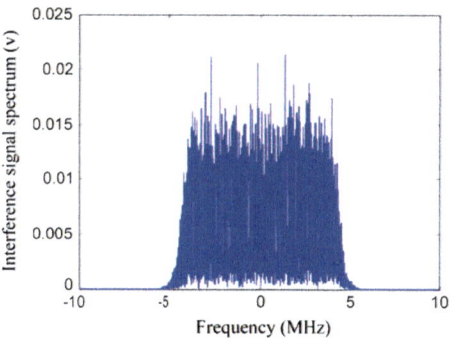

polarization narrow band white noise jamming, with jamming band width of 8 M, and jamming signal of $j(t) = P_J \cdot I_J(t) \cdot e^{j\varphi_J(t)}$ where P_J determines jamming power, and amplitude $I_J(t)$ is subject to Rayleigh distribution and phase $\varphi_J(t)$ subject to $[0, 2\pi]$ even distribution, as well as interference/noise ratio in main lobe direction of the lobe is 30 dB. Figure 7.14 is jamming signal spectrum obtained by simulation, and Fig. 7.15 gives time domain distribution of jamming signal, with jamming signal power of 12.23 dBw. The target echo is in 3000–3200th resolution element, pulse width is 10 μs, target polarization scattering matrix is $S = \begin{bmatrix} 1 & 0.3j \\ 0.3j & 0.9 \end{bmatrix}$, echo signal power is 0 dBw, and input signal to interference ratio is $SIR_i = -12.23$ dB; at this moment, distance information of the target is completely interfered and suppressed, and target information cannot be detected.

Figure 7.16 gives time domain distribution of receive signal amplitude for a distance element obtained when the antenna main lobe sweeps the interference direction; at this moment, envelope of the receive signal is irregular but approximately subject to Gaussian distribution profile; Figs. 7.17, 7.18 and 7.19 give output signal of the receiver at three antenna scanning positions, and Fig. 7.20 shows spatial virtual polarization filter results. It can be seen that, before virtual polarization filter, echo will be flooded in interference signal at individual azimuth angles, and the target cannot be detected effectively; it can be known through

Fig. 7.15 Space-time
distribution of blanketing
interference

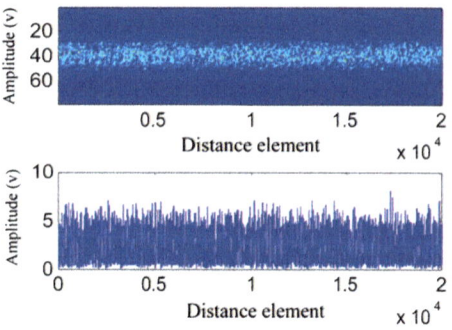

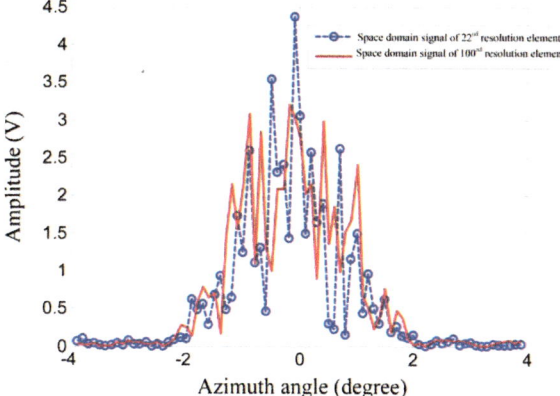

Fig. 7.16 Time domain distribution of interference signal

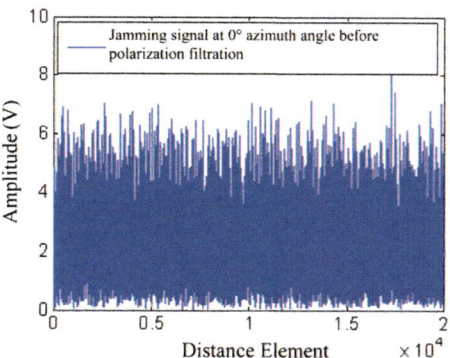

Fig. 7.17 Interference signal received in main lobe center

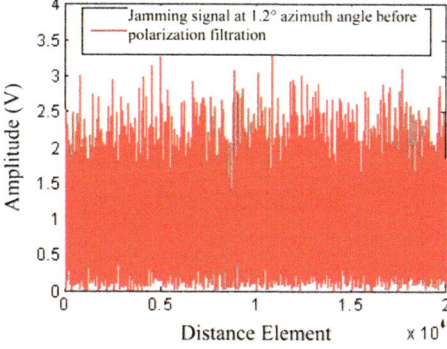

Fig. 7.18 Jamming signal received by the antenna at 1.2° scanning angle

calculation that input signal to interference ratio is $SIR_{in} = -5.18$ dB. After spatial virtual polarization filter, interference signal of individual distance elements can be suppressed effectively, and the target signal is shown, input signal to interference ratio $SIR_o = 8.72$ dB, and improvement factor EIF reaches 13.9037 dB. Through value calculation, Fig. 7.21 gives interference suppression improvement factor

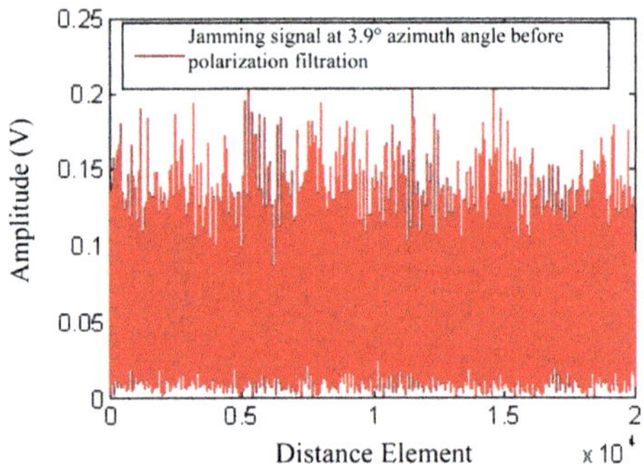

Fig. 7.19 Jamming signal received by the antenna at 3.9° scanning Angle

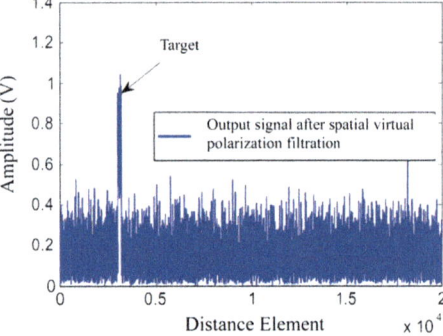

Fig. 7.20 Output signal after spatial virtual polarization filter

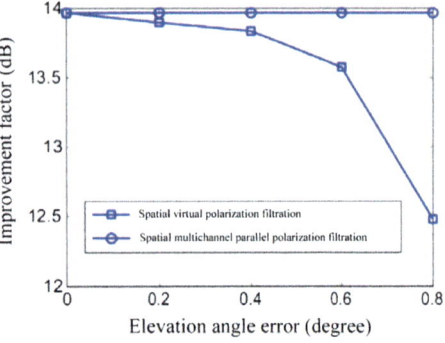

Fig. 7.21 Spatial virtual multichannel parallel polarization filter performance

versus elevation angle error profile alternative relation after polarization filter. When elevation angle measurement error is increased to 0.8°, the improvement factor is declined by about 1.4 dB, indicating that elevation angle error will reduce performance of interference suppression. Through use of spatial multichannel polarization filter, it is possible to improve such condition, so as to ensure effectiveness of interference suppression, and eliminate influence of elevation angle on interference polarization suppression.

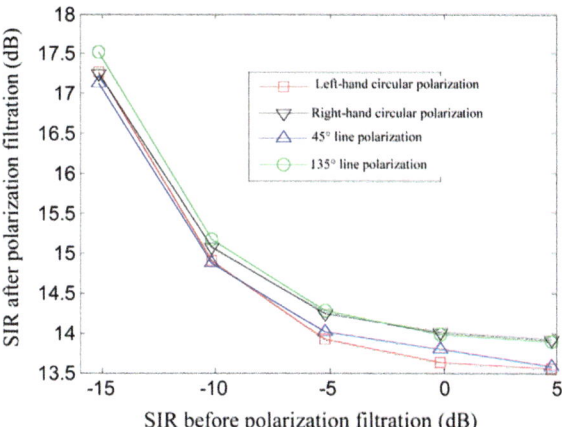

Fig. 7.22 Suppression performance of spatial virtual polarization filter against different polarization jamming

To analyze suppression performance of spatial virtual polarization filter algorithm against different polarization interference, target echo parameter is fixed during experiment, and left-hand circular and right-hand circular polarization, 45° linear polarization and 135° linear polarization are taken for interference polarization, respectively, as well as interference signal power is adjusted to set signal to interference ratio SIR to 4.8205, −0.1795, −5.1795, −10.1795, −15.1795 dB. Through simulation, Fig. 5.22 shows relationship of SIR improvement factor *EIF* in individual interference conditions with *SIR* before filter. It can be seen that *SIR* improvement factor can be up to above 13 dB (Fig. 7.22).

7.4 The Validity Demonstration of Polarization Filtering Within Polarization Estimated Error

In recent years, with gradually in-depth radar polarization theoretical study and significant improvement in technical level of radar apparatus, polarization filter plays increasingly important role in the field of anti-interference technology. Scholars designed several kinds of polarization filters for different applications, including single notch polarization filter (SPC), multi-notch polarization filter (MLP), adaptive polarization filter (APC), frequency-domain polarization filter, interference suppression polarization filter (ISPF), optimum polarization filter (OPC), multilook polarization whitening filter (MPWF), SINR polarization filter [162, 164–166]. In condition where the interference echo polarization is different from the target, the polarization filter can suppress interference effectively. Through selectivity of different incident wave in polarization field, it is possible to improve receive quality of the signal. Most of these works suppose that interference direction is in main lobe direction, ignore influence of the antenna polarization characteristics, and usually do not take account of influence of channel amplitude and phase error, channel noise and polarization measurement algorithm on polarization filter's performance, so it is intuitively thought that polarization filter efficiency only depends on polarization estimation accuracy. During evaluation of the polarization filter, theoretical analysis is carried out from the prospect of difference between real polarization and estimated value, it is overlooked that polarization filter object is not subject to real polarization, but is subject to receive polarization. Specifically, output signal of the receive channel directly characterizes amplitude and phase characteristics of electromagnetic wave signal under influence of antenna polarization characteristics, direction characteristics, channel amplitude and phase characteristics, noise characteristics and polarization algorithm and other factors. The orthogonal polarization dual-channel output signal can be estimated directly as incident wave polarization, i.e., optimum estimation of electromagnetic wave polarization in current conditions, which corresponds to steady solution of polarization estimation algorithm. However, output of polarization channel cannot be taken unbiased estimation of real polarization, i.e., polarization information obtained according to channel output signal contains certain polarization error, and polarization filter vector also is built based on this error; therefore such error is compensated during polarization filter, i.e., because polarization error resulted from such factors as channel inconsistency and antenna polarization characteristic will not influence effectiveness of polarization filter, polarization estimation accuracy does not directly restrict key problem determining polarization filter performance. From the prospect of polarization filter, this section proves the above-mentioned conclusion, gives detailed analysis, which provides guidance in improving existing single polarization radar, increasing polarization measurement and anti-jamming ability.

7.4.1 Polarization Estimation Based on Orthogonal Polarization Channel

By selecting horizontal and vertical polarization (h, v) as polarization base, electric field vector of an incident wave (interference signal) can be represented as $\boldsymbol{h}_J = [\,h_{JH} \quad h_{JV}\,]^T$, $\|\boldsymbol{h}_J\| = 1$ at this polarization base, interference signal can be represented at port of the radar's receive antenna:

$$\boldsymbol{e}_J(t) = \boldsymbol{h}_J J(t) \tag{7.4.1}$$

where $J(t)$ is modulation signal of oppressive interference, approximated to zero mean value white noise, and whose power spectrum density is σ_J^2.

Supposing that co-polarization normalization direction pattern of the antenna is $g_H(\theta)$, cross polarization normalization direction pattern is $g_V(\theta)$, and antenna peak gain is G_r, antenna spatial polarization vector is written as:

$$\boldsymbol{g}(\theta) = G_r \cdot \begin{bmatrix} g_H(\theta) \\ g_V(\theta) \end{bmatrix} \tag{7.4.2}$$

Specifically, for horizontal polarization antenna, polarization purity at center position of the antenna's main lobe is up to maximum value. The above-mentioned equation can be approximately written as $\boldsymbol{g}(\theta) = G_r \cdot \begin{bmatrix} 1 \\ 0 \end{bmatrix}$; for vertical polarization antenna, $\boldsymbol{g}(\theta) = G_r \cdot \begin{bmatrix} 0 \\ 1 \end{bmatrix}$; polarization vector is function of θ in other spatial direction θ.

Supposing that $\boldsymbol{\Gamma}$ is multiplying error coefficient matrix introduced when orthogonal polarization dual-way amplitude and phase, $\boldsymbol{\Gamma} = \mathrm{diag}[r_H, r_V]$. $r_H = A_H e^{j\varphi_H}$ and $r_V = A_V e^{j\varphi_V}$ are amplitude error and phase error of horizontal and vertical polarization channel, respectively.

Therefore, actual receive voltage of interference signal entering horizontal polarization channel is:

$$v_H(t) = r_H \cdot P_J \cdot \frac{k_{RF} G_r}{L_R} \cdot \boldsymbol{g}^{\mathrm{T}}(\theta) \cdot \boldsymbol{h}_J \cdot J(t) + n_H(t) \tag{7.4.3}$$

where P_S is interference signal power, k_{RF} is radio amplification coefficient, and $L_R = \frac{\lambda^2}{(4\pi R)^2 \cdot 10^L}$ is loss coefficient taking account of loss of electromagnetic wave space propagation and measuring system antenna feed line and devices. When receiving horizontal polarization and vertical polarization signal of the signal source, respectively, such parameter can be thought identical. $n_H(t)$ is noise in the horizontal polarization channel, and is subject to normal distribution, i.e., $n_H \sim N(0, \sigma_m^2)$.

Similarly, actual receive voltage of vertical polarization channel is as follows:

$$v_V(t) = r_V \cdot P_J \cdot \frac{k_{RF}^* G_r}{L_R^*} \cdot \boldsymbol{g}^{\mathrm{T}}(\theta) \cdot \boldsymbol{h}_J \cdot J(t) + n_V(t) \tag{7.4.4}$$

According to Eqs. (7.4.3) and (7.4.4), it can be known that when orthogonal polarization channel receives interference signal, it modulates receive channel characteristics, radio link characteristics, antenna spatial polarization characteristics, amplitude and phase characteristics of interference, and receive channel noise level and other factors. Therefore, for the purpose of analyzing performance of polarization estimation error, it can be started from the above-mentioned factors, and it is required to comprehensively considering polarization estimation algorithm, find factors of restricting polarization estimation accuracy, and give detailed estimation.

7.4.1.1 Influence of Antenna Polarization Characteristics

The pre-conditions of studying active interference polarization filter or false target interference polarization discrimination remain consistent, i.e., supposing that the jammer measurement radar is placed in electric axis direction of the polarization measurement radar, vice versa. Actually, when altitude of the jammer changes and radar beam direction deviates from the normal direction in real radar defense-attack confrontation, the defense and attack parties are usually deviated from counterpart's electric axis direction, so original supposition has certain irrationality. Especially, for the mechanical scanning radar, jamming signal sent by the covering jammer from its side lobe is always deviated from electric axis direction.

In non-electric axis direction, electromagnetic waves received by antennas which are orthogonal to each other do not only receive antenna gain modulation, but also receive modulation of antenna spatial polarization characteristic vector. They are not kept orthogonal to each other strictly, but also keep certain relativity. In such condition, jamming signal generated by the jammer has different characteristics when it incomes along the electric axis direction.

When the jamming signal is situated in main lobe of the radar antenna, Eqs. (7.4.3) and (7.4.4) can be written as follows:

$$V_H(t) = r_H \cdot P_S \cdot \frac{k_{RF} G_r}{L_R} \cdot h_{JH} \cdot J(t) + n_H(t) \tag{7.4.5}$$

$$V_V(t) = r_V \cdot P_S \cdot \frac{k_{RF} G_r}{l_R} \cdot h_{JV} \cdot J(t) + n_V(t) \tag{7.4.6}$$

Usually, jamming signal has significant advantage with respect to its power, that is to say, jammer-to-noise ratio, so ratio between output signal of orthogonal polarization channel at element where the jamming occupies is approximately calculated as follows:

$$\left\langle \frac{V_V(t)}{V_H(t)} \right\rangle \approx \frac{r_V \cdot P_S \cdot \frac{k_{RF} G_r}{L_R} \cdot h_{JV} \cdot J(t)}{r_H \cdot P_S \cdot \frac{k_{RF} G_r}{L_R} \cdot h_{JH} \cdot J(t)} \tag{7.4.7}$$

If amplitude and phase error between channels does not exists, the above-mentioned equation is further written as follows:

$$\left\langle \frac{V_V(t)}{V_H(t)} \right\rangle = \frac{h_{JV}}{h_{JH}} \tag{7.4.8}$$

represents ensemble average Obviously, polarization state of jamming signal can be estimated from Eq. (7.4.8) and restriction conditions $\|\boldsymbol{h}_J\| = 1$, and complex polarization ratio of the jamming signal is calculated as follows:

$$\rho_J = \left\langle \frac{V_V(t)}{V_H(t)} \right\rangle = \frac{h_{JV}}{h_{JH}} = tg\gamma_J e^{j\phi_J} \tag{7.4.9}$$

Actually, polarization characteristics of the antenna (which can also be referred to as polarization purity) changes with observation direction. Supposing that initial polarization vector of the radar's horizontal polarization antenna $\boldsymbol{h} = \begin{bmatrix} 1 & 0 \end{bmatrix}^T$, its polarization purity gradually decreases in different observation space, and cross polarization component increases; supposing polarization angle γ increases linearly, it can be represented as follows:

$$\gamma_H(\theta) = K_{\text{Polar}} \cdot |\theta|, \ \theta \in [-\theta_0/2, +\theta_0/2] \tag{7.4.10}$$

$$\gamma_V(\theta) = \frac{\pi}{2} - K_{\text{Polar}} \cdot |\theta|, \ \theta \in [-\theta_0/2, +\theta_0/2] \tag{7.4.11}$$

where $K_{\text{Polar}} > 0$, which is rate of change of polarization angle of the antenna. If K_{Polar} is bigger, it indicates that the antenna polarization changes rapidly, and also indicates that spatial polarization characteristics of the antenna is more obvious. The polarization angle $\phi_H(\theta) = -\phi_V(\theta)$.

When jamming signal is near side lobe of the radar antenna, it is possible to obtain following equations:

$$
\begin{aligned}
V_H(t) &= r_H \cdot P_S \cdot \frac{k_{RF}G_r}{L_R} \cdot [\cos(\gamma(\theta))h_{JH} + \sin(\gamma(\theta))h_{JV} \cdot \exp(j\phi_H)] \cdot J(t) + n_H(t) \\
&= r_H \cdot P_S \cdot \frac{k_{RF}G_r}{L_R} \cdot [\cos(K_{\text{Polar}} \cdot |\theta|)h_{JH} + \sin(K_{\text{Polar}} \cdot |\theta|)h_{JV} \cdot \exp(j\phi_H(\theta))] \\
&\quad + n_H(t)
\end{aligned}
\tag{7.4.12}
$$

$$
\begin{aligned}
V_V(t) &= r_V \cdot P_S \cdot \frac{k_{RF}G_r}{L_R} \cdot \left[\cos\left(\frac{\pi}{2} - K_{\text{Polar}} \cdot |\theta|\right)h_{JH} + \sin\left(\frac{\pi}{2} - K_{\text{Polar}} \cdot |\theta|\right)h_{JV} \cdot \exp(j\phi_V(\theta))\right] \\
&= r_V \cdot P_S \cdot \frac{k_{RF}G_r}{L_R} \cdot [\sin(K_{\text{Polar}} \cdot |\theta|)h_{JH} + \cos(K_{\text{Polar}} \cdot |\theta|)h_{JV} \cdot \exp(j\phi_V(\theta))] \\
&\quad + n_V(t)
\end{aligned}
\tag{7.4.13}
$$

According to antenna theory, it can be known that co-polarization and cross polarization vectors are orthogonal to each other under spherical coordinate system in any observation position. So, polarization of the antenna any two times of observation forms a group of complete bases of the polarization space.

So, in the elements where interference occupies, ratio between output signal of orthogonal polarization channel is approximately calculated as follows:

$$\left\langle \frac{V_V(t)}{V_H(t)} \right\rangle = \frac{r_V \cdot P_S \cdot \frac{k_{RF}G_r}{L_R} \cdot [\sin(K_{\text{Polar}} \cdot |\theta|)h_{JH} + \cos(K_{\text{Polar}} \cdot |\theta|)h_{JV} \cdot \exp(j\phi_V(\theta))]}{r_H \cdot P_S \cdot \frac{k_{RF}G_r}{L_R} \cdot [\cos(K_{\text{Polar}} \cdot |\theta|)h_{JH} + \sin(K_{\text{Polar}} \cdot |\theta|)h_{JV} \cdot \exp(j\phi_H(\theta))]}$$
$$= \frac{\tan(K_{\text{Polar}} \cdot |\theta|)h_{JH} + h_{JV} \cdot \exp(j\phi_V(\theta))}{h_{JH} + \tan(K_{\text{Polar}} \cdot |\theta|)h_{JV} \cdot \exp(j\phi_H(\theta))}$$

$$(7.4.14)$$

Polarization of the interference signal is written in the form of polarization ratio. For any fixed θ, $\tan(K_{\text{Polar}} \cdot |\theta|) = m_p$, $\rho_J = \frac{h_{JV}}{h_{JH}}$, the above-mentioned equation can be written as follows:

$$\left\langle \frac{V_V(t)}{V_H(t)} \right\rangle = \frac{\tan(K_{\text{Polar}} \cdot |\theta|) + \rho_J \cdot \exp(j\phi_V(\theta))}{1 + \tan(K_{\text{Polar}} \cdot |\theta|)\rho_J \cdot \exp(j\phi_H(\theta))}$$
$$= \frac{m_p + \rho_J \cdot \exp(j\phi_V(\theta))}{1 + m_p \cdot \cdot \rho_J \cdot \exp(j\phi_H(\theta))}$$

$$(7.4.15)$$

It can be seen from the above-mentioned equation that polarization estimation value output by the polarization channel shows nonlinear change with respect to real polarization, which is mainly influenced by two factors: the first factor is polarization angle of the antenna in interference incidence direction, and the second one is phase characteristic of the antenna in interference incidence direction. This indicates that interference signal entered from the side lobe is declined after being subject to two polarization channel receive power, compared to that entered from main lobe; polarization of output signal is modulated by antenna polarization characteristics, the polarization characteristics are varied, and polarization amplitude and phase produce estimation error, i.e., polarization directly estimated by orthogonal polarization channel output obviously cannot be considered as unbiased estimation of interference polarization. Figure 7.23 gives law of change in polarization estimation error versus antenna characteristics and output voltage.

Fig. 7.23 Law of change in polarization estimation error versus antenna characteristics and output voltage

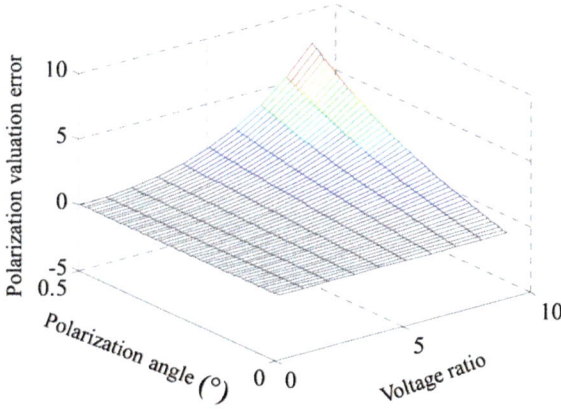

7.4.1.2   Influence of Polarization Channel Characteristics

To simplify analysis in ideal conditions, array element response, array element position disturbance, cross coupling, and signal wave front distortion of the radar receive/transmit antenna are ignored, but in a real system, the above-mentioned factors inevitably cause significant influence on receive signal. These errors can be represented by "Amplitude and phase error". Although amplitude and phase errors are corrected in engineering, but residual amplitude and phase error still exists. Therefore, influence of amplitude and phase error on polarization estimation and polarization filter performance is a problem that should be considered in practical application.

Under influence of amplitude and phase error, actual receive voltage of interference signal entered from main lobe for horizontal polarization channel is as follows:

$$
\begin{aligned}
V_H(t) &= A_H e^{j\varphi_H} \cdot P_S \cdot \frac{k_{RF} G_r}{L_R} \cdot \boldsymbol{g}^T(\theta) \cdot \boldsymbol{h}_J \cdot J(t) + n_H(t) \\
&= A_H e^{j\varphi_H} \cdot P_S \cdot \frac{k_{RF} G_r}{L_R} \cdot h_{JH} \cdot J(t) + n_H(t)
\end{aligned}
\tag{7.4.16}
$$

Actual receive voltage of interference signal for vertical polarization channel is as follows:

$$
\begin{aligned}
V_V(t) &= A_V e^{j\varphi_V} \cdot P_S \cdot \frac{k_{RF} G_r}{L_R} \cdot \boldsymbol{g}^T(\theta) \cdot \boldsymbol{h}_J \cdot J(t) + n_V(t) \\
&= A_V e^{j\varphi_V} \cdot P_S \cdot \frac{k_{RF} G_r}{L_R} \cdot h_{JV} \cdot J(t) + n_V(t)
\end{aligned}
\tag{7.4.17}
$$

Therefore, ratio between orthogonal polarization channel output signal is approximately calculated as follows:

$$
\begin{aligned}
\left\langle \frac{V_V(t)}{V_H(t)} \right\rangle &= \frac{A_V e^{j\varphi_V} \cdot P_S \cdot \frac{k_{RF} G_r}{L_R} \cdot h_{JV} \cdot J(t) + n_V(t)}{A_H e^{j\varphi_H} \cdot P_S \cdot \frac{k_{RF} G_r}{L_R} \cdot h_{JH} \cdot J(t) + n_H(t)} \\
&= \frac{A_V e^{j\varphi_V}}{A_H e^{j\varphi_H}} \cdot \frac{h_{JV}}{h_{JH}} = \xi_A \cdot \exp(j\phi_\xi) \cdot \frac{h_{JV}}{h_{JH}}
\end{aligned}
\tag{7.4.18}
$$

where, $\xi_A = \frac{A_V}{A_H}$ is amplitude inconsistency error between channels, and $\phi_\xi = \varphi_V - \varphi_H$ is phase inconsistency error between channels.

Thus, polarization of signal output by the polarization channel modulates a 年 amplitude error item and a phase error item for real polarization characteristics.

7.4.1.3   Influence of Receiver's Noise

Electric vector of interference incident wave is written $h = [j_h \quad j_v]^T$ at (h, v). After being received by orthogonal polarization dual-channel measurement system, the channel output forms a 2D complex vector, which is written as $x = h + n$, where n is noise vector of the measuring system. If clutter and interference are not considered, n usually represents output noise of two-way channel receiver. In actual condition, output vector of the measuring system is often taken to estimate incidence signal polarization directly [163], i.e.,

$$\hat{h} = \frac{x}{\|x\|} = \frac{h + n}{\|h + n\|} = \frac{h + \tilde{h}}{\|h + \tilde{h}\|} \tag{7.4.19}$$

Supposing that ϑ is vector angle between real polarization g_h of the incoming wave and estimated polarization $\hat{g}_h$, g_h and $\hat{g}_h$ are Stokes subvector for incoming wave polarization and estimated polarization. Supposing that $J_h = [1, g_h^T]^T$ and $\hat{J}_h = [1, \hat{g}_h^T]^T$ are Stokes vector for real polarization and estimated polarization incoming wave, then $J_h = R(h \otimes h^*)$ and $\hat{J}_h = R(\hat{h} \otimes \hat{h}^*)$, where the superscript "*" represents conjugate, and R is a quasi-unitary matrix.

h and $\hat{h}$ represents phase descriptor of real polarization and estimated polarization. For $\hat{h} = \frac{h + \tilde{h}}{\|h + \tilde{h}\|}$, $\tilde{h}$ is noise vector of the measuring system. g_h and $\hat{g}_h$ are 3D real vector, and meet $\|g_h\| = \|\hat{g}_h\| = 1$.

Through normalization of estimated incoming wave polarization, it is possible to obtain "normalized" estimated incoming wave vector $\hat{h}_{\text{uni}}$, and matching coefficient between h and $\hat{h}_{\text{uni}}$ is m_p, thus:

$$m_p = \frac{1}{2}(1 + \cos \vartheta) = \frac{\left|h^H \hat{h}_{\text{uni}}\right|^2}{\|h\|^2 \left\|\hat{h}_{\text{uni}}\right\|^2} \tag{7.4.20}$$

$\hat{h}_{\text{uni}} = \frac{\hat{h}}{\|\hat{h}\|} = \frac{h + \tilde{h}}{\|h + \tilde{h}\|}$ is substituted into the above-mentioned equation to obtain

$$\sin^2 \frac{\vartheta}{2} = \frac{\left\|\tilde{h}\right\|^2 - \left\|\tilde{h}^H h\right\|^2}{1 + h^H \tilde{h} + \tilde{h}^H h + \left\|\tilde{h}\right\|^2} = \frac{\left\|\tilde{h}\right\|^2 - \left\|\tilde{h}^H h\right\|^2}{\left\|\hat{h}\right\|^2} \tag{7.4.21}$$

If signal-to-noise ratio of the measuring system is very high, i.e. $\|\tilde{h}\| \ll \|h\| = 1$, then $\|\hat{h}\| \approx 1$, and the Eq. (5.7.8) can be approximately written as follows:

$$\sin^2\frac{\vartheta}{2} = \tilde{h}^H\tilde{h} - \tilde{h}^H hh^H\tilde{h} = \tilde{h}^H Q\tilde{h} \qquad (7.4.22)$$

where, $Q = I_{2\times2} - hh^H$, in which $I_{2\times2} = \mathrm{diag}\{1,1\}$, and it can be known that Q is a non-negative definite Hermite matrix.

Unitary similarity variant $Q = U^H\Lambda U$ of Q is obtained by taking $h = \begin{bmatrix} \cos\gamma & \sin\gamma e^{j\phi} \end{bmatrix}^T$ and substituting it into Q expressions, in which $\Lambda = \mathrm{diag}\{1,0\}$, $U = \begin{bmatrix} \sin\gamma e^{j\phi} & -\cos\gamma \\ \cos\gamma & \sin\gamma e^{-j\phi} \end{bmatrix}$ is unitary matrix; And supposing that $b = U\tilde{h}$ is unitary transformation vector of measurement error, then Eq. (5.5.9) can be written as $\sin^2\frac{\vartheta}{2} = b^H\Lambda b$, and based on the above-mentioned supposition of $\|\tilde{h}\| \ll 1$, then $\vartheta \ll 1$. At this moment, the above-mentioned equation can be further written approximately as follows:

$$\vartheta \approx 2\sqrt{b^H\Lambda b} \qquad (7.4.23)$$

According to the above-mentioned analysis, it can be known that $\tilde{h}$ is zero mean value complex Gaussian vector, so $b = U\tilde{h}$ is a zero mean value complex Gaussian vector, and its covariance matrix is $R_b = UR_{\tilde{h}}U^H$. Error angle ϑ between estimated polarization and real polarization depends on noise vector of the measuring system, and polarization parameter of interference signal. Figure 7.24 gives a statistical histogram for probability density distribution of polarization error angle under two kinds of signal-to-noise ratio level, and estimated times is 2×10^4. The statistical result proves the above-mentioned conclusion.

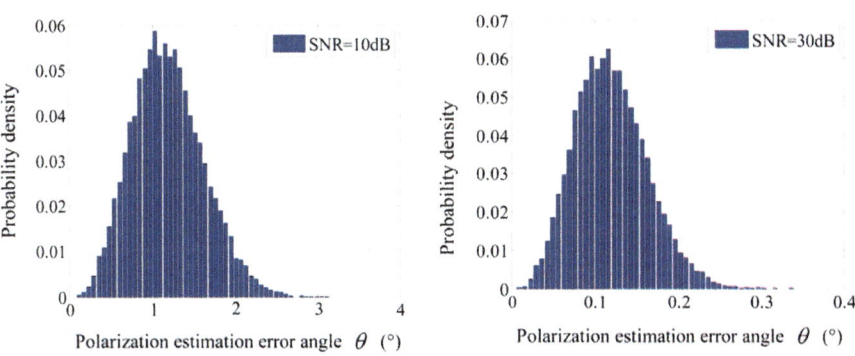

Fig. 7.24 Probability density distribution of polarization estimation error angle θ

7.4.1.4 Influence of Polarization Estimation Algorithm

In actual radar, statistical characteristic in the time domain for output of the measuring system is often used to describe polarization state of the interference, but according to time frequency invariability of polarization state, signal polarization also can be finished. To compare accuracy of polarization measurement in two processing field, the following gives comparison between polarization signal frequency domain measurement principle and measurement performance.

According to stokes parameters of signal polarization, time domain polarization parameters [167] can be obtained by following method.

$$\bar{\gamma} = arctg\left(\sqrt{\frac{\bar{g}_{op} - \bar{g}_1}{\bar{g}_{op} + \bar{g}_1}}\right) \tag{7.4.24}$$

$$\bar{\delta} = arctg\left(\frac{\bar{g}_3}{\bar{g}_2}\right) \tag{7.4.25}$$

where, $\bar{g}_0 = \left\langle |E_h(t)|^2 \right\rangle + \left\langle |E_v(t)|^2 \right\rangle$, $\bar{g}_1 = \left\langle |E_h(t)|^2 \right\rangle - \left\langle |E_v(t)|^2 \right\rangle$, $\bar{g}_2 = 2\mathrm{Re}\left\langle E_h(t)E_v^*(t) \right\rangle$, $\bar{g}_3 = -2\mathrm{Im}\left\langle E_h(t)E_v^*(t) \right\rangle$, $\bar{g}_0 = (\bar{g}_1^2 + \bar{g}_2^2 + \bar{g}_3^2)^{\frac{1}{2}}$.

Similarly, time of electric field vector of plane resonance single-color electromagnetic wave propagated along z axis is represented as follows [167]:

$$E(z,t) = \begin{bmatrix} E_H(z,t) \\ E_V(z,t) \end{bmatrix} = E(t) \cdot e^{j\omega t} \begin{bmatrix} \cos \gamma \\ \sin \gamma \cdot \exp(j\eta) \end{bmatrix} \cdot e^{j(\Omega t - kz)} \tag{7.4.26}$$

where, Ω is signal angle frequency, and $E(t)$ indicates electric field intensity which varies with time.

For single frequency signal whose amplitude remains unchanged and polarization keeps constant, absolute phase should be neglected, and its time function is solved as per following equation:

$$\begin{aligned} E_H(t) &= E_h(t) \cdot e^{j\Omega t} \\ E_V(t) &= E_v(t) \cdot e^{j(\Omega t + \eta)} \end{aligned} \cdot \tag{7.4.27}$$

where, $\rho_s = \frac{E_v(t)}{E_h(t)} = tg\gamma$ represents amplitude of polarization ratio between two polarization channels, and $\eta = \arg[E_V(t)] - \arg[E_H(t)]$ represents phase difference of two polarization channel.

Two ways of polarization signal is sampled, and supposing that the sampling period is T_s, corresponding digital frequency is $\omega_0 = \Omega T_s$, and after sampling, digital signal output is as follows:

$$\boldsymbol{E}(n) = \begin{bmatrix} E_H(n) \\ E_V(n) \end{bmatrix} = \begin{bmatrix} E \cos \gamma \cdot \exp(j\omega_0 n) \\ E \sin \gamma \cdot \exp[j(\omega_0 n + \eta)] \end{bmatrix} \tag{7.4.28}$$

where N is number of samples, $n = 1, 2, \ldots, N$, FFT is carried out for two ways of signal to obtain:

$$E(e^{j\omega}) = \begin{bmatrix} E_h(e^{j\omega}) \\ E_v(e^{j\omega}) \end{bmatrix} = \begin{bmatrix} E \cos \gamma \cdot \sum_{m=1}^{N} 2\pi\delta(\omega - \omega_0) \\ E \sin \gamma e^{j\eta} \cdot \sum_{m=1}^{N} 2\pi\delta(\omega - \omega_0) \end{bmatrix} \tag{7.4.29}$$

So, amplitude ratio between two ways of signal spectrum is taken as estimation of the polarization ratio amplitude:

$$\hat{\rho}_s = \left| \frac{E \sin \gamma e^{j\varphi} \cdot \sum_{m=1}^{N} 2\pi\delta(\omega - \omega_0)}{E \cos \gamma \cdot \sum_{m=1}^{N} 2\pi\delta(\omega - \omega_0)} \right| = tg\gamma = \rho_s \tag{7.4.30}$$

$$\hat{\eta} = \arg\left[E_v(e^{j\omega}) \right] - \arg\left[E_h(e^{j\omega}) \right] = \eta \tag{7.4.31}$$

Therefore, through time domain and frequency domain processing of two ways of output signal, it is possible to complete effective estimation of signal polarization state. Figures 7.25 and 7.26 gives performance curve for polarization parameters by conducting time domain and frequency domain processing of the polarization signal, respectively, and it can be seen obviously that frequency domain processing method is equivalent to relevant accumulation of time domain through FFT transformation, so as to improve signal/noise ratio (SNR), and its estimation accuracy is higher than time domain processing method.

Fig. 7.25 Time domain estimation performance of polarization state

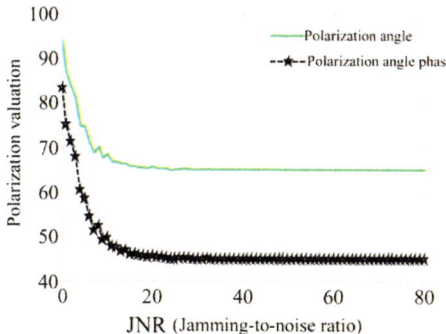

Fig. 7.26 Frequency domain
estimation performance of
polarization state

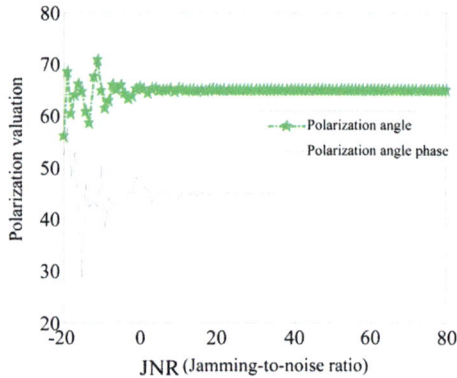

7.4.2   Validity Analysis of Polarization Filtering

Usually, polarization filter efficiency depends on polarization estimation accuracy.
If no estimation error exists, the residual interference power is 0. This conclusion
neglects an important factor, i.e., object of polarization filter is not real polarization,
but receive polarization. Specifically, output signal of the receive channel directly
characterizes amplitude and phase characteristics of electromagnetic wave under
such factors as antenna polarization characteristics, direction characteristics,
channel amplitude and phase characteristics, noise characteristics and polarization
estimation algorithm. Through analysis set forth above, the polarization estimation
error can be equivalent to combination of two kinds of multiplying error and one
kind of summing error, of which multiplying error can be represented by amplitude
and phase error in a unified manner. Whether this factor will influence effectiveness
of polarization filter, the following gives detailed analysis.

Supposing that real polarization of the interference is $\boldsymbol{h} = \begin{bmatrix} \cos\gamma & \sin\gamma e^{j\phi} \end{bmatrix}^{\mathrm{T}}$,
and certain estimation error in polarization estimation resulted from antenna
polarization characteristics and channel's non-ideal characteristics, polarization of
channel output signal is deviated from real polarization, which can be represented
as $\hat{\boldsymbol{h}} = \begin{bmatrix} \cos\hat{\gamma} & \sin\hat{\gamma}e^{j\hat{\phi}} \end{bmatrix}^{\mathrm{T}} = \begin{bmatrix} \cos(\gamma + \Delta\gamma) \\ \sin(\gamma + \Delta\gamma)e^{j(\phi + \Delta\phi)} \end{bmatrix}$, and supposing that polarization filter vector $\boldsymbol{H_r} = \begin{bmatrix} \cos\gamma_r \\ \sin\gamma_r e^{j\phi_r} \end{bmatrix}$, and because estimation error is considered
when estimating polarization filter vector, following relationship exists:

$$\begin{cases} \gamma_r = \gamma + \Delta\gamma + \frac{\pi}{2} \\ \phi_r = -(\phi + \Delta\phi) \end{cases} \tag{7.4.32}$$

i.e., orthogonal polarization $\boldsymbol{H_r} = \hat{\boldsymbol{h}}_\perp = \left[-\sin\hat{\gamma} \quad \cos\hat{\gamma}e^{-j\hat{\phi}} \right]^T$, it can be known that Stokes subvector $\hat{\boldsymbol{g}}_{h\perp} = \Lambda_{12}\hat{\boldsymbol{g}}_h$, so $\Lambda_{12} = \mathrm{diag}\{-1,-1,1\}$. Because of restriction by polarization estimation accuracy, $\hat{\boldsymbol{h}}_\perp$ is usually not orthogonal to real polarization $\boldsymbol{h}$ of interference in a strict manner. Reference considers that some interference signal will leak to the radar receiver. This part of residual interference power is as follows:

$$P_r = \left| \hat{\boldsymbol{h}}_\perp{}^T \boldsymbol{h} \right|^2 = \frac{1}{2}\hat{\boldsymbol{J}}_\perp{}^T \boldsymbol{U}_4 \boldsymbol{J}_h = \frac{1}{2}\left(1 + \hat{\boldsymbol{g}}_{h\perp}{}^T \Lambda_3 \boldsymbol{g}_h\right)$$
$$= \frac{1}{2}\left(1 - \hat{\boldsymbol{g}}_h{}^T \hat{\boldsymbol{g}}_h\right) = \frac{1}{2}\left(1 - \cos\vartheta\right) \tag{7.4.33}$$

where $\hat{\boldsymbol{J}}_\perp = \boldsymbol{R}\left(\hat{\boldsymbol{h}}_\perp \otimes \hat{\boldsymbol{h}}_\perp^*\right) = \left[1 \quad \hat{\boldsymbol{g}}_{h\perp}{}^T\right]^T$ is Stokes vector corresponding to $\hat{\boldsymbol{h}}_\perp$, $\boldsymbol{U}_4 = \mathrm{diag}\{1,1,1,-1\}$, $\Lambda_3 = \mathrm{diag}\{1,1,-1\}$.

The above-mentioned equation indicates that residual interference power only depends on difference between interference polarization estimation and real interference polarization. When both polarization equal, the residual power $P_r = 0$. However, in actual system, polarization filter vector is not real polarization $\boldsymbol{h}$ filter, and as shown in Fig. 7.27, estimated optimum interference suppression polarization vector $\hat{\boldsymbol{h}}_\perp$ should filters channel output signal $\hat{\boldsymbol{h}}J(\mathrm{t})$.

For incidence signal, orthogonal polarization channel output has been not suitable for being taken as optimum estimation of electromagnetic wave polarization, but is more like an approximate solution, due to modulation of spatial polarization characteristics of radar receive antenna, and influence of element coupling effect, channel noise and amplitude and phase error. Therefore, in traditional analysis of

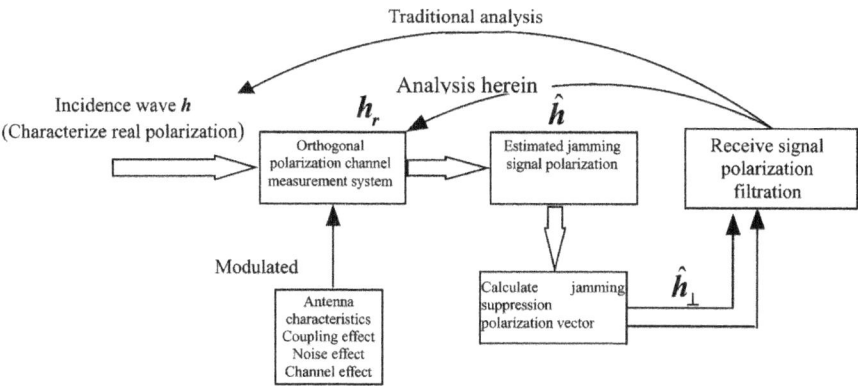

Fig. 7.27 Open loop model of polarization filter in real polarization radar

polarization filter performance, polarization with estimated error is usually used to build filter vector, while filter of incidence "real polarization" does not meet actual signal process flow. A "output polarization h_r" is added in open loop model for polarization filter of the whole actual polarization radar, and polarization filter performance is judged through three quantities including incoming wave real polarization, output polarization, estimated polarization, and the evaluation rule is not built purely from estimated polarization and incoming wave real polarization. Therefore, the Eq. (7.4.33) can be rewritten as follows:

$$P_r = \left| \hat{\boldsymbol{h}}_\perp{}^{\mathrm{T}} \boldsymbol{h}_r \right|^2 = \left| \hat{\boldsymbol{h}}_\perp{}^{\mathrm{T}} \hat{\boldsymbol{h}} + \hat{\boldsymbol{h}}_\perp{}^{\mathrm{T}} \boldsymbol{n} \right|^2 \tag{7.4.34}$$

The above-mentioned equation including two items, of which the first item is estimated polarization obtained based on channel output, which has certain estimation error, and error modeling and analysis is specifically analyzed in Sect. 7.4.1; the second one is noise vector. It can be known by separately analyzing the first item that output signal can be represented as follows after known filter:

$$
\begin{aligned}
E_1(t) &= J(t) \cdot \hat{\boldsymbol{h}}_\perp^{\mathrm{T}} \hat{\boldsymbol{h}} \\
&= J(t) \cdot \left[\cos(\gamma + \Delta\gamma + \tfrac{\pi}{2}) \quad \sin(\gamma + \Delta\gamma + \tfrac{\pi}{2}) e^{j(\phi + \Delta\phi)} \right] \cdot \begin{bmatrix} \cos(\gamma + \Delta\gamma) \\ \sin(\gamma + \Delta\gamma) e^{-j(\phi + \Delta\phi)} \end{bmatrix} \\
&= J(t) \cdot \begin{bmatrix} \cos(\gamma + \Delta\gamma + \dfrac{\pi}{2}) \cos(\gamma + \Delta\gamma) + \sin(\gamma + \Delta\gamma + \dfrac{\pi}{2}) e^{j(\phi + \Delta\phi)} \\ \times \sin(\gamma + \Delta\gamma) e^{-j(\phi + \Delta\phi)} \end{bmatrix} \\
&= 0
\end{aligned}
$$

$$\tag{5.5.35}$$

It can be known that output power of the first item is approximated to 0.

The second item is $\left| \hat{\boldsymbol{h}}_\perp{}^{\mathrm{T}} \boldsymbol{n} \right|^2 = \sigma_n^2$, being equivalent to noise power output after polarization filter.

It can be known from the above-mentioned analysis that polarization estimation error and polarization filter vector calculation are influenced by each other. The polarization filter vector is based on polarization estimation error, and polarization estimation error is compensated at polarization filter vector calculation, or polarization estimation error is compensated during polarization filter. Therefore, polarization estimation error does not directly restrict effectiveness of polarization filter in the actual processing.

7.5 Electronic Deception Jamming Countermeasure Method Based on Spatial Polarization Characteristics of Mono-Pulse Radar Antenna

7.5.1 False Target Jamming Polarization Estimation Method Based on Sum-Dif Beam Property

Supposing that amplitude comparison sum-dif is composed of two antennas and receive branch A and B, the antenna beam is totally consistent, and phase center of these two antennas is deviated from the line of vision axis at an angle θ_0. Their normalization direction function can be written as follows:

$$\begin{cases} \boldsymbol{F}_A = F(\theta - \theta_0) \cdot \boldsymbol{h}_A(\theta) \\ \boldsymbol{F}_B = F(\theta + \theta_0) \cdot \boldsymbol{h}_B(\theta) \end{cases} \tag{7.5.1}$$

where, $\boldsymbol{h}_A(\theta) = [\,1 \quad \rho_A(\theta)\,]^T$, $\boldsymbol{h}_B(\theta) = [\,1 \quad \rho_B(\theta)\,]^T$ is antenna spatial polarization vector, and $\|\boldsymbol{h}_A\| = 1$, $\|\boldsymbol{h}_B\| = 1$, $\rho_A(\theta)$ and $\rho_B(\theta)$ are polarization ratio of two antenna elements, respectively.

Supposing that the target direction is θ, and waveform influence is neglected, radar transmit signal can be written as follows:

$$\boldsymbol{E}_T = F(\theta - \theta_0) \cdot \boldsymbol{h}_A(\theta) + F(\theta + \theta_0) \cdot \boldsymbol{h}_B(\theta) \tag{7.5.2}$$

Supposing that the jammer polarization in θ direction is $\boldsymbol{h}_J$, signal received by the jammer can be represented as follows:

$$E_J = \boldsymbol{h}_J^T [F(\theta - \theta_0)\boldsymbol{h}_A + F(\theta + \theta_0)\boldsymbol{h}_B] \tag{7.5.3}$$

where $\boldsymbol{h}_J = [\,\cos \varepsilon_J \quad \sin \varepsilon_J \cdot \exp(j\eta_J)\,]^T$ is Jones vector form in interference polarization state, ε_J represents amplitude relation (equivalent to polarization angle) of orthogonal polarization component in interference signal, η_J represents phase difference in orthogonal polarization components. It can be seen that $E_J = \boldsymbol{h}_J^T [F(\theta + \theta_0)\boldsymbol{h}_A + F(\theta - \theta_0)\boldsymbol{h}_B]$ represents matching degree between interference polarization and radar antenna, and can be considered a power factor, determining output power of radar transmit signal in the jammer.

Through DRFM sampling of the jammer, polarization of the jammer transmit signal is identical to polarization method of the interference antenna. With respect to false target interference signal, it can be known from the antenna receive theory that signal received by radar's branch A is represented as follows:

$$A(t) = F(\theta - \theta_0)\boldsymbol{h}_A^T \chi_J E_J \cdot \boldsymbol{h}_J \cdot J(t) \tag{7.5.4}$$

where $J(t)$ is time domain waveform of interference signal, and χ_J is amplification coefficient.

Similarly, signal received by the branch B can be represented as follows:

$$B(t) = F(\theta + \theta_0)\mathbf{h}_B^T \chi_J E_J \cdot \mathbf{h}_J.J(t) \tag{7.5.5}$$

After two ways of receive signal is subject to orthogonal mode T, they become summation and difference signal, of which summation signal can be represented as follows:

$$\sum(t) = A(t) + B(t)$$

$$= \chi_J E_J J(t) \begin{bmatrix} F(\theta - \theta_0) + F(\theta + \theta_0) \\ F(\theta - \theta_0)\rho_A(\theta) + F(\theta + \theta_0)\rho_B(\theta) \end{bmatrix}^T \begin{bmatrix} \cos \varepsilon_J \\ \sin \varepsilon_J \cdot \exp(j\eta_J) \end{bmatrix} \tag{7.5.6}$$

Difference signal is represented as follows:

$$\Delta(t) = A(t) - B(t)$$

$$= \chi_J E_J J(t) \begin{bmatrix} F(\theta - \theta_0) - F(\theta + \theta_0) \\ F(\theta - \theta_0)\rho_A(\theta) - F(\theta + \theta_0)\rho_B(\theta) \end{bmatrix}^T \begin{bmatrix} \cos \varepsilon_J \\ \sin \varepsilon_J \cdot \exp(j\eta_J) \end{bmatrix} \tag{7.5.7}$$

To simplify the expression, it is possible to suppose

$$\begin{cases} F_1(\theta) = F(\theta - \theta_0) + F(\theta + \theta_0) \\ F_2(\theta) = F(\theta - \theta_0)\rho_A(\theta) + F(\theta + \theta_0)\rho_B(\theta) \\ F_3(\theta) = F(\theta - \theta_0) - F(\theta + \theta_0) \\ F_4(\theta) = F(\theta - \theta_0)\rho_A(\theta) - F(\theta + \theta_0)\rho_B(\theta) \end{cases}, \text{and } \chi' = \chi_J E_J$$

Then, the Eqs. (7.5.6) and (7.5.7) can be rewritten in the form of matrix:

$$\begin{bmatrix} \sum(t) \\ \Delta(t) \end{bmatrix} = \begin{bmatrix} F_1(\theta) & F_2(\theta) \\ F_3(\theta) & F_4(\theta) \end{bmatrix} \cdot \begin{bmatrix} \chi'J(t)\cos \varepsilon_J \\ \chi'J(t)\sin \varepsilon_J \cdot \exp(j\eta_J) \end{bmatrix}. \tag{7.5.8}$$

It can be seen from the equation above that summation and difference channel echo does not only modulate co-polarization component of the signal, but also contains signal component generated under influence of the antenna cross polarization gain and the cross polarization component of signal.

Because element $\chi'J(t)\cos \varepsilon_J$ and $\chi'J(t)\sin \varepsilon_J \exp(j\eta_J)$ in the column matrix represents orthogonal polarization component of interference signal $J(t)$, respectively, $\chi'J(t)\cos \varepsilon_J$ can be replaced by $J_H(t)$, and $\chi'J(t)\sin \varepsilon_J \exp(j\eta_J)$ is replaced by $J_V(t)$.

Then, the Eq. (7.5.8) can be simplified as follows:

$$\begin{bmatrix} \sum(t) \\ \Delta(t) \end{bmatrix} = \begin{bmatrix} F_1(\theta) & F_2(\theta) \\ F_3(\theta) & F_4(\theta) \end{bmatrix} \begin{bmatrix} J_H(t) \\ J_V(t) \end{bmatrix} \tag{7.5.9}$$

Receive voltage signal of the summation and difference channels can be divided into M distance resolution elements, and 2 voltage values of the same distance element are formed to vector $\boldsymbol{r}(t_k)$, $k = 1, \ldots, M$. Polarization signal decomposition is to multiply voltage vector $\boldsymbol{r}(t_k) = \begin{bmatrix} \sum(t_k) \\ \Delta(t_k) \end{bmatrix}$ of each distance element by inverse matrix of the antenna polarization characteristic matrix $\boldsymbol{G}$ on the premise that signal angle of arrival θ and antenna polarization characteristic $\boldsymbol{G}(\theta) = \begin{bmatrix} F_1(\theta) & F_2(\theta) \\ F_3(\theta) & F_4(\theta) \end{bmatrix}$ have been known, so as to obtain two ways of orthogonal polarization signal.

Mathematically, orthogonal polarization echo of the signal can be obtained according to inverse operation of the matrix, and with respect to each sampling time t_k or distance resolution element, orthogonal polarization component of the interference can be written as follows:

$$\begin{bmatrix} J_H(t_k) \\ J_V(t_k) \end{bmatrix} = \begin{bmatrix} F_1(\theta) & F_2(\theta) \\ F_3(\theta) & F_4(\theta) \end{bmatrix}^{-1} \begin{bmatrix} \sum(t_k) \\ \Delta(t_k) \end{bmatrix} \tag{7.5.10}$$

Precondition for this method is that

$$\boldsymbol{G}(\theta) = \begin{bmatrix} F(\theta - \theta_0) + F(\theta + \theta_0) & F(\theta - \theta_0)\rho_A(\theta) + F(\theta + \theta_0)\rho_B(\theta) \\ F(\theta - \theta_0) - F(\theta + \theta_0) & F(\theta - \theta_0)\rho_A(\theta) - F(\theta + \theta_0)\rho_B(\theta) \end{bmatrix}$$

matrix is invertible. By taking co-polarization direction pattern function of the summation beam as $F_\Sigma = F(\theta - \theta_0) + F(\theta + \theta_0)$ and cross polarization direction pattern function of the difference beam as $F_\Delta = F(\theta - \theta_0) - F(\theta + \theta_0)$, $F(\theta - \theta_0)\rho_A(\theta) + F(\theta + \theta_0)\rho_B(\theta)$ and $F(\theta - \theta_0)\rho_A(\theta) - F(\theta + \theta_0)\rho_B(\theta)$ can be considered as cross polarization direction pattern function of the summation beam and difference beam, respectively.

Due to restriction by processing level, each antenna element cannot be totally consistent with characteristics of the receive branch, so $\rho_A(\theta) \neq \rho_B(\theta)$, i.e. no ideal polarization antenna exists, enabling the above-mentioned antenna characteristic matrix $\boldsymbol{G}(\theta)$ to be an invertible matrix. Figure 7.28 gives polarization direction pattern of sum-dif beams obtained by Grasp9.0 electromagnetic calculation, and it can be seen that there is obvious difference between summation and difference beam polarization direction patterns, and cross polarization direction pattern of summation beam is identical to difference beam with respect to their structure, i.e. a zero point exists at main lobe's center position, so that cross polarization receive power reaches minimum value or local minimum at maximum co-polarization position. Therefore, this method is tenable, whose advantage is to make use of antenna's non-ideal factors.

Fig. 7.28 Polarization direction pattern of summation and difference beams

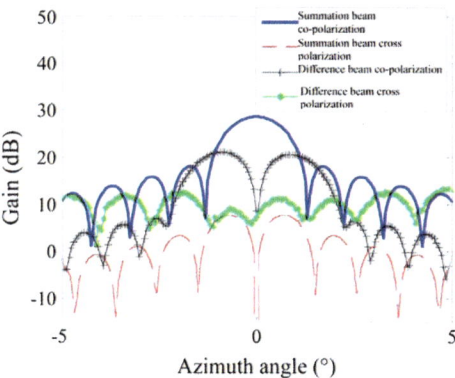

According to estimated orthogonal polarization channel echo, polarization ratio of interference signal can be obtained by following equation:

$$\rho_J = \left\langle \frac{J_V(t_k)}{J_H(t_k)} \right\rangle \tag{7.5.11}$$

where $\langle \cdot \rangle$ represents geometric mean value. Meanwhile, restriction condition $\|\rho_J\| = 1$ is considered to estimate polarization state of the interference signal.

For full polarization monopulse radar with polarization diversity, its receive signal model can be represented as follows:

$$\begin{bmatrix} \Sigma_H(t) \\ \Sigma_V(t) \\ \Delta_H(t) \\ \Delta_V(t) \end{bmatrix} = \begin{bmatrix} F(\theta - \theta_0) + F(\theta + \theta_0) & 0 \\ 0 & F(\theta - \theta_0) + F(\theta + \theta_0) \\ F(\theta - \theta_0) - F(\theta + \theta_0) & 0 \\ 0 & F(\theta - \theta_0) - F(\theta + \theta_0) \end{bmatrix} \begin{bmatrix} J_H(t) \\ J_V(t) \end{bmatrix} \tag{7.5.12}$$

The above-mentioned equation can be further written as follows:

$$\begin{bmatrix} \Sigma_H(t) \\ \Sigma_V(t) \\ \Delta_H(t) \\ \Delta_V(t) \end{bmatrix} = \begin{bmatrix} F(\theta - \theta_0) + F(\theta + \theta_0)J_H(t) \\ F(\theta - \theta_0) + F(\theta + \theta_0)J_V(t) \\ F(\theta - \theta_0) - F(\theta + \theta_0)J_H(t) \\ F(\theta - \theta_0) - F(\theta + \theta_0)J_V(t) \end{bmatrix} \tag{7.5.13}$$

It can be known from the Eq. (7.5.13) that factors of antenna direction pattern and signal angle of arrival are not considered in full polarization measurement mechanism, and optimum estimation of signal polarization state can be obtained by echo signal of two ways of orthogonal polarization channel. However, starting point for the method proposed in this section skillfully makes use of difference between cross polarization direction patterns of summation and difference beam, and it is not required to process orthogonal polarization antenna and dual-polarization channel, so as to reduce quantity and complexity of polarization information processing equipment.

7.5.2 Target Return Signal Polarization Estimation Method Based on Sum-Dif Beam Property

The radar transmit wave can be represented as follows:

$$E_t(t) = A \cdot \boldsymbol{h}_{tm} s_m(t) \tag{7.5.14}$$

where, M is number of transmit pulses, $\boldsymbol{h}_{tm}$ is Jones vector representation of transmit antenna polarization, and $\|\boldsymbol{h}_{tm}\| = 1$, $s_m(t)$ is transmit waveform function, which is usually represented in the form of rectangular pulse or linear modulation frequency pulse. A represents amplitude.

After the transmit signal is scattered from the target, its receive echo can be represented as follows:

$$E_s(t) = \chi \cdot \boldsymbol{h}_{tm}{}^T \boldsymbol{S} \boldsymbol{h}_{tm} s_m(t - \tau), \quad m = 1, \ldots, M \tag{7.5.15}$$

where $\boldsymbol{S} = \begin{bmatrix} S_{HH} & S_{HV} \\ S_{VH} & S_{VV} \end{bmatrix}$ is target polarization scattering matrix, χ represents signal amplitude, which is a value determined jointly by the radar receiver's processing gain and elements in radar equation (except for target scattering cross-section). $\chi = \frac{k_{RF}}{16\pi^2 R^4 L_R} \sqrt{\frac{P_t}{4\pi L_t}}$, where k_{RF} is radio frequency amplification coefficient, R is distance from the radar to the target, P_t is radar transmit power, L_t and L_R are comprehensive loss of transmit and receive signal, respectively.

Waveform influence is neglected, and target scattering is considered only, then target echo signal received by branch A can be represented as follows:

$$A(t) = \chi_s \left(F(\theta - \theta_0)^2 \boldsymbol{h}_A^T \boldsymbol{S} \boldsymbol{h}_A + F(\theta + \theta_0) F(\theta - \theta_0) \boldsymbol{h}_A^T \boldsymbol{S} \boldsymbol{h}_B \right) \tag{7.5.16}$$

Signal received by branch B can be represented as follows:

$$B(t) = \chi_s \left(F(\theta + \theta_0)^2 \boldsymbol{h}_B^T \boldsymbol{S} \boldsymbol{h}_B + F(\theta + \theta_0) F(\theta - \theta_0) \boldsymbol{h}_B^T \boldsymbol{S} \boldsymbol{h}_A \right) \tag{7.5.17}$$

To simplify the expression, suppose $F(-) = F(\theta - \theta_0)$ and $F(+) = F(\theta + \theta_0)$, so the summation signal can be represented as follows:

$$\sum(t) = A(t) + B(t)$$
$$= \chi_s \left[F(-)^2 \boldsymbol{h}_A^T \boldsymbol{S} \boldsymbol{h}_A + F(+)F(-)\left(\boldsymbol{h}_A^T \boldsymbol{S} \boldsymbol{h}_B + \boldsymbol{h}_B^T \boldsymbol{S} \boldsymbol{h}_A\right) + F(+)^2 \boldsymbol{h}_B^T \boldsymbol{S} \boldsymbol{h}_B \right]$$
$$(7.5.18)$$

The difference signal can be represented as follows:

$$\Delta(t) = A(t) - B(t)$$
$$= \chi_s \left[F(-)^2 \boldsymbol{h}_A^T \boldsymbol{S} \boldsymbol{h}_A + F(+)F(-)\left(\boldsymbol{h}_A^T \boldsymbol{S} \boldsymbol{h}_B - \boldsymbol{h}_B^T \boldsymbol{S} \boldsymbol{h}_A\right) - F(+)^2 \boldsymbol{h}_B^T \boldsymbol{S} \boldsymbol{h}_B \right]$$
$$(7.5.19)$$

Supposing $F(-)\boldsymbol{h}_A + F(+)\boldsymbol{h}_B = \boldsymbol{E}_S$, the Eq. (7.5.18) can be written as follows:

$$\sum(t) = \chi_s \left(F(-)\boldsymbol{h}_A^T + F(+)\boldsymbol{h}_B^T \right) \boldsymbol{S} \cdot \boldsymbol{E}_S \qquad (7.5.20)$$

The Eq. (7.5.19) can be rewritten as follows:

$$\Delta(t) = \chi_s \left(F(-)\boldsymbol{h}_A^T - F(+)\boldsymbol{h}_B^T \right) \boldsymbol{S} \cdot \boldsymbol{E}_S \qquad (7.5.21)$$

The common item $\boldsymbol{S}\boldsymbol{E}_S$ in the above-mentioned equation can be considered as orthogonal polarization component of target echo, written as $\boldsymbol{S}\boldsymbol{E}_S = \begin{bmatrix} s_h(t) \\ s_v(t) \end{bmatrix}$, so receive signal of real target in the summation and difference channel is written in the form of measurement equation matrix:

$$\begin{bmatrix} \sum(t) \\ \Delta(t) \end{bmatrix} = \begin{bmatrix} F(-)\boldsymbol{h}_A^T + F(+)\boldsymbol{h}_B^T \\ F(-)\boldsymbol{h}_A^T - F(+)\boldsymbol{h}_B^T \end{bmatrix} \begin{bmatrix} s_h(t) \\ s_v(t) \end{bmatrix} \qquad (7.5.22)$$

Thus, according to polarization characteristic matrix of the antenna and signal angle of arrival, the orthogonal polarization component of the target echo can be estimated by means of inverse matrix operation. This is identical to the method by which false target interference polarization is estimated, set forth in Sect. 7.5.1.

$$\begin{bmatrix} F(-)\boldsymbol{h}_A^T + F(+)\boldsymbol{h}_B^T \\ F(-)\boldsymbol{h}_A^T - F(+)\boldsymbol{h}_B^T \end{bmatrix} = \begin{bmatrix} F(\theta - \theta_0) + F(\theta + \theta_0) & F(\theta - \theta_0)\rho_A + F(\theta + \theta_0)\rho_B \\ F(\theta - \theta_0) - F(\theta + \theta_0) & F(\theta - \theta_0)\rho_A - F(\theta + \theta_0)\rho_B \end{bmatrix}$$
$$= \begin{bmatrix} F_1(\theta) & F_2(\theta) \\ F_3(\theta) & F_4(\theta) \end{bmatrix}$$
$$(7.5.23)$$

It can be known from the Eqs. (7.5.18), (7.5.19) that real target echo modulates square characteristics of the radar transmit antenna's polarization direction pattern and polarization scattering characteristics of the target. While interference signal echo only modulates polarization characteristics of the jammer antenna and primary characteristic of the radar receive antenna polarization direction pattern, which is different from the real target echo obviously. Therefore, discrimination parameters can be built by difference in polarization estimation value between both echoes, so as to identify and suppress active multiple false target jamming.

7.5.3 Multi-false Target Jamming Discrimination Method Based on Polarization Similarity

7.5.3.1 Polarization Characteristics of False Target Jamming

Multiple target interference is an important deception jamming form, whose intent is to provide many false targets at different distance from real target to the enemy's radar, so that the radar cannot distinguish true or false target or delays the time of identifying true target due to difficulty to identify the target. By taking multiple false target as example, waveform of each pulse is similar to the radar transmit signal waveform extremely. Its matching filter output and target echo has no difference in time domain and frequency domain. Therefore, the false jamming is usually used for self-defense jamming of the target, so as to coordinate with the target. Through transfer delay, several false targets are formed around the real target. Figure 7.29 simulates filter output result matching the true and false target echo. It can be seen that power of the false target to the radar receiver antenna should be bigger than that of the real target echo.

At horizontal and vertical polarization bases, active false target jamming signal at radar receive antenna port can be represented as follows:

$$e_J(t) = h_{J1}J_1(t) \tag{7.5.24}$$

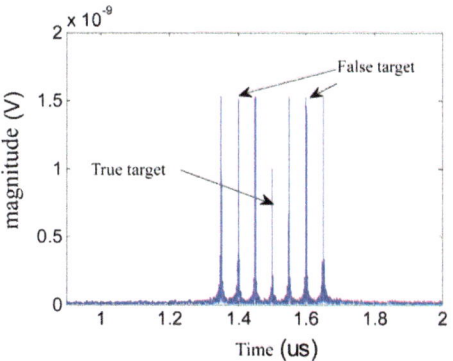

Fig. 7.29 Filter output of real and false target jamming

where $J_1(t)$ is modulation signal of false target jamming, which can be in any waveform. Its characteristic should be similar to modulation characteristics of target scattering wave for the purpose of protecting it from being identified by the radar from time domain and frequency domain; $\boldsymbol{h}_{J1} = [h_{JH1}, h_{JV1}]^{\mathrm{T}}$ is polarization form of current jamming signal, $\|\boldsymbol{h}_{J1}\| = 1$.

Generally, the jamming antenna is circular polarization to obtain jamming effect. At left- and right-hand orthogonal polarization base $(\hat{l}, \hat{r})$, false target signal transferred by the jammer can be represented as follows:

$$\boldsymbol{e}_{false}(t) = \begin{bmatrix} E_L(t) \\ E_R(t) \end{bmatrix} = \begin{bmatrix} 1 \\ 0 \end{bmatrix} \delta(t) = \begin{bmatrix} 1 \\ 0 \end{bmatrix} \zeta(t) \exp(j2\pi f_0 t + j\varphi(t)) \tag{7.5.25}$$

where $\delta(t)$ is modulation form of the active false target, which can simulate time domain, frequency domain and Droplet information of the radar target distinctly, so it is hard to distinguish conventional time domain and frequency domain discrimination methods; $\zeta(t)$ is modulation function containing target scattering echo amplitude characteristic information; f_0 is transmit signal carrier frequency; $\varphi(t)$ is phase modulation function of jamming signal.

Because the radar usually adopts linear polarization working method, the active false target polarization base is transformed to radar polarization base $\hat{h}, \hat{v}$ by use of polarization base transformation formula, which can be represented as follows:

$$\boldsymbol{e}_{false}(t) = \frac{1}{\sqrt{2}} \begin{bmatrix} 1 & 1 \\ j & -j \end{bmatrix} \begin{bmatrix} E_L(t) \\ E_R(t) \end{bmatrix} = \frac{1}{\sqrt{2}} \begin{bmatrix} 1 \\ j \end{bmatrix} \delta(t) \tag{7.5.26}$$

Because circular characteristics does not change due to change in jammer altitude, jamming signal polarization state will not be affected even if the jammer's false target signal amplitude and phase characteristics change, i.e., $\delta(t)$ has complex modulation. At this moment, $\boldsymbol{h}_{false} \approx \frac{1}{\sqrt{2}} \begin{bmatrix} 1 \\ j \end{bmatrix}$.

If the jamming antenna is also horizontal, vertical or 45° line polarization form, its horizontal and vertical polarization base is written as $(\hat{x}, \hat{y})$. Due to change in the jammer altitude, horizontal polarization direction of the radar antenna has a deviation angle ϑ. Therefore, polarization of the jamming signal can be represented at the polarization base where the radar is situated:

$$\boldsymbol{e}_{false}(t)^{\mathrm{H}} = \begin{bmatrix} \cos\vartheta & -\sin\vartheta \\ \sin\vartheta & \cos\vartheta \end{bmatrix} \begin{bmatrix} 1 \\ 0 \end{bmatrix} \delta(t) = \begin{bmatrix} \cos\vartheta \\ \sin\vartheta \end{bmatrix} \delta(t) \tag{7.5.27}$$

$$\boldsymbol{e}_{false}(t)^{\mathrm{V}} = \begin{bmatrix} \cos\vartheta & -\sin\vartheta \\ \sin\vartheta & \cos\vartheta \end{bmatrix} \begin{bmatrix} 0 \\ 1 \end{bmatrix} \delta(t) = \begin{bmatrix} -\sin\vartheta \\ \cos\vartheta \end{bmatrix} \delta(t) \tag{7.5.28}$$

$$e_{false}(t)^{45} = \begin{bmatrix} \cos\vartheta & -\sin\vartheta \\ \sin\vartheta & \cos\vartheta \end{bmatrix} \frac{1}{\sqrt{2}} \begin{bmatrix} 1 \\ 1 \end{bmatrix} \delta(t) = \frac{1}{\sqrt{2}} \begin{bmatrix} \cos\vartheta - \sin\vartheta \\ \cos\vartheta + \sin\vartheta \end{bmatrix} \delta(t) \quad (7.5.29)$$

It can be known from the above-mentioned analysis that polarization characteristics of active false target jamming only depends on polarization state of the jammer antenna and altitude of the jammer. This is decided by working principle of the active transfer type false target jammer, and transfer time delay between different false targets is decided by setting of the jammer. The polarization state of the false target jamming is transmitted by the same jamming antenna. Although the jamming signal may fluctuate, ratio between orthogonal polarization components (i.e. polarization ratio) does not change, polarization state is relatively stable, measurement value of polarization vector for different false targets is very similar, strong relativity exists, and polarization "similarity metric" is approximately equal to 1. But polarization state of target echo is irrelevant to the jammer antenna, which is a quantity jointly decided by the radar transmit antenna and target scattering characteristics. If the polarization state is decomposed in a Euclidean space, polarization characteristics of the active false target necessarily keeps a distance from polarization of real target in the Euclidean space, and the "polarization similarity" is weak. For this reason, as a judgment basis, it is possible to identify true and false target by use of the above-mentioned polarization estimation method and through calculating "polarization similarity metric" of two neighboring target echoes, so as to realize rapid and effective suppression.

7.5.3.2 Active Multiple False Target Jamming Discrimination Technology Based on Polarization Similarity

As discussed in the introduction, the identification method of conventional false target polarization has higher requirements for sensors, and it is required to have high-accuracy polarization measurement ability, and to estimate polarization scattering matrix of the target, and equivalent scattering matrix of false target (four components), i.e., it is required to obtain absolute measurement value. This accuracy is restricted by several kinds of physical factors of the sensor and electromagnetic environment, and the value cannot be accurately obtained, and it is more difficult to ensure that it still meets reciprocity and singularity and several other judgment natures, so it is possible to cause the polarization discrimination algorithm to be failed. It is worth to mention that signal measurement at orthogonal polarization channel reflects relative measurement result, and non-ideal measurement factor has the same influence on orthogonal polarization channel of the target and interference. So even if there is coupling error in echo polarization vector obtain, it is still possible to carry out balance and comparison through "similarity metric" between polarization valuation vectors, so as to take it as a kind of sounder judgment quantity.

7.5.3.3   Design of False Target Jamming Discrimination Parameters

From the prospect of model identification, such three physical quantities as included angle cosine, relevant coefficient and Euclidean distance can reflect variation information among polarization vectors, and similarity. Difference between included angle cosine and relevant coefficient is not big, and is similar, and Euclidean distance can reflect measurement deviation of the polarization vector. By making use of this nature, "similarity metric" between vectors can be used as polarization vector index, and it is required to measure and calculate included angle cosine, correlation coefficient and Euclidean distance of polarization vector of neighboring targets (Fig. 7.30).

The included angle cosine is to solve implementation of two vector included angle cosine concept in multiple spaces in geometry. Polarization estimation value of each target to be identified is represented as a model vector in 2D space, and included angle between 2 model vectors (i.e. polarization vector valuation) is ζ_{pol} in a polarization space. The bigger the similarity between 2 polarization estimation values, the smaller the included angle between 2 vectors is, and the included cosine coefficient is closer to 1. Included angle cosine value between polarization estimated values for targets to be discriminated after polarization measurement can be obtained by following equation.

$$
\begin{aligned}
\zeta_{pol} &= \frac{h(k)^H h(k+1)}{\|h(k)\| \cdot \|h(k+1)\|} \\
&= \frac{j_H(k) \cdot j_H(k+1) + j_V(k) \cdot j_V(k+1)}{\sqrt{(|j_H(k)|^2 + |j_V(k)|^2)(|j_H(k+1)|^2 + |j_V(k+1)|^2)}}
\end{aligned}
\tag{7.5.30}
$$

where, $\|\cdot\|$ indicates Feobenius normal number of polarization vector, $h(k) = [j_H(k) \quad j_V(k)]^T$ and $h(k+1) = [j_H(k+1) \quad j_V(k+1)]^T$ are 2D complex vector, which indicates measurement value of kth and $k+1$th echo polarization vector; ζ_{pol} is referred as "polarization measurement similarity". It can be known that $|\zeta_{pol}| \le 1$, and ζ_{pol} value is closer to 1, indicating that difference between two target polarization is smaller.

Fig. 7.30 Included angle between polarization vectors

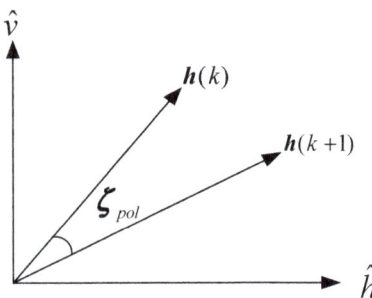

Correlation coefficient is initially used to measure degree of correlation between variables, and then to measure degree of similarity to the variables to be measured in cluster analysis. After estimating orthogonal polarization component of target or false target jamming through measurement algorithm, it is required to calculate and compare correlation characteristic difference between neighboring echo polarization, so as to discriminate real target and active false target jamming. The "polarization correlation coefficient" R_{pol} expression is defined as follows:

$$R_{pol} = \frac{(j_H(k) - \bar{j}_H) \cdot (j_H(k+1) - \bar{j}_H) + (j_V(k) - \bar{j}_V) \cdot (j_V(k+1) - \bar{j}_V)}{\sqrt{(|j_H(k) - \bar{j}_H|^2 + |j_V(k) - \bar{j}_V|^2)} \cdot \sqrt{(|j_H(k+1) - \bar{j}_H|^2 + |j_V(k+1) - \bar{j}_V|^2)}}$$

$$(7.5.31)$$

where, $\bar{j}_H$ represents mean value of horizontal polarization component in polarization estimation value of several targets, and $\bar{j}_V$ represents mean value of vertical polarization component in polarization estimation value of several targets.

With respect to active false target, polarization vector of neighboring echoes has strong correlation, and the correlation coefficient will approach to 1, indicating that these two targets are false target. Otherwise, polarization vector of neighboring echoes has weak correlation, and the correlation coefficient is less than 1, indicating that one target is a real target, and it is required to carry out discrimination of next pair of target.

Euclidean distance similarity reflects degree of intimacy between study objects, which can be used to describe degree of similarity for polarization estimation value of neighboring targets. The Euclidean distance expression is defined as follows:

$$D_{pol} = \sqrt{(j_H(k) - j_H(k+1))^2 + (j_V(k) - j_V(k+1))^2} \qquad (7.5.32)$$

With respect to false target, polarization estimation value of neighboring targets is very close to each other, and at this moment, Euclidean distance $D_{pol} = 0$; otherwise, polarization estimation value of neighboring targets has certain difference, i.e., $D_{pol} \neq 0$. With increase in difference in polarization estimation value, Euclidean distance D_{pol} increases, indicating one target is a real radar target, so as to carry out discrimination of next pair of targets.

7.5.3.4 Design of Polarization Discrimination Algorithm and Its Performance Analysis

Polarization similarity metric, polarization correlation coefficient, and polarization Euclidean distance portray effective parameter for difference in polarization characteristics of real and false targets. It is possible to distinguish real and false targets with different polarization state only if relative estimation value of the target in orthogonal polarization channel is obtained, while it is not required to accurately measure absolute value of complete target polarization scattering matrix. The following gives active false target discrimination algorithm which takes $\Theta_1 = \zeta_{pol}$, $\Theta_2 = R_{pol}$, $\Theta_3 = D_{pol}$ as discrimination parameter set, whose discrimination expression is as follows:

$$l_\Theta = \begin{cases} \{S(|\Theta_1 - 1| \leq Th)\&S(|\Theta_2 - 1| \leq Th)\&S(|\Theta_3| \leq Th)\} = 1, & \text{Active false target} \\ \{S(|\Theta_1 - 1| \leq Th)\&S(|\Theta_2 - 1| \leq Th)\&S(|\Theta_3| \leq Th)\} = 0 & \text{Radar target} \end{cases}$$

$$(7.5.33)$$

where, $S(x) = \begin{cases} 1, & \text{Logic relation is tenable} \\ 0, & \text{Logic relation is tenable} \end{cases}$ represents that the logic relation is tenable; symbol "&" represents logic and operation; Th represents uniform judgment gate limit. Specific discrimination process for active false targets is shown in Fig. 7.31.

In conditions where measurement noise influence is neglected, Sects. 7.5.3.1 and 7.5.3.2 analyze polarization characteristics of active false target and radar target. Based on this, this section gives polarization discrimination algorithm process of active false target, and the following analyzes performance of discrimination algorithm in noise condition. With respect to radar adopting single polarization tracking mechanism, measurement data of target polarization state that is estimated by the above-mentioned method can be represented as follows:

$$\begin{bmatrix} \Sigma_S(t) \\ \wedge_\Lambda(t) \end{bmatrix} = \begin{bmatrix} A \cdot F(\theta - \theta_0)^2 + A \cdot F(\theta + \theta_0)^2 & A \cdot F(\theta - \theta_0)^2 \rho_A(\theta) + A \cdot F(\theta + \theta_0)^2 \rho_B(\theta) \\ A \cdot F(\theta - \theta_0)^2 - A \cdot F(\theta + \theta_0)^2 & A \cdot F(\theta - \theta_0)^2 \rho_A(\theta) - A \cdot F(\theta + \theta_0)^2 \rho_B(\theta) \end{bmatrix}$$
$$\times \begin{bmatrix} s_h(t) \\ s_v(t) \end{bmatrix} + \begin{bmatrix} n_\Sigma(t) \\ n_\Delta(t) \end{bmatrix}$$

$$(7.5.34)$$

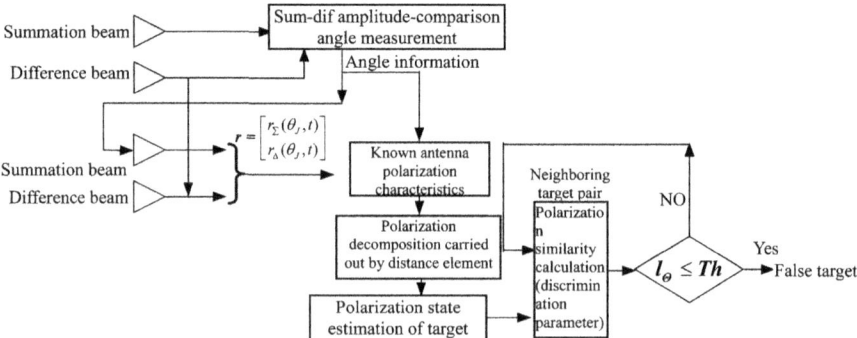

Fig. 7.31 Polarization discrimination process flow for active false target

Meanwhile, false target jamming signal received by summation and difference channel is written in the form of measurement equation matrix:

$$\begin{bmatrix} \sum_J(t) \\ \Delta_J(t) \end{bmatrix} = \begin{bmatrix} F(\theta - \theta_0) + F(\theta + \theta_0) & F(\theta - \theta_0)\rho_A(\theta) + F(\theta + \theta_0)\rho_B(\theta) \\ F(\theta - \theta_0) - F(\theta + \theta_0) & F(\theta - \theta_0)\rho_A(\theta) - F(\theta + \theta_0)\rho_B(\theta) \end{bmatrix}$$
$$\cdot \begin{bmatrix} j_h(t) \\ j_v(t) \end{bmatrix} + \begin{bmatrix} n_\Sigma(t) \\ n_\Delta(t) \end{bmatrix}$$

$$(7.5.35)$$

where $n_\Sigma(t)$ and $n_\Delta(t)$ are summation and difference channel noise, respectively, and any time t can be taken as random variable which is subject to zero mean value normal distribution, i.e., meeting $n_\Sigma(t) \sim N(0, \sigma_s^2)$ and $n_\Delta(t) \sim N(0, \sigma_D^2)$.

Therefore, the measurement error can be represented as follows:

$$\begin{bmatrix} \delta s_h(t) \\ \delta s_v(t) \end{bmatrix} = \begin{bmatrix} A \cdot F(\theta - \theta_0)^2 + A \cdot F(\theta + \theta_0)^2 & A \cdot F(\theta - \theta_0)^2 \rho_A(\theta) + A \cdot F(\theta + \theta_0)^2 \rho_B(\theta) \\ A \cdot F(\theta - \theta_0)^2 - A \cdot F(\theta + \theta_0)^2 & A \cdot F(\theta - \theta_0)^2 \rho_A(\theta) - A \cdot F(\theta + \theta_0)^2 \rho_B(\theta) \end{bmatrix}^{-1}$$
$$\times \begin{bmatrix} n(t) \\ n_\Delta(t) \end{bmatrix}$$

$$(7.5.36)$$

$$\begin{bmatrix} \delta j_h(t) \\ \delta j_v(t) \end{bmatrix} = \begin{bmatrix} F(\theta - \theta_0) + F(\theta + \theta_0) & F(\theta - \theta_0)\rho_A(\theta) + F(\theta + \theta_0)\rho_B(\theta) \\ F(\theta - \theta_0) - F(\theta + \theta_0) & F(\theta - \theta_0)\rho_A(\theta) - F(\theta + \theta_0)\rho_B(\theta) \end{bmatrix}^{-1}$$
$$\times \begin{bmatrix} n_\Sigma(t) \\ n_\Delta(t) \end{bmatrix}$$

$$(7.5.37)$$

At this moment, target echo power of summation channel can be represented as follows:

$$\left| \sum(t) \right|^2 = A^2 \cdot F(\theta - \theta_0)^4 \left| S_{HH} + 2S_{HV}\rho_A(\theta) + S_{VV}\rho_A^2(\theta) \right|^2$$
$$+ A^2 \cdot F(\theta + \theta_0)^4 \left| S_{HH} + 2S_{HV}\rho_B(\theta) + S_{VV}\rho_B^2(\theta) \right|^2$$
$$+ 2A^2 \cdot F(\theta - \theta_0)^2 F(\theta + \theta_0)^2 \left| S_{HH} + 2S_{HV}\rho_A(\theta) + S_{VV}\rho_A^2(\theta) \right|$$
$$\times \left| S_{HH} + 2S_{HV}\rho_B(\theta) + S_{VV}\rho_B^2(\theta) \right|$$

$$(7.5.38)$$

Target signal power in difference channel is as follows:

$$
\begin{aligned}
|\Delta(t)|^2 &= A^2 \cdot F(\theta - \theta_0)^4 |S_{HH} + 2S_{HV}\rho_A(\theta) + S_{VV}\rho_A^2(\theta)|^2 \\
&\quad + A^2 \cdot F(\theta + \theta_0)^4 |S_{HH} + 2S_{HV}\rho_B(\theta) + S_{VV}\rho_B^2(\theta)|^2 \\
&\quad - 2A^2 \cdot F(\theta - \theta_0)^2 F(\theta + \theta_0)^2 |S_{HH} + 2S_{HV}\rho_A(\theta) + S_{VV}\rho_A^2(\theta)| \\
&\quad \times |S_{HH} + 2S_{HV}\rho_B(\theta) + S_{VV}\rho_B^2(\theta)|
\end{aligned}
\tag{7.5.39}
$$

At this moment, signal/noise ratio in the summation and difference channels is as follows, respectively:

$$
\begin{aligned}
\mathrm{SNR}_\Sigma &= \frac{|\Sigma(t)|^2}{\sigma_s^2} \\
&\approx \frac{A^2 \cdot \left[\begin{array}{c} F(\theta - \theta_0)^4 + F(\theta + \theta_0)^4 \\ + 2A^2 \cdot F(\theta - \theta_0)^2 F(\theta + \theta_0)^2 \end{array}\right] |S_{HH} + 2S_{HV}\rho_A(\theta) + S_{VV}\rho_A^2(\theta)|^2}{\sigma_s^2}
\end{aligned}
\tag{7.5.40}
$$

$$
\begin{aligned}
\mathrm{SNR}_\Delta &= \frac{|\Delta(t)|^2}{\sigma_D^2} \\
&\approx \frac{A^2 \cdot \left[\begin{array}{c} F(\theta - \theta_0)^4 + F(\theta + \theta_0)^4 \\ - 2A^2 \cdot F(\theta - \theta_0)^2 F(\theta + \theta_0)^2 \end{array}\right] |S_{HH} + 2S_{HV}\rho_A(\theta) + S_{VV}\rho_A^2(\theta)|^2}{\sigma_D^2}
\end{aligned}
\tag{7.5.41}
$$

It can be seen from the above-mentioned equations that performance of polarization estimation depends on angle measurement accuracy θ, and signal/noise ratio SNR in the channel and false target jamming INR. It can be known from analysis given in Sect. 7.5.1 that signal strength of active false target is higher than real target, so jamming/noise ratio received by the false target in summation and difference channel is bigger than signal/noise ratio, i.e. INR > SNR; therefore, it is not required to analyze expression of INR in detail.

7.5.4 Discrimination Experiment and Result Analysis

According to algorithm principle given in Sect. 7.5.1, first carry out simulation analysis for polarization estimation method. Supposing that the target is in center direction of the antenna beam and polarization scattering matrix $S_t = \begin{bmatrix} 1 & 0.3j \\ 0.3j & 0.9 \end{bmatrix}$ of the target, co-polarization and cross-polarization direction pattern of the antenna and difference beam are shown as the figure. Where signal/noise ratio is 30 dB, results obtained by 50 times of polarization estimation for target echo are shown in Fig. 7.32, in which times of estimation is determined by number of echo pulses, and at this moment, the echo pulse is 50. It can be seen that vertical polarization component in the echo component is dominated, which is co-polarization of the radar antenna is vertical polarization, and the echo is dominated by co-polarization component (vertical) of the antenna. Due to variable polarization effect of the target, polarization ratio of the echo is about 1.414, polarization angle is about 55°, and close to 60° linearly polarization state. Figure 7.33 gives performance curve for polarization estimation of the target echo. When signal/noise ratio SNR is bigger than 20 dB, the algorithm tends to stable, and the measurement accuracy is high. Figures 7.34 and 7.35 give interference signal polarization estimation results. At this moment, jamming/noise ratio INR = 25 dB, jamming polarization is left-hand circular polarization. With respect to 50 times of echo pulse measurement, estimation amplitude of orthogonal polarization component is approximated to 0.707, and matches with real jamming polarization, and polarization estimation accuracy increases with increase in jamming/noise ratio.

Fig. 7.32 Polarization estimation/signal to noise ratio of target echo 30 dB

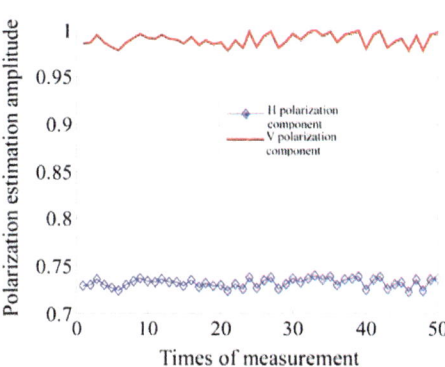

Fig. 7.33 Performance curve of target echo polarization estimation

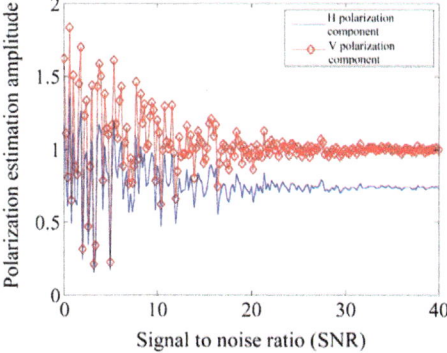

Fig. 7.34 Polarization estimation/signal to noise ratio of jamming echo 30 dB

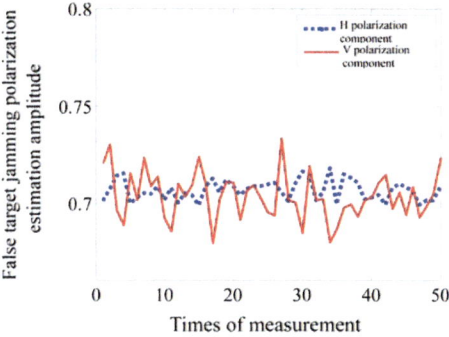

Fig. 7.35 Performance curve of jamming echo polarization estimation

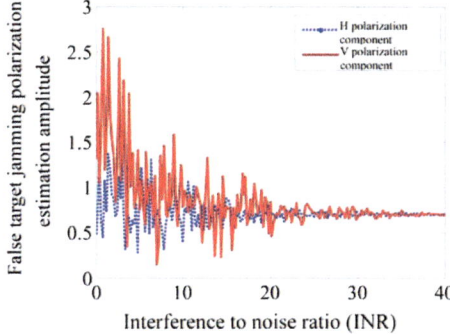

Fig. 7.36 Amplitude of real and false target polarization estimation

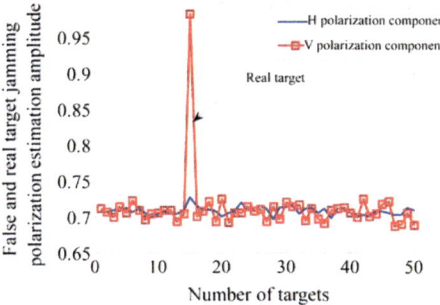

Figure 7.36 gives orthogonal polarization component estimation distribution result of 50 targets. It can be seen that H component of horizontal polarization for 50 targets is very close to each other, and V component of vertical polarization of 15th target is obviously different from other targets, so it is possible to initially judge the real target. Through the method given in Sect. 7.5.3.2, polarization similarity of every two neighboring targets is calculated, so that 49 pairs of measurement values shown in Fig. 7.37. It can be seen that polarization similarity between most targets is approximated to 1, and similarity of 14th and 15th pair of

Fig. 7.37 Polarization
similarity of neighboring
targets

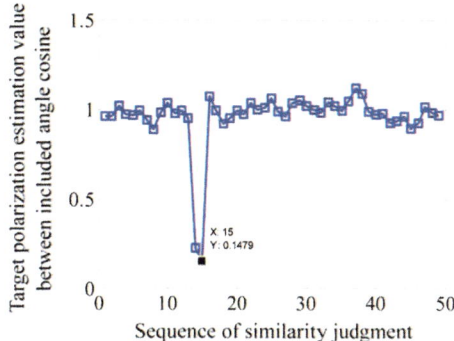

targets approximated to 0.2, indicating that polarization similarity and correlation
between 14th and 15th targets and between 15th and 16th targets are poor, so that it
indicates polarization of the 15th target is significantly different from polarization of
other targets. It can be taken as a basis for detecting the real target, while other 49
targets can be taken as false target. Meanwhile, distribution of polarization
Euclidean distance among 50 dense targets is shown in Fig. 7.38, and most of 49
polarization Euclidean distance calculation values are distributed nearby 0 value,
while distance value of 14th and 15 pairs of targets is about 1.6, which indicates
that polarization estimation value of the 15th target is significantly different from
polarization of other targets, and can be taken as a basis for judging the real target,
while other targets can be initially judged as false targets. Monte-Carlo simulation
experiment is carried out by making use false target discrimination algorithm and
processing flow given in this section, so as to obtain correct discrimination probability of active false target jamming versus SNR/INR curve, as shown in Fig. 7.39.
Number of simulations is 1000, and discrimination threshold Th = 0.2. When the
false targets are detected as the target, INR is usually high, and at this moment,
SNR ≈ INR > 10dB. It can be seen from the false target discrimination curve that
when SNR/INR is bigger than 15 dB, the false targets can be discriminated cor-
rectly; while SNR/INR is low, the discrimination probability can still reach above

Fig. 7.38 Polarization
Euclidean distance of
neighboring target

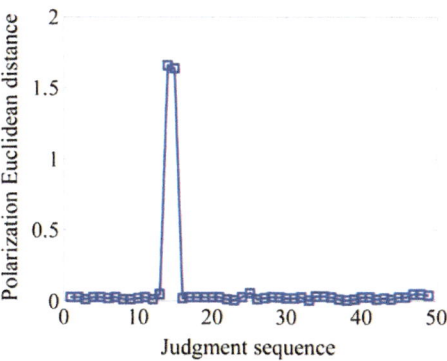

Fig. 7.39 Identification performance curve of false targets

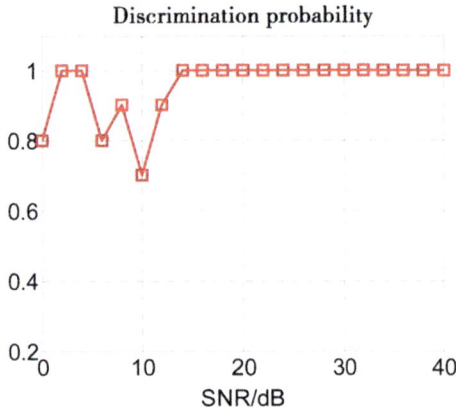

70%, indicating that discrimination performance of false target jamming is excellent, and the discrimination method is effective.

To sum up, multiple-target jamming has deceptive and blanketing advantage, and shows excellent jamming effect for tracking mechanism radars adopting different working means. Existing false target polarization discrimination algorithm mainly utilizes characteristic difference between equivalent scattering matrix of active false target and scattering matrix of radar target polarization. It is required that radar sensor has orthogonal polarization transmit and receive ability or variable polarization receive/transmit ability, and it is necessary to measure four components of polarization scattering matrix of targets. However, due to restriction by various factor in actual application, existing radar sensor cannot obtain complete and accurate polarization scattering information of the targets or jamming signal, so that the method of identify and suppressing targets and jamming by such natures as reciprocity and singularity of polarization scattering matrix is failed. This chapter ingeniously utilizes characteristics of cross polarization direction pattern for summation and difference beam of conventional single-polarization tracking radar to study a new polarization measurement method, design sound characteristic parameters used to identify false targets, and gives false target jamming identification algorithm and processing flow. This method does not require absolute value of accurate measurement of complete target polarization scattering information, but only requires relative estimation value of the target's orthogonal polarization component, being able to effectively distinguish real and false targets, and reduce calculation and processing difficulty and being easier to achieve. By this technology, it is not required to add polarization channel, but only required to obtain full polarization direction pattern and angle of arrival in advance, and improve signal processing means, so that it is hopeful to enhance multi-false-target anti-jamming ability of the single polarization tracking radar.

References

1. Zhaowen, Zhuang, Xiao Shunping, and Wang Xuesong. 1999. *Radar polarization information processing and application*, 1. Beijing: National Defense Industry Press.
2. Peikang, Huang, Yin Hongcheng, and Xu smallsword. 2005. *Radar target characteristics*, 3. Beijing: Electronic Industry Press.
3. Xuesong, Wang. 1995. Broadband polarization information processing research. Ph.D. thesis, Graduate School of National University of Defense Technology, Changsha, 5.
4. Giuli, Dino. 1986. Polarization diversity in radars. *Proceedings of IEEE* 74 (2): 245–269.
5. Boerner, W.M. 1992. *Direct and inverse methods in radar polarimetry*. Netherlands: Kluwer Academic Publishers.
6. Yongzhen, Li. 2004. Research on statistical characteristics and treatment of transient polarization. Doctoral thesis, Graduate School of National University of Defense Technology, Changsha, 12.
7. Mott, H. 1986. *Polarization in antenna and radar*. New York: Wiley-Interscience.
8. Extraction and recognition of the characteristics of polarimetric domain in the middle segment of the trajectory. Doctoral dissertation, Graduate School of National University of Defense Technology, Changsha, 2006, 10.
9. Giuli, Dino, L. Facheris, and M. Fossi. 1990. Simultaneous scattering matrix measurement through signal coding. In *Proceedings of IEEE 1990 international radar conference*. Arlington, VA, USA, 258–262.
10. Giuli, Dino, and M. Fossi. 1993. Radar target scattering matrix measurement through orthogonal signals. *Proceedings of IEEE, Pt. F* 140 (4): 233–242.
11. Poelman, A.J. 1981. Virtual polarisation adaptation: A method for increasing the detection capability of a radar system through polarisation-vector processing. *Proceedings of IEE, Pt. F* 128 (5): 261–270.
12. Poelman, A.J., and J.R.F. Guy. 1984. Multinotch logic-product polarisation suppression filters: A typical design example and its performance in a rain clutter environment. *Proceedings of IEEE, Pt. F* 131 (4): 383–396.
13. Poelman, A.J. 1983. Polarisation-vector translation in radar systems. *IEE Proceedings* 130 (2): 161–165.
14. Poelman, A.J., and J.R.F. Guy. 1984. Nonlinear polarization-vector translation in radar systems: A promising concept for Real-time polarization-vector signal processing via a single-notch polarization suppression filter. *IEE Proceedings* 131 (5): 451–464.
15. Poelman, A.J., and C.J. Hilgers. 1991. Effectiveness of multinotch logic-product polarisation filters in radar for countering rain clutter. *IEE Proceedings-F* 138 (5): 427–436.
16. Poelman, A.J. 1976. Cross correlation of orthogonally polarized backscatter components. *IEEE Transaction on Aerospace and Electronic Systems* 12 (12): 647–682.
17. Stapor, D.P. 1995. Optimal receive antenna polarization in the presence of interference and noise. *IEEE Transaction on Antennas and Propagation* 43 (5): 473–477.

© National Defense Industry Press, Beijing and Springer Nature Singapore Pte Ltd. 2019
H. Dai et al., *Spatial Polarization Characteristics of Radar Antenna*,
https://doi.org/10.1007/978-981-10-8794-3

18. Guoyi, Zhang. 2002. Research on polarization anti-interference technique of high frequency ground wave radar. Doctoral Dissertation, Graduate School of Harbin Institute of Technology, Harbin, 10.

19. Xuesong, Wang, Dai Hai, and Xu Zhenhai. 2004. Performance evaluation and selection of polarization filters. *Progress in Natural Science* 14 (4): 442–448.

20. Xue-song, Wang, Chang Yu-liang, Dai Da-hai, et al. 2007. Band characteristics of SINR polarization filter. *IEEE Transaction on Antennas and Propagation* 55 (4): 1148–1154.

21. Novak, L.M., M.B. Sechtin, and M.J. Cardullo. 1989. Studies of target detection algorithms that use polarimetric radar data. *IEEE Transaction on Aerospace and Electronic Systems* 25 (2): 150–165.

22. Novak, L.M., M.C. Burl, and W.W. Irving. 1993. Optimal polarimetric processing for enhanced target detection. *IEEE Transaction on Aerospace and Electronic Systems* 29 (1): 234–243.

23. Maio, A.D. 2002. Polarimetric adaptive detection of range-distributed targets. *IEEE Transaction on Signal Processing* 50 (9): 2152–2158.

24. Garren, David A., Anne C. Odom, Michael K. Osborn, et al. 2002. Full-polarization matched-illumination for target detection and identification. *IEEE Transaction on Aerospace and Electronic Systems* 38 (3): 824–835.

25. Xuesong, Wang, Li Yongzhen, Xu Zhenhai, et al. 2000. Research on polarization detection of high resolution radar signals. *Acta Electronic Journal* 28 (12): 15–18.

26. Yongzhen, Li, Wang Xuesong, Xu Zhenhai, et al. 2001. Detection of high resolution polarimetric targets based on the radial accumulation of strong scattering points. *Acta* 29 (3): 307–310.

27. Yongzhen, Li, Wang Xuesong, and Shunping Xiao. 2005. The weak target detection algorithm based on ISVS. *Acta Electronic Journal* 33 (6): 1028–1031.

28. Zhenhai, Xu, Wang Xuesong, and Shunping Xiao. 2004. Polarization sensitive array signal detection: Partial polarization case. *Acta* 32 (6): 938–941.

29. Yonghu, Zeng, Wang Xuesong, and Shunping Xiao. 2005. Based on polarization filtering in time frequency domain high resolution polarization radar signal detection. *Journal of Electronics* 33 (3): 524–526.

30. Songhua. 1993. High range resolution MMW radar target recognition theory and application. Ph.D. thesis, National University of Defense Technology Department of electronic technology, Changsha.

31. Yonghu, Zeng. 2004. Polarization radar time-frequency analysis and target recognition research. Ph.D. thesis, Institute of Electronic Science and engineering, National University of Defense Technology, Changsha, 6.

32. Fuller, D.F., A.J. Terzuoli, P.J. Collins, et al. 2004. Approach to object classification using dispersive scattering centres. *IEE Proceedings on Radar and Sonar Navigation* 151 (2): 85–90.

33. Emre, E., and C.P. Lee. 2000. Polarimetric classification of scattering centers using M-ary Bayesian decision rules. *IEEE Transaction on Aerospace and Electronic Systems* 36 (3): 738–749.

34. Lee, Jong-See, Ernst Krogager, Thomas L. Ainsworth, et al. 2006. Polarimetric analysis of radar signature of a manmade structure. *IEEE Geoscience and Remote Sensing Letters* 3 (4): 555–559.

35. Karnychev, V., A.K. Valery, P.L. Leo, et al. 2004. Algorithms for estimating the complete group of polarization invariants of the scattering matrix (SM) based on measuring all SM elements. *IEEE Transaction on Geoscience and Remote Sensing* 42 (3): 529–539.

36. Duquenoy, Mickael, Jean Philippe Ovarlez, Laurent Ferro-Famil, et al. 2006. Study of dispersive and anisotropic scatterers behaviour in radar imaging using time-frequency analysis and polarimetric coherent decomposition. In *Proceedings of ICR*, 180–185.

37. Touzi, R., W.M. Boerner, J.S. Lee, et al. 2004. A review of polarimetry in the context of synthetic aperture radar: Concepts and information extraction. *Canadian Journal of Remote Sensing* 30 (3): 380–407.

38. Dahai, Dai. 2003. POLSAR image simulation and Research on target detection and classification method. Master's degree thesis, Graduate School of National University of Defense Technology, Changsha, 11.

39. Kim, K.T., S.W. Kim, and H.T. Kim. 1998. Two-dimensional ISAR imaging using full polarisation and super-resolution processing techniques. *IEE Proceedings on Radar, Sonar Navigation* 145 (4): 240–246.

40. Dahai, Dai, Wang Xuesong, and Xiao Shunping. 2007. Full polarimetric ISAR superresolution imaging based on two dimensional CP-GTD model. *Natural Science Progress* 17 (9): 131–140.

41. Dahai, Dai. 2008. Polarimetric radar imaging and target feature extraction. Doctoral dissertations, Graduate School of National University of Defense Technology, Changsha, 6. Cohen, M.N., E.S. Sjoberg. 1985. Intrapulse polarization agile radar. In *Advances in radar techniques*, ed. J. Clarke, Peter Peregrinus Ltd.

42. Beide, Wang. 1996. Progress in radar polarization research in recent three years. *Modern Radar* 18 (2): 1–14.

43. http://www.les.com.cn/nanjing/part/les/database/weng/2nd/nmd.htm.

44. Dai. 2000. Bowei polarimetric synthetic aperture radar system and polarization information processing research. Doctoral dissertation, Institute of electronics Chinese, Beijing.

45. Kecheng, Liu, et al. 1989. *Antenna principle*. Changsha: National University of Defense Technology press.

46. Xuesong, Wang, et al. 2001. Statistical modeling and analysis of the influence of antenna polarization error on antenna receiving power. *Natural Science Progress* 11 (11): 1210–1215.

47. Dai, H.Y., X.S. Wang, J. Luo, Y.Z. Li, and S.P. Xiao. 2010. Spatial polarization characteristics and scattering matrix measurement of orthogonal polarization binary array radar. *Science in China (Series F)* 53 (12): 2687–2695.

48. Jia, Luo, and Wang Xuesong Xiao Shunping. 2007. The polarization characteristics of the real measured antenna. *Journal of Radio Wave Science* 22 (Sup): 373–376.

49. Jia, Luo, Wang Xue-song, et al. 2007. Spacial polarization characteristics of antenna. In *2007 1st Asian and Pacific conference on synthetic aperture radar proceedings (APSAR-2007)*. Nov. 5–9, Huangshan, China, 139–144.

50. McGrath, D., T. Schuneman, N. Shively. 2003. Polarization properties of scanning arrays. In *2003 IEEE international symposium on phased array systems and technology*, Oct 2003, 295–299.

51. Schrank, Hal, et al. 1993. Design of offset-parabolic-reflector antennas for low cross-pol and low sidelobes. *IEEE Antennas and Propagation Magazine* 35 (6): 47–50.

52. Alan Rudge, W., et al. 1978. Offset-parabolic-reflector antennas: A review. *Proceedings of the IEEE* 66 (2): 1592–1681.

53. Kildal, Per-Simon, et al. 1988. Losses, sidelobes, and cross polarization caused by feed-support structs in reflector antennas: Design curves. *IEEE Transaction on Antennas and Propagation* 36 (2): 182–190.

54. Ludwig, A.C. 1973. The definition of cross polarization. *IEEE Transaction on Antennas and Propagation*, 117–120.

55. Roy, J.E., et al. 2001. Generalization of the ludwig-3 definition for linear copolarization and cross-polarization. *IEEE Transaction on Antennas and Propagation* 49 (6): 1006–1010.

56. Damavandi, Nader. 1997. Cross polarization characteristics of annular ring microstrip antennas. *IEEE*, 1878–1881.

57. Hanle, E. 1992. 3-D polarimetry for target acquisition and classification with electronically steered planar array systems. In *IEEE international conference on radar*, Brighton, 222–225.

58. Hanle, E. 1995. Adaptive chaff suppression by polarimetry with planar phased arrays at off-broadside. In *IEEE international conference on radar*, 108–112.

59. Hanle, E. 1995. Polarimetry at off-broadside directions with planar arrays compensation methods in radar and communications. In *Workshop on radar polarimetry*, Nantes, March 1995, 21–23.

60. Hanle, E. 1994. Polarimetric suppression of cluster at off-broadside directions. In *International conference on radar*, Paris, 222–225.

61. Worms, J.G. 1995. About the influence of polarization agile jammers to adaptive antenna array. In *IEEE proceedings on radar* 95, USA, 619–623.

62. Jinlin, N., et al. 2002. Influence of unit cross polarization on the performance of adaptive array. *Journal of electronics and information* 24 (1): 97–101.

63. Fossi, M., M. Gherardelli, P. Girrnino, et al. 1986. Experimental results of dual-polarization behaviour of ground clutter. In *Record of CIE 1986 international conference on radar*, Nanjing.

64. Livingstone, C.E., A.L. Gray, R.K. Hawkins, et al. 1988. CCRS C/X-Airborne synthetic aperture radar: An R and D tool for the ERS-1 Time Frame. In *Proceedings of the 1988 IEEE national radar conference*, 15–21.

65. Livingstone, C.E., T.I. Lukowski, M.T. Rey, et al. 1989. CCRS/DREO synthetic aperture radar polarimetry-status report. In *IGARSS'89, Proceedings of the international geoscience and remote sensing symposium*. Vancouver, Canada, 10–14.

66. Abou-El-Magd, A.M., V. Chandrasekar, V.N. Bringi, and W. Strapp. 2000. Multiparameter radar andin situ aircraft observation of graupel and hail. *IEEE Transactions on Geoscience and Remote Sensing* 38 (1): 570–578.

67. Agrawal, A.P., and W.M. Boerner. 1989. Redevelopment of Kennaugh's target characteristic polarization state theory using the polarization transformation ratio formalism for thecoherent case. *IEEE Transactions on Geoscience and Remote Sensing* 27: 2–14.

68. Al-Jumily, K.J., R.B. Charlton, and R.G. Humphries. 1991. Identification of rain and hail with circular polarization radar. *Journal of Applied Meteorology* 30: 1075–1087.

69. Aydin, K., Y. Zhao, and T.A. Seliga. 1989. Rain-induced attenuation effects on C-banddual-polarization meteorological radars. *IEEE Transactions on Geoscience and Remote Sensing* 27: 57–66.

70. Bader, M.J., S.A. Clough, and G.P. Cox. 1987. Aircraft and dual polarization radar observations of hydrometeors in light stratiform precipitation. *Quarterly Journal of the Royal Meteorological Society* 103: 269–280.

71. Balakrishnan, N., and D.S. Zrnić. 1990. Use of polarization to characterize precipitation and discriminate large hail. *Journal of the Atmospheric Sciences* 47: 1525–1540.

72. Doviak, R.J., and D.S. Zrnić. 1993. *Doppler radar and weather observations*, 2nd ed. San Diego, CA: Academic Press.

73. Gorgucci, E., G. Scarchilli, V. Chandrasekar, and V.N. Bringi. 2000. Measurement of mean raindrop shape from polarimetric radar observations. *Journal of Atmospheric Science* 57: 3406–3413.

74. Hendry, A., Y.M.M. Antar, and G.C. McCormick. 1987. On the relationship between the degree of preferred orientation in precipitation and dual polarization radar echo characteristics. *Radio Science* 22: 37–50.

75. Cloude, S.R. 1985. Radar target decomposition theorems. *Electronics Letters* 21 (1): 22–24.

76. Cloude, S.R., and E. Pottier. 1996. A review of target decomposition theorems in radar polarimetry. *IEEE Transactions on Geoscience and Remote Sensing* 34 (2): 498–517.

77. Cloude, S.R., and E. Pottier. 1997. An entropy based classification scheme for land applications of polarimetric SAR. *IEEE Transactions on Geoscience and Remote Sensing* 35 (1): 68–78.

78. Lombardo, P. 2002. Optimal classification of polarimetric SAR images using segmentation. In *Proceedings of IEEE on radar conference*, Long Beach, CA, USA, 8–13.

79. Qong, M. 2004. Scattering mechanism identification based on the rotation and eccentric angles of polarimetric SAR data. In *Proceedings of international geoscience and remote sensing symposium (IGARSS'04)*, Anchorage, AK, USA, 3054–3057.

80. Xu, J.Y., et al. 2002. Using cross-entropy for polarimetric SAR image classification. In *Proceedings of international geoscience and remote sensing symposium (IGARSS'02)*, Toronto, Canada, 1917–1919.

81. Junyi, Xu, Yang Jian, and Peng Yingning. 2005. A new method for classification of remote sensing images of dual band polarization radar. *Chinese Science (series E)* 35 (10): 1083–1095.

82. Jin, Y.Q., and F. Chen. 2002. Polarimetric scattering indexes and information entropy of the SAR imagery for surface monitoring. *IEEE Transactions on Geoscience and Remote Sensing* 40 (11): 2502–2506.

83. Yaqiu, Jin, and Chen Fei. 2003. Application of SAR image polarization scattering index and the information entropy in surface recognition. *Natural Science Progress* 13 (2): 174–178.

84. Plombardo, Dpastina, and Tbucciarelli. 2001. Adaptive Polarimetric target detection with coherent radar Part II: Detection against Non-Gaussian background. *IEEE Transactions on Aerospace and Electronic Systems* 37 (4): 1207–1220.

85. Maio, A.D. 2002. Polarimetric adaptive detection of ranged-distributed target. *IEEE Transactions on Signal Processing* 50 (9): 2152–2159.

86. Park, H.R., J. Li, and H. Wang. 1995. Polarization-space-time domain generalized likelihood ratio detection of radar targets. *Signal Processing* 41: 153–164.

87. Park, H.R., Y.K. Kwag, and H. Wang. 2003. An efficient adaptive Polarimetric processor with an embedded CFAR. *ETRI Journal* 25 (3): 171–178.

88. Pastina, D., Plombardo, and TBucciarelli. 2001. Adaptive Polarimetric target detection with coherent radar Part I: Detection against Gaussian background. *IEEE Transactions on Aerospace and Electronic Systems* 37 (4): 1194–1206.

89. Plombardo, DPastina, and TBucciarelli. 2001. Adaptive Polarimetric target detection with coherent radar Part II: Detection against Non-Gaussian background. *IEEE Transactions on Aerospace and Electronic Systems* 37 (4): 1207–1220.

90. Hughes, P.K. 1983. A high-resolution radar detection strategy. *IEEE Transactions on Aerospace and Electronic Systems* 19 (5): 663–667.

91. Gerlach, Karl, and M.J. Steiner. 1999. Adaptive detection of range distributed. *IEEE Transactions on Signal Processing* 47 (7): 1844–1851.

92. G, Alfano, A. De Maio, and A. Farina. 2004. Model-based adaptive detection of range-spread targets. *IEE Proceedings-Radar, Sonar and Navigation* 151 (1): 2–10.

93. De Maio, Antonio. 2002. Polarimetric Adaptive detection of range-distributed targets. *IEEE Transactions on Signal Processing* 50 (9): 2152–2159.

94. Wenfeng, Sun, He Songhua, Guo Guirong, et al. 1999. Adaptive distance unit accumulation detection method and its application. *Acta Electronic Journal* 27 (2): 111–113.

95. Zhenhai, Xu, Wang Xuesong, Zhou Ying, et al. 2001. High resolution polarimetric radar target detection algorithm based on PWF fusion. *Acta* 29 (12): 1620–1623.

96. Yongzhen, Li, Wang Xuesong, Li Jun, et al. 2001. High resolution polarization detection based on Stokes vector. *Modern Radar* 23 (1): 52–58.

97. Yongzhen, Li, Wang Xuesong, and Xiao Shunping. 2000. Detection of high resolution polarimetric target based on nonlinear accumulation. *Infrared and Millimeter Wave Journal* 19 (4): 307–312.

98. Zhenhai, Xu, Wang Xuesong, Xiao Shunping, and Zhuang Zhaowen. 2004. Filtering performance of polarization sensitive array: Completely polarized case. *Chinese Journal of Electronics* 32 (8): 1310–1313.

99. Zhenhai, Xu, Wang Xuesong, Xiao Shunping, and Zhuang Zhaowen. 2004. Filtering performance of polarization sensitive array: Interference case. *Journal on Communications* 25 (10): 8–15.

100. Zhenhai, Xu, Wang Xuesong, Xiao Shunping, Zhuang Zhaowen. 2001. Target detection of high resolution and full polarization based on PWF and scattering points. In *IEEE RADAR*, Beijing, 376–379.
101. Zhenhai, Xu, Wang Xuesong, Xiao Shunping, Zhuang Zhaowen. 2003. Joint spectrum estimation of polarization and space. In *IEEE ICNNSP*, Nanjing, 1285–1289.
102. Stapor, D.P. 1995. Optimal receive antenna polarization in the presence of interference and noise. *IEEE Transactions on Antennas and Propagation* 43 (5): 473–477.
103. Santalla, V., M. Vera, A.G. Pino. 1993. A method for polarimetric contrast optimization in the coherent case. In *Antennas and propagation society international symposium*, 1288–1291.
104. Yang, J., Y. Yamaguchi, and W.M. Boerner. 2000. Numerical methods for solving the optimal problem of contrast enhancement. *IEEE Transactions on Geoscience and Remote Sensing* 38 (2): 965–971.
105. Yang, J., G.W. Dong, Y.N. Peng, et al. 2004. Generalized optimization of polarimetric contrast enhancement. *IEEE Transactions on Geoscience and Remote Sensing* 42 (3): 171–174.
106. Yongzhen, Li, Wang Xuesong, Wang Tao, et al. 2004. Study on polarization identification of active decoys. *National Defense Science Journal* 26 (3): 83–88.
107. Yongzhen, Li, Wang Xuesong, and Xiao Shunping. 2004. Polarization identification algorithm for true and false targets basded on IPPV. *Modern Radar* 26 (9): 38–42.
108. Yongzhen, Li, Xiao Shunping, and Wang Xuesong. 2005. The discrimination ability of active false target polarization for ground-based defense radar. *System Engineering and Electronic Technology* 27 (7): 1164–1168.
109. Tao, Wang, Wang Xuesong, and Xiao Shunping. 2006. Research on polarization identification of random modulated single polarized active false targets. *Natural Science Progress* 16 (5): 611–617.
110. Longfei, Shi, Wang Xuesong, and Xiao Shunping. 2009. Repeater decoy interference polarization discrimination. *China series F: Information Science* 39 (4): 468–475.
111. Jinliang, Li Wang Xuesong, and Li Yongzhen. 2008. The polarization characteristics of normal space oriented chaff. *Journal of Electric Wave Science* 23 (3). 1–7.
112. QingFu, Lai, Li Jinliang, Feng Dejun, and Wang Xuesong. 2010. Research on the dual polarization statistical properties of ships and chaff. *Journal of Radio Science* 25 (6): 1079–1084.
113. Qingpu, Liu, and Shen Yunchun. 1996. Chaff cloud polarization identification scheme performance analysis. *System Engineering and Electronic Technology* 11 (5): 1–7.
114. Yunchun, Shen, Xie Junhao, and Liu Qingpu. 1995. Identification of chaff cloud new scheme. *System Engineering and Electronics Technology* 17 (4): 11–14.
115. Luojia, Wang Xuesong, Li Yongzhen, Xiao Shunping, and Dai Huanyao. 2008. A new method for estimating the signal polarization state of the incoming wave. *National Defense Science Journal* 30 (5): 56–61.
116. Huanyao, Dai, Wang Xuesong, Li Yongzhen, Luo Jia, and Xiao Shunping. 2011. Spatial polarization characteristics of orthogonal polarization two element array radar and scattering matrix measurement method. *China Science (series F)* 41 (8): 945–954.
117. Lufei, Ding. 2004. *Radar principle*. Xi'an: Xi'an Electronic and Science University press.
118. Jia, Luo, Wang Xuesong, Li Yongzhen, and Xiao Shunping. 2008. Characterization and analysis of the polarization characteristics of antenna. *Journal of Radio Science* 23 (4): 620–628.
119. Zuji, Zhang, Jinlin, et al. 2005. *Radar antenna technology*. Beijing: Electronic Industry Press.
120. Warren Stutzman, L., and A. Gary Thiele. 1997. *Antenna theory and design*, 12. Hoboken: Wiley.
121. Kraus, John D., Ronald J. Marhefka. 2006. *Antennas: For all applications*, 3rd ed.

122. Wanzheng, Lu. 2004. *Antenna theory and technology*. Xi'an Electronic and Science University press: Xi'an.
123. Qixiao, Ye. 2006. *Handbook of applied mathematics*. Beijing: Science Press.
124. Kezhong, Yang, Yang Zhiyou, and Zhang Rirong. 1993. *New technology of modern antenna*. Beijing: People's Posts and Telecommunications Publishing House.
125. Chongcan, Zhu, Huang Jingxi, and Lu Shu (eds.). 1996. *Antenna*. Wuhan: Wuhan University press.
126. GRASP9—General reflector and antenna farm analysis software. http://www.ticra.com/script/site/page.asp?artid=33.
127. GRASP9. http://www.vi-re.com/products2.asp?id=21.
128. Huanyao, Dai, Luo Jia, and Wang Xuesong. 2009. Study on the characteristics of spatial instantaneous polarization of parabolic reflection antenna. *Chinese Journal of Radio Science* 24 (1): 126–131.
129. Jia, Luo, Wang Xuesong, and Xiao Shunping. 2009. A novel polarimetric scattering matrix measurement method of radar target. *Signal Processing* 25 (6): 868–873.
130. Bray Matthew, G., and H. Werner Douglas. 2002. Optimization of thinned aperiodic linear phased arrays using genetic algorithms to reduce grating lobes during scanning. *IEEE Transaction on Antennas and Propagation* 50 (12): 1732–1742.
131. Nan, Wang, Xue Zhenghui, and Liu Ruixiang. 2006. The characters of time domain radiated field of ultra wide band ultra low side lobe phased array antenna. *Acta Electronica Sinica* 34 (9): 1605–1609.
132. Xiaohu, Du, Li Jianxin, and Zheng Xueyu. 2002. Design of X-band dual polarization active phased array antenna. *Modern Radar* 5 (9): 67–69.
133. Zi-sen, Qi, GuoYing, and Wang Bu-hong. 2007. Performance analysis of MUSIC for conformal array. In *2007 international conference on wireless communications, networking and mobile computing (WICOM07)* Shanghai, 168–171.
134. Zisen, Qi, Guo Ying, and Wang Buhong. 2008. Performance analysis of MUSIC for conformal array. *Journal of Electronics and Information Technology* 30 (11): 2674–2677.
135. Xueyu, Zheng, and Wan Changning. 1995. Wide-band wide-angle scanning phased array antenna. *Chinese Journal of Radio Science* 10 (1): 33–38.
136. Jianying, Li, and Liang Changhong. 1999. Unit characteristic of rectangular waveguide end-slot finite phased arrays. *Acta Electronica Sinica* 27 (12): 102–104.
137. fenn, Alan, A. Guy Thiele, and A. Benelike Mrrik. 1982. Moment method analysis of finite rectangular waveguide phase arrays. *IEEE Transaction on Antennas and Propagation* 30 (4): 554–564.
138. Peiguo, Liu, Mao Junjie. 2004. *Radio and antenna*. Changsha: National University of Defense Technology press.
139. Dai, H.Y., Y.Z. Li, X.S. Wang, and J. Zhao. 2009. Polarization property of scanning slot phased array. *IEEE Asia-Pacific Conference on Synthetic Aperture Radar (APSAR) Xi'an* 10: 567–570. xi'an.
140. Dai, Huanyao, Y.Z. Li, and X.S. Wang. 2010. Spatial polarization characteristics of electronically scanning dipole phased arrays antenna. *Journal of National University of Defense Technology* 32: 1–84. 89.
141. Huanyao, Dai, Li Yongzhen, Xue Song, and Wang Xuesong. 2011. High frequency simulation analysis of spatial polarization characteristic of phased array antenna. *Chinese Journal of Radio Science* 26 (2): 316–322.
142. Brockett, T., and Y. Rahmat-Samii. 2008. A novel portable bipolar near-field measurement system for millimetre-wave antennas: Construction, development, and verification. *IEEE Magazine on Antennas and Propagation* 50 (5): 121–130.
143. Laitinen, T.A, S. Ranvier, J. Toivanen, J. Ilvonen, L. Nyberg, J. Krogerus, C. Icheln, and P. Vainikainen. 2009. Research activities on small antenna measurements at Helsinki University of Technology. In *IEEE international workshop on antenna technology*, 2–4 March 2009, 1–4.

144. Hirose, M., S. Kurokawa, and K. Komiyama. 2007. Antenna measurements by one-path two-port calibration using radio-on-fiber extended port without power supply. *IEEE Transactions on Instrumentation and Measurement* 56 (2): 397–400.

145. Shang Junping, Fu, and Deming, Yu. Ding. 2006. Study of fast measurement method of phased array antennas based on the control of circular bit shift. *Journal of Microwaves* 22 (6): 1–4.

146. Huili, Zheng, and Mao Naihong. 1996. A study of radar antenna testing method. *Chinese Journal of Radio Science* 11 (3): 46–51.

147. Hu Hongfei, Fu, and Tan Jizhao Demin. 2003. A modification of sampling space selection criterion for near-field antenna measurements. *Radar Science and Technology* 1 (1): 51–64.

148. Di, Li, and Wang Hua. 2005. Research of measurement technique for phased array scan pattern in intermediate range. *Modern Radar* 27 (7): 48–50.

149. Dai, H.Y, Y. Liu, Y.Z. Li, X.S. Wang, S.P. Xiao. 2010. New far field measurement method for antenna polarization characteristics based on calibrator target. In *2010 IEEE international radar conference*, Arlington, Virginia, USA.

150. Huanyao, Dai, Liu Yong, Li Yongzhen, and Wang Xuesong. 2009. Far field measurement for spatial polarization characteristics of antenna. *Journal of Applied Sciences* 27 (6): 606–611.

151. Huanyao, Dai, Liu Yong, and Wang Xuesong. 2010. Antenna polarization characteristic measurement method based on target scattering character of calibrator. *Journal of Microwaves* 26 (1): 5–11.

152. Huanyao, Dai, Li Jinliang, and Liu Yong. 2010. Effect of polarization basis mismatching on antenna polarization measurement and calibration method. *Journal of Microwaves* 26 (4): 32–36.

153. Huanyao, Dai, and Li Jinliang. 2010. Measurement and calibration of spatial instantaneous polarization characteristics of practical radar antenna. *Modern Radar* 32 (7): 83–86.

154. Xianrong, Shu, and He Bingfa. 2008. Query on reciprocity between ARP and ATP. *Journal of Microwaves* 24 (6): 43–46.

155. Jia, Luo, Wang Xuesong, and Xiao Shunping. 2009. A novel polarimetric scattering matrix measurement method of radar target. *Signal Processing* 25 (8): 868–873.

156. Dai, H.Y., X.S. Wang, and Y.Z. Li. 2012. Main-lobe jamming suppression method of using spatial polarization characteristics of antenna. *IEEE Transaction on Aerospace and Electronic Systems* 48 (3): 2167–2179.

157. Dai, H.Y., X.S. Wang, Y. Liu, Y.Z. Li, and S.P. Xiao. 2012. Novel research on main-lobe jamming polarization suppression technology. *Science in China (Series F)* 55 (2): 368–376.

158. Dai, Huanyao, Yongzhen Li, Wang Xuesong, and Liu Yong. 2012. A new method of main lobe interference suppression of polarization. *Chinese Science (series F)* 42 (4): 460–466.

159. Huanyao, Dai, Li Yongzhen, and Liu Yong. 2011. Spatial null phase shift interference suppression polarization filter design for single polarized radars. *Systems Engineering and Electronics* 33 (2): 290–295.

160. Mao, X.P., and Y.T. Liu. 2007. Null phase-shift polarization filtering for high-frequency radar. *IEEE Transactions on Aerospace and Electronic System* 43 (4): 1397–1407.

161. Xingpeng, Mao, Liu Yongtan, and Deng Weibo. 2008. Frequency domain null phase-shift multi notch polarization filter. *Acta Electronica Sinica* 36 (3): 537–542.

162. Xuesong, Wang, Wang Liandong, and Xiao Shunping. 2004. Theoretical performance analysis of adaptive polarization filters. *Acta Electronica Sinica* 32 (8): 1326–1329.

163. Guoyi, Zhang, and Liu Yongtan. 2001. Polarization suppression of multidisturbance in HF ground wave radar. *Acta Electronica Sinica* 29 (9): 1206–1209.

164. Yunfu, Yang, Tao Ran, and Wang Yue. 2007. Analysis of filter characteristics of SINR optimal polarization: Partially polarized case. *Progress in Natural Science* 17 (3): 370–378.

165. Longfei, Shi, Wang Xuesong, and Xiao Shunping. 2006. The iterative filtering scheme and its performance analysis of APC. *Journal of Electronics and Information Technology* 28 (9): 1560–1564.

166. Guoyi, Zhang, Tan Zhongji, and Wang Jiantao. 2006. Modification of polarization filtering technique in HF ground wave radar. *Journal of Systems Engineering and Electronics* 17 (4): 737–742.

167. Pace, P.E., D.J. Fouts, S. Ekestrrm, et al. 2002. Digital false-target image synthesizer for countering ISAR. *IEE Proceedings Radar Sonar and Navigation (S0956-375X)* 149 (5): 248–257.

168. Sharpin, D.L., and J.B.Y. Tsui. 1995. Analysis of linear amplifier analog-digital convertor interface in a digital microwave receiver. *IEEE Tranaction on AES (S0018-9251)* 31 (1): 248–255.

169. Yanbin, Zhang, Gao Meiguo, and Yuan Qi. 2009. Digital realization of multiple artificial target jamming on velocity. *Transactions of Beijing Institute of Technology* 29 (1): 59–62.

170. Jiaqi, Liu, Liu Jin, Dan Mei, Feng Dejun, and Wang Guoyu. 2008. Simulation research on multiple-face targets jamming against missile defense guidance radar. *Journal of System Simulation* 20 (3): 557–561.

171. Hongya, Liu. 2008. Methods to recognize false target generated by digital-image-synthesiser. In *2008 international symposium on information science and engineering*, ISISE 2008, v1, 71–75.

172. Kostis, and G. Theodoros. 2009. Angular glint effects generation for false naval target verisimility requirements. *Measurement Science and Technology* 20 (10): 832–837.

173. Bin, Rao, Liu Yi, Xiao Shun-Ping, Wang Xue-Song. 2009. Conservation-law based discrimination method of exoatmosphere range false targets. In *IET international radar conference 2009*, n 551.

174. Li, Yuan, Gaohuan Lv, Huilian Chen. 2008. A new technology of multi-false targets deception against chirp waveform inverse synthetic aperture radar. In *2008 9th international conference on signal processing*, ICSP 2008, 2477–2480.

Printed by Printforce, the Netherlands